FROM NATURAL HISTORY TO
THE HISTORY OF NATURE

Figure 1 - Buffon near age 53, by Drouais

From Natural History to the History of Nature:

Readings from Buffon and His Critics

Edited, translated, and with introductions by
John Lyon and Phillip R. Sloan

UNIVERSITY OF NOTRE DAME PRESS
NOTRE DAME

University of Notre Dame Press
Notre Dame, Indiana 46556
undpress.nd.edu

Copyright © 1981 by University of Notre Dame
Published in the United States of America

Paperback edition published in 2017

Library of Congress Cataloging in Publication Data

Main entry under title:

From natural history to the history of nature.

   Bibliography: p.
   1. Natural history—History—Addresses, essays, lectures. 2. Buffon, Georges Louis Leclerc, comte de, 1707-1788—Addresses, essays, lectures. 3. Naturalists—France—Biography—Addresses, essays, lectures. I. Lyon, John. II. Sloan, Phillip R.
QH15.F76   500   81-1320
                                         AACR2
ISBN: 978-0-268-15974-0 (paper)
ISBN: 978-0-268-00955-7 (hardcover)

Contents

PREFACE   *ix*

ACKNOWLEDGMENTS   *xiii*

1. INTRODUCTION   1

PART I: THE FORMATIVE YEARS

2. Buffon's Preface to the *Vegetable Staticks* of
   Stephen Hales (1735). Translated by Phillip R.
   Sloan.   35

3. Buffon's Preface to Isaac Newton's *Fluxions*
   (1740) (selected). Translated by John Lyon.   41

4. Buffon's Moral Arithmetic (1730, 1777)
   (selected). Translated by John Lyon.   51

5. Buffon on Newton's Law of Attraction, (1749),
   (selected). Translated by Phillip R. Sloan.   77

PART II: THE EMERGENCE OF THE *NATURAL HISTORY*

6. The "Initial Discourse" to Buffon's *Histoire
   naturelle* (1749). Translated by John Lyon.   89

7. The "Second Discourse" and "Proofs of the
   Theory of the Earth" from Buffon's *Histoire
   naturelle* (1749) (selected).   131

   a. "Second Discourse: The History and
      Theory of the Earth." Translated by
      J.S. Barr.   134

   b. "Proofs of the Theory of the Earth,"
      (1749) Article I. Translated by J.S.
      Barr.   151

8. Buffon on the Generation of Animals (1749) (selected). ... 165

   a. "Of Reproduction in General" (1749) (selected). Translated by J. S. Barr. ... 170

   b. "Of Nuturition and Growth" (1749) (selected). Translated by J.S. Barr. ... 181

   c. "Experiments on the Method of Generation" (1749) (selected). Translated by J.S. Barr. ... 187

   d. "Reflections on the Preceding Experiments" (1749) (selected). Translated by J.S. Barr. ... 201

PART III: THE FIRST RECEPTIONS

9. The *Journal de Trévoux* Reviews (1749-50). Translated by John Lyon. ... 213

10. The *Journal des savants* Reviews (1749) (selected). Translated by John Lyon. ... 231

11. The *Nouvelles ecclésiastiques* Reviews (1750). Translated by John Lyon. ... 235

12. The *Bibliothèque raisonée* Reviews (1750-51). Translated by John Lyon. ... 253

13. The Sorbonne's Condemnation of the *Histoire naturelle* (1751). Translated by John Lyon. ... 283

14. Buffon on Hypotheses: The Haller Preface to the German Translation of the *Histoire naturelle* (1750). Translated by Phillip R. Sloan. ... 295

15. Haller on Buffon's Theory of Generation (1751) (selected). Translated by Phillip R. Sloan. ... 311

16. Malesherbes' *Observations* on Buffon's Natural History (1749, 1798), (selected). Translated by John Lyon. ... 329

PART IV: BUFFON IN RETROSPECT: HIS STYLE AND GLORY

17. Hérault de Séchelles' Visit to Buffon
(1785). Translated by John Lyon.   *349*

LIST OF ILLUSTRATIONS   *387*

INDEX OF NAMES AND SUBJECTS   *391*

Preface

We have had to make several significant decisions concerning the content of the following collection of readings and translations. The finest available collection of readings from Buffon, the masterful <u>Oeuvres philosophiques de Buffon</u>, edited by Jean Piveteau (Paris: Presses Universitaires de France, 1954), served as our beginning point. Other collections include the lesser Buffon: <u>Morceaux choisis</u>, edited by A.M. Petitjean (Paris: Gallimard, 1939), and the recent unorthodox collection of texts in <u>Un autre Buffon</u>, edited by Jacques-Louis Binet and Jacques Roger (Paris: Hermann, 1977). In addition to these, there are now available three recent collections from the body of the William Smellie English translation: the selections from the <u>Natural History, General and Particular</u>, edited by Frank N. Egerton, (History of Ecology Series; New York: Arno Press, 1977); and the selections from the <u>Natural History of the Birds</u> and the <u>Natural History of Oviparous Quadrupeds</u>, reprinted under the editorship of Keir B. Sterling (Biologists and their World Series; New York: Arno Press, 1978). In light of the available selections from the <u>Natural History</u> itself, it was our conclusion that the kind of collection most needed was one which could supply a greater context to Buffon's work, and give some explanation of the importance of his thought in the history of science and in the intellectual culture of the Enlightenment. By choosing initially a selection from writings which seemed to lay some of the conceptual foundations for the <u>Natural History</u>, our intent was to reveal some of the critical methodological and conceptual developments evident in Buffon's non-biological writings that might give a key to the enigmatic features of his mature thought. Through a selection of reviews by his contemporaries, we sought a more concrete insight into the novel aspects of his works as his immediate peers perceived them.

Commentators have frequently discussed in recent years the great impact of the <u>Natural History</u> on the thought of the Enlightenment. Otis Fellows, in an assessment of Buffon's general importance, saw in him nearly the ideal <u>philosophe</u> ("Buffon's Place in the Enlightenment," <u>Studies in Voltaire and the Eighteenth Century</u> 25 [1963], 603-29), and has extended such claims in his recent collaborative biography with Stephen F. Milliken, Buffon (New York: Twayne, 1972). Daniel Mornet's somewhat surprising disclosure in 1910 that Buffon's Natural History was more frequently represented in eighteenth century French libraries than any other scientific or philosophical text with the exception of Pierre Bayle's Dictionnaire (D. Mornet, "Les Enseignements des bibliothèques privées [1750-80]," Revue d' historie littéraire de la France 17 [1910], 449-96), does not itself indicate anything about the nature of Buffon's actual impact, nor does it warrant an automatic inference from simple possession to critical reading of the text.

Yet questions still remain. Was Buffon simply a great "popularizer" of natural history? Were his scientific speculations held in high regard by his scientific peers, or dismissed as the hypotheses of a facile amateur? Did Buffon warrant the sort of respect which his contemporaries accorded to Leibniz, Diderot, Voltaire, Wolff, Hume and Montesquieu? Is his current location in the footnotes of intellectual history warranted? We believe that the seminal importance of Buffon in the development of modern natural science can be inferred from the texts which we have collected and translated here.

It has not been our intent to supply a comprehensive selection of texts from the Natural History itself. We have consequently included only those selections from that work which are most directly relevant to the subject matter of the reviews and commentaries which we have thought most centrally important to an understanding of the enduring significance of the man and his work. Consequently, our collection does not closely replicate collections of Buffon's writings available in any language, and is intended in part as a complement and supplement to those collections. Ideally a set of new translations from the original texts of the Natural History will be made in the future that will also include selections from the important Daubenton articles that have never appeared in the English language. Our collection does not intend to supply this deficiency.

With the exception of the "Premier discours", which has only recently been translated in its entirety for the first time (by one of the editors of this collection), and appears here in a revised form, selections from the body of the Natural History itself have been made from the rare J.S. Barr edition of 1792. While this is, on some points, still not an ideal translation, it is the only English translation which remains closely faithful in detail to the French original. To increase the value of the Barr translations, we have not hesitated to make minor alterations and restorations after comparison with the original texts. The more commonly encountered William Smellie translation, first issued in 1781-85 and often reprinted, is in many respects a more literate translation, but suffers from repeated editorial alteration and omissions which make it a text which must be used with care on finer points of interpretation. The more literal William Kenrick translation of 1775 includes several texts not present in the other editions, particularly texts from the important Suppléments and Histoire naturelle des mineraux, but suffers in many places from an incompetence that does not warrant its reprinting.

Because it has been our primary intent to provide a set of Buffon readings and commentary at a reasonable price for use by scholars and students of eighteenth century science and intellectual history generally, editorial commentary and critical appartus have been kept to a minimum. Readability and reliability have taken precedence over the demands of producing scholarly "critical" editions of these texts. For students interested in a deeper exploration of the Buffon literature, the logical beginning point is with the magnificent critical bibliography by E. Genet-Varcin and Jacques Roger appended to the Piveteau edition of the Oeuvres philosophiques de Buffon.

In selecting from a primary literature as large as that surrounding Buffon's work, decisions on selections have been made that may be controversial to some, for instance our failure to publish selections from the Abbé Joseph LeLarge de Lignac's polemical Lettres à un Amériquain sur l'histoire naturelle (Hamburgh, 1751). Considerations of space and the diffuse content of this five-volume work dictated its omission. One obvious lacuna in our work is the absence of reviews from the English-language periodical literature. To our surprise, we have been able to locate no substantive reviews in the British periodical press in any proximity to the original publication date, and no evidence for such reviews can be obtained from either the Genet-Varcin and Roger bibliography, nor the one serious study of Buffon's reception among the British, Stephen F. Milliken's unpublished doctoral dissertation, "Buffon and the British," (Columbia University, 1965). Our sole evidence of serious notice of Buffon's work in Great Britain in the 1750's has been the appearance of an anomymous translation of part of the Premier discours, without reference to authorship or source, in the Universal Magazine (volume 9 [July-December, 1751], pp. 49-57).

We considered, and then abandoned, the idea of including in the final section the Eloges by Condorcet, Vicq-d'Azyr and Georges Cuvier, in favor of the presentation of a complete text of Hérault de Séchelles' Voyage à Montbard, a text which has been surrounded by controversy, but one which we feel gives a plausible picture of Buffon, one consistent with our own reading of his writings.

Our decision has been to chose reviews of the initial three volumes of the Natural History that appeared together in 1749, rather than a selection spanning a wider span of time. The later reviews, while often substantive, lack the freshness of those which confronted Buffon's work for the first time. Those dealing with the highly important Époques de la nature of 1778, on the other hand, seem best located with an eventual critical edition of that work, which has never appeared in complete form in English.

Unknown to us until after preparations had been made to publish these texts, our title had been used previously in other languages by Wolf Lepenies, first in his <u>Das Ende der Naturgeschichte</u> (Munich: C. Hanser, 1976), chapter 8, and more recently in his article, "De l'histoire naturelle a l'histoire de la nature," <u>Dix-huit siècle</u> 11 (1979), 175-84. Although his intent in his use of this title and interpretation of the issues is quite different than our own, a common root in some of Kant's reflections on these questions is obviously involved.

We should also note that, though we have sought to achieve agreement on points of interpretation among ourselves, we have not succeeded in doing so. Consequently, rather than arbitrarily suppress one interpretation, some of our differences have been allowed to stand, in the hope that a fruitful dialogue might ensue.

Those who have attempted to translate material such as we have worked with here, that is, those who have tried to move the complex meanings of such texts across barriers of time, space, language and mental <u>gestalten</u>, know the difficulties of translation, and also come to a deeper realization of their own intellectual and spiritual limitations. It is chastened by this dual experience that we send forth this volume.

                                                  Phillip R. Sloan
                                                  John Lyon
                                                  Notre Dame, Indiana
                                                  January, 1981

Acknowledgments

The many difficulties in assembling documents for a volume such as this have placed us in the debt to many individuals. Particularly helpful among these was Professor Paul L. Farber of Oregon State University, who provided us with copies of the scarce J.S. Barr translation for inclusion, and gave helpful encouragement and suggestions on several points. We also acknowledge our debt to Professor John Neubauer of the University of Pittsburgh, John Greene of the University of Connecticut, and Ms. Margaret Humphries. The libraries of the University of Oklahoma, the University of Chicago, the University of Cincinnati, Oregon State University, Harvard University, and Yale University provided copies of most of the materials which we have translated here. Our thanks to the Journal of the History of Biology and D. Reidel Publishing Company, Dordrecht, Holland, for permission to reprint a revised version of the translation of Buffon's "Initial Discourse." With the exception of articles appearing in the English editions of the Natural History, the "Initial Discourse" is, to the best of our knowledge, the only item in this volume to have appeared in English. Sister Elaine Abels, R.C.S.J., and Dr. James B. McCormick of Replica Rara Ltd. provided technical assistance and the use of replicated instruments for historical microscopy, and supplied valuable references for exploring the Buffon-Needham experiments. Additional technical assistance on microscopy was also obtained from Fr. James McGrath of Notre Dame University. Our research was assisted by grants from the O'Brien Fund of the University of Notre Dame and the Lilly Foundation of Indianapolis. Professors Bernard Doering and Robert Nuner of Notre Dame, and Michael and Teresa Marcy of St. Mary's College provided helpful assistance on matters of translation. Any gaucheries of translation which remain rest solely on our shoulders.

The preparation of this unusually complicated manuscript was graciously and patiently facilitated by Cheryl Reed and Sandra DeWulf, with help from Janet Wright and Amy Kizer. We owe them a special debt of thanks. John Scanlon and Richard Carnell deserve particular thanks for the regular and conscientious performance of their work as our student assistants.

Finally, and most specially, a debt of appreciation is due to our respective families, who have in many ways encouraged this project. To them this book is dedicated.

1. Introduction
Phillip R. Sloan and John Lyon

> It is evident, that the knowledge of natural objects as they are at present, would still leave the desire for knowledge of them as they have been in former times, and of the series of changes they have undergone in order to attain their present condition in every locale. The history of nature, which we still almost wholly lack, would teach us the changes of the earth's form, and likewise those which the earth's creatures (plants and animals) have undergone through natural changes, and their alterations which have thence taken place away from the original form of the stem genus. This presumably would trace back a great many apparently different species to races of one and the same genus, and thus convert the presently greatly extended formal system of the description of nature into a physical system for the understanding.
>
> Immanuel Kant[1]

I

Natural history, as an inquiry concerned with the manifold of concrete "things of nature"--the rocks, animals, plants and fossils that man has found around him--is an inquiry that, in an organized form, dates at least from the works of Aristotle. In spite of its antiquity, however, there have been certain critical periods when, as a disciplined inquiry, it underwent significant transformation in content, in basic aims, and in the underlying concepts that have served to redirect its focus. Such a period of transformation began in the middle of the eighteenth century, and from this a set of issues and questions emerged that governed its inquiry for the next century.

From this period of conceptual change, natural history finally emerged as a coherent set of disciplines standing in their own right, and separated in methodology, approach, and even fundamental epistemology from the experimental sciences on one hand, and the abstract mathematical disciplines of mechanics and astronomy on the other.

At the end of the seventeenth century, an age which had seen the remarkable transformation of the physical sciences and mathematics, there had also been established a set of fundamental epistemological assumptions, a consensus on the guiding role of mathematics in the inquiry into nature, and some general conclusions about the relation of God to the world, which had defined a dominant view on the subject matter and domain of inquiry of natural history. This definition, clearly delimited first by Bacon, had made "natural history" a science concerned with the description of the works of nature, one providing the inductive foundation of true "natural philosophy," and having, except for its role in natural theology, little theoretical function of its own.[2] John Harris, in his influential <u>Lexicon Technicum</u>, well summarizes this conception of natural history at the opening of the eighteenth century:

> <u>Natural History</u> is a Description of any of the Natural Products of the Earth, Water or Air, such as Beasts, Birds, Fishes, Metals, Minerals, Fossils, together with such <u>Phaenomena</u> as at any time appear in the material world; such as Meteors &c.[3]

By the end of the eighteenth century, however, a remarkable change had taken place in natural history. Historical geology and cosmology, raised as options in the seventeenth century by Descartes, Steno, Burnet, Whiston and Woodward, but thwarted in their further development by factors we shall examine shortly, were by the end of the century in active development as a result of the works of Buffon, James Hutton, William Herschel, LaPlace, Lamarck, Kant, Lambert and Johann Blumenbach. Many of these same individuals, joined by Christoph Girtanner, Diderot, and Lacépède, were actively reflecting on the duration of organic species, and the possibility of their historical transformation. Kant's distinction in 1775 between the mere description of nature (<u>Naturbeschreibung</u>), and a genuine historical understanding of nature in its temporal development (<u>Naturgeschichte</u>),[4] was symptomatic of a fundamental shift in consciousness. Studies of plant and animal distribution patterns in relation to geographical change, historical cosmology and geology, the renewed interest in the descriptive study of the stages of embryological development, the conceptualization of comparative anatomy, studies on plant and animal hybridization, all begin to take shape as discrete disciplines, narrowing their focus,

Introduction

sharpening their inquiries, and probing in ever-increasingly technical ways into the variety of biological phenomena.

This transformation in natural history, marking the historical root of modern evolutionary biology, biogeography, ecology, physical anthropology, historical geology and cosmology, is in many respects as great an intellectual event as the scientific revolution of the seventeenth century. Apart from its narrower scientific importance, the new blend of Enlightenment philosophy, empirical inquiry, philosophic naturalism and materialism, and historical thinking, provided a great rational alternative to the physical sciences, with profound inplications for the philosophical directions of the nineteenth century. Quality, process, historicity, and concreteness are elevated in these sciences above mathematical abstraction, quantification, mechanism, and rigorous, deductive analysis. To the sciences of the "abstract" were now opposed the sciences of the "concrete," which by their concern with concrete, historical and vital process, were to bear increasingly on questions of ethics, politics and religion. The science of Newton, Clairaut, and La Grange seems remote from the thought of Schelling, Herder, Hegel, Feuerbach, Goethe, Comte, Marx and Darwin. That of Buffon, Diderot, D'Holbach, Kant, Blumenbach, Hutton, Kielmeyer and Haller does not, and this difference in tone, content, focus, and even underlying epistemology reveals a fundamental break of this group of sciences, and of philosophy more generally, away from the physicist's paradigm of seventeenth century science. The fact that many seminal thinkers in the natural history tradition--Buffon, Linnaeus, and later Cuvier--seemed to draw more direct inspiration from the works of Aristotle than from Newton is only a superficial indication of the change in underlying ontology and epistemology taking place. Natural history was no longer to be an inquiry dedicated to the collection of facts, but a science concerned most broadly with the "history of nature." The category of "nature" itself, which for seventeenth century science had functioned as an inert, divinely ordered system of bodies in mathematically describable motions,[5] had become a <u>vital</u>, almost teleological entity, historically changing, and endowed with self-actuating and self-realizing powers which were presumably sufficient to explain the origin of organic beings and even the apparent miraculous order that had led seventeenth century naturalists into paeans over intelligent design.

Once this "history of nature" was no longer confined to the historical development of mountain ranges, plants and animals, but included in its development "human nature" as well, the consequences for a larger set of critical issues in ethics and politics necessarily became the subject of profound reflection.

Rousseau, writing at the brink of this development, perceived well some of the possible implications when he wrote in the <u>Discours sur l'inégalité</u> of 1755:

It is in this slow succession of things that [man] will see the solution to an infinite number of problems of ethics and politics which the philosophers cannot resolve. He will sense that, the human race of one age [is not] the human race of another . . . . In a word, he will explain how the soul and human passions, altering imperceptibly, change their nature so to speak; why our needs and our pleasures change their objects in the long run; why, original man vanishing by degrees, society no longer offers to the eyes of the wise man anything except an assemblage of artifical men and factitious passions which are the work of all these new relations and have no true foundation in [original] nature.6

In presenting this collection of primary documents, we have done so with the intent of disclosing at least some of the critical stages in this reorientation of natural history as they center around what we see as the pivotal work, Buffon's monumental Historie naturelle, générale et particulière. By collecting together several inaccessible and little known texts, revealing both Buffon's early intellectual development and the initial reactions his works generated, we can see the interplay of epistemological, empirical and metaphysical questions that gave Buffon's work such obvious importance in the Enlightenment, and rendered it of possibly critical significance for the formation of the problems that were to lie at the center of the great intellectual struggles of the nineteenth century.

II

The main details of Buffon's life have been told sufficiently in recent years to require only a summary.7 Born in 1707 in the provincial capital of Dijon, France, to the family of a minor parliamentary official, Buffon, like so many of the great philosophes, received his early education under the Jesuits, in his case at the Collège des Godrans in Dijon. While a student there, Buffon demonstrated a keen interest in mathematics, and through these interests began in 1727 a long-lasting correspondence with the Swiss mathematician, Gabriel Cramer, an enigmatic figure in Enlightenment science, who served as an intermediary for an interacting group of significant scientist-philosophers, including, eventually, Pierre de-Maupertuis, Charles Bonnet, Jean Sénebrier, Louis Bourguet, Samuel Koenig, Jean Jallabert, John and Daniel Bernoulli, and

Introduction

Etienne Condillac. Significant for some of our later discussion is the fact that Cramer was also the editor of the works of Johann and Jakob Bernoulli, as well of as the philosophical and mathematical correspondence between Leibnitz and Johann Bernoulli. He also edited a Swiss edition of Christian Wolff's Elementa matheseos universae.

In 1728, Buffon moved from Dijon to the University of Angers, where evidence suggests he may have studied medicine and mathematics, and it is from this date that it can with certainty be determined that he first read Newton's Principia. The student years at Angers ended suddenly in 1730, when Buffon was forced to leave France as the result of a near-fatal duel with an English student. Travelling in the company of two Englishmen, the Duke of Kingston, and the Duke's tutor, an obscure physician, entomologist, and Fellow of the Royal Society of London by the name of Dr. Nathan Hickman, Buffon spent the period between November of 1730 and the spring of 1732 travelling in Switzerland, southern France, and Italy. Letters written in this period reveal that Buffon visited leading mathematicians, including Gabriel Cramer of Geneva and Fr. Guido Grandi of Pisa, and seems to have been in consultation with several other natural scientists at Montpellier and other centers of scientific learning on his route. If we can believe Condorcet, whose late éloge seems to be the only source of this claim,[8] it was on this journey that Buffon became actively interested in geological processes, and resolved to devote his life to the study of natural philosophy.

On his return to France in 1732, Buffon became immersed in Parisian scientific and social circles, and began the enduring practice of dividing his residence between Paris and his maternal ancestral estate at Montbard, which he had managed to wrest from his father as the outcome of an acrimonious legal suit created by his father's remarriage. This division of residence was a practice he would continue until his death in 1788, and served, at least in his early years, to associate Buffon both with Parisian and provincial scientific and intellectual circles.

In this period appeared Buffon's first scientific writing, a treatise on gambling probabilities, which was only published in its complete version in 1777, incorporated into his Essais d'arithmétique morale. It was this early work on probability theory that resulted in Buffon's somewhat surprising nomination and subsequent election to the prestigious Académie des Sciences in 1734 as an adjoint-méchanicien. The broader range of Buffon's scientific interests is revealed, however, by his translation in the following year of the important Vegetable Staticks of the English natural philosopher Stephen Hales, a work which attempted to follow the quantitative experimentalism of Newton's Opticks into plant physiology and pneumatic chemistry.

In 1740, Buffon advanced his reputation as a mathematician by publishing his translation of Newton's work on the calculus, the Fluxions. This had appeared for the first time in print in 1736, published by Buffon's correspondent, Peter Colson, and Buffon added to the work a historical preface discussing the priority dispute between Newton and Leibnitz over the discovery of the calculus, in which he sided generally with Newton.

It is in this same period that we also see Buffon making one of those great career changes which have proved so important in the history of biology. As a product more of political preferment than obvious scientific credentials, Buffon was appointed to succeed to the recently vacated directorship of the Jardin du Roi in 1739, and was elected to the position of associate Botanist at the Académie des Sciences, filling the place left open by Bernard de Jussieu, who was advanced to that of full pensionnaire.

Through this event, Buffon, although with his main background in mathematics and physics, embarked on a career which was to make him in a short time one of the foremost natural historians in all of Europe, rivalled in dominance over the field only by his contemporary Linnaeus.

By the early 1740's, Buffon had begun work on what was destined to become the great Histoire naturelle, générale et particulière, avec la description du cabinet du Roi. How the work was initially conceived is unclear. As late as April of 1744, Buffon had written to Gabriel Cramer that he was simply at work on an ". . . historical catalog of an immense cabinet of natural curiosities which I have put in order."[9] It is only clear with its publication how far beyond a simple "catalog" the work in fact extended. In 1745, Buffon was joined in his project by the physician and anatomist Louis-Jean-Marie Daubenton, and this collaboration was undoubtedly at least one key factor in the expansion of the focus of the work.

The scope of the original project, as announced in the original publisher's prospectus of the work,[10] was enormous and even, it would seem, naive. In a mere fifteen volumes, Buffon and Daubenton proposed to treat all of geology, zoology, botany and minerology, an extensive undertaking when compared to the schematic taxonomies of the three kingdoms that had to that date been published by Linnaeus in his Systema naturae, but nothing approaching the forty-five volumes that were eventually to appear under Buffon's direct or indirect supervision, never reaching botany.

The Histoire naturelle began its public career with the publication of three volumes in 1749 which clearly demonstrated Buffon's intent to write something more than a catalog. Introducing the first volume was a general programme statement that constituted a Discours de la méthode for Enlightenment natural history. Like its Cartesian predecessor, which was obviously not far from Buffon's mind when composing this document, the new method in natural history was also to be illustrated by its application to specific subject areas, in this case, geology, cosmology, biological theory, and anthropology.

Introduction

From available documents it is evident that Buffon had originally intended to treat all the warm-blooded quadrupeds in two volumes, due to appear, and apparently written in large part, in 1750. Undoubtedly reflecting the remarkable response and controversy surrounding the first three volumes, however, there was an evident strategic alteration of the original program, and the fourth volume, appearing in 1753, began a much more extensive species by species discussion of the quadrupeds which would occupy twelve volumes, and would only be completed in 1767. For each animal discussed, Buffon's intent was to write an eloquent description and discussion of the animal in its natural habitat, describing the details of its life history, food, and other "ecological" and economic aspects, and Daubenton would then add an article giving an anatomical description of the animal.

This fruitful and highly significant collaboration between Buffon, the mathematical physicist turned natural historian, and Daubenton, the physician turned comparative anatomist, was one of the most significant events in the history of Enlightenment biology, and had ramifications for the way zoology was practiced for the subsequent biological tradition.

For reasons that seem primarily to reduce to personal tensions created between Buffon and Daubenton, related in a large degree to Buffon's authorization of a reprint edition of the Histoire naturelle which omitted Daubenton's anatomical descriptions [Paris: Panckoucke, 1769-1770], this collaboration ended in 1767.[11] The subsequent nine-volume Histoire naturelle des oiseaux (1770-1783), originally intended to occupy only a single tome, was undertaken with a variety of collaborators, and there was never again the attempt to combine Buffon's descriptions with comparative anatomy. This was the only other group of organisms treated with any completeness in the published Histoire naturelle. The Histoire naturelle des quadrupèdes ovipares (2 vols., 1788-1789) the Histoire naturelle des poissons (5 vols., 1798-1803), and the Histoire naturelles des cétacés (1 vol., 1804), once intended to comprise two volumes of the original project, were only brought to completion after Buffon's death by his understudy, Bernard de Lacépède, apparently with the use of Buffon's notes.

Mineralogy and electrical and magnetic theory were only dealt with late in Buffon's life in the Histoire naturelle des mineraux (5 vols., 1783-1788), and much of the remainder of the original project, including the natural history of the invertebrates and microscopic animals, and the whole domain of botany, would never appear, except insofar as we can interpret the prolific "Natural Histories" of various organic groups by naturalists such as Lamarck, Olivier, Maundyt, Brugière, Daubenton, Bonaterre, Cuvier, Valenciennes and Duméril as attempts to complete this research program. The seven volumes of Suppléments à l'histoire naturelle (1774-1789) also served to fill out the project, and it is in this context that Buffon's synthetic and possibly most significant work, the Époques de la

Nature, appeared in 1778, finally bringing together in one work his main ideas on historical cosmology, geology and biology.

Buffon's personal vanity, accurately captured by Hérault de Séchelles, his prestigious position in the court of Louis XVI, and his personal sense of the great significance of his own work, coupled with his general contempt for the work of most of his contemporaries, served to place Buffon in a special kind of intellectual independence. He was neither a man to create students nor disciples in any ordinary sense, nor one who can be easily located in the main traditions of the French scientific establishment of the latter eighteenth century. His locus of power as the sole and autocratic director of the Jardin du Roi gave him a freedom from the more orthodox institution of scientific certification, the Académie des Sciences, and often placed him in opposition to Condorcet and D'Alembert, the leading lights of the Académie.

His consistent work habits, divided between the periods of intensive private work alone at Montbard, where all his creative scientific work seems to have been carried out, and periods of administrative work at Paris, also served to insulate him, more than one might expect, from ordinary networks of scientific interaction. Buffon's critical intellectual connections lie more in the direction of Louis Bourguet, Cramer, Montesquieu, the Bernoullis, Maupertuis, Du Châtelet and Samuel Koenig than they do with Condorcet, Condillac, Diderot, D'Alembert, or LaPlace. Even in biology, careful analysis has revealed the fundamental philosophical and conceptual divergence of Buffon's thought from that of Daubenton, Lamarck, Adanson and even Lacépedè.[12] In this closer affinity of Buffon's thought to Berlin and Geneva, rather than to Paris and London, we may perhaps finally see the solution to the enigmatic problem of the unity and coherence of Buffon's thought.

In 1771, Buffon was formally made a count by Louis XV, and by that date he was the most eminent natural historian in Europe outside his rival and contemporary Linnaeus, honored as "the Pliny and the Aristotle of France."[13] By that date the Histoire naturelle had been translated into all the main Western European languages, although the first English edition would only appear in 1775, a delay caused at least in part, it seems, by English sympathies with Linnaeus, to whom Buffon was openly hostile.

His last years were marred by the torment of the stone, vividly described by Hérault de Séchelles, and he died of kidney failure on April 16, 1788, only a year before the beginning of the cataclysmic events that would have carried such consequences for Buffon's life-style. He was thus spared the defamation of his name and monuments by sans-culottes, the paradoxical Jacobin preference for Linnaeus, and the guillotining of his only son in 1793.

It is reported that thousands lined the streets at his funeral procession, and perhaps no tribute would have been of

Introduction

greater pleasure to him than the epigram penned by the Marquis de Caraccioli, which closed his obituary notice in a German scientific periodical:

> Hic silet Naturae Lingua.[14]

III

Received tradition has painted an image of Buffon as a thinker whose scientific and philosophical thought was either inconsistent and incoherent, or else subject to radical change on fundamental issues. This tradition is well summarized by Isidore Geoffroy St. Hilaire's remark in 1859 that "Buffon, from one part to the other of the <u>Natural History</u>, completely changes opinion and language."[15] St. Hilaire thus proposed three periods in Buffon's intellectual development, each period supposedly characterized by a particular focus on the question of the permanence of biological species. This periodization is often implicit in subsequent writings on Buffon. The first time span, running from 1749 to 1756, was a period when Buffon, except for some inconsistent statements in the "Premier discours," was still hypothetically under the influence of a fixist view of species. A second period dating from 1761 to 1766, is allegedly characterized by Buffon's advocacy of a broad transformism; and a third stage, extending from 1765 to 1778, supposedly saw Buffon return to a reaffirmation of the fixity of species.

On a more penetrating and interesting level, Jacques Roger has more recently reaffirmed the claim of some kind of fundamental change in Buffon's thought, albeit without accepting such a tidy periodization:

> Buffon was particularly sensitive to the disorder that appeared to rule nature. . . . He found fault with classifiers, especially Linnaeus, for trying to imprison nature within an artificial system, since man cannot even hope to understand nature completely. Only in mathematics is there evident truth because that particular science is man-made. Physics deals only with the probable. . . .
> As time went on, Buffon's ideas changed. [Later] he seems to admit that man is actually capable of ascertaining fundamental laws of nature. . . .[16]

This assertion of fundamental intellectual change, presumably involving alterations in Buffon's epistemology, his concept of species, geological theory, and conception of historical process, nevertheless presents some curious anomalies. It is clear that there is some development and elaboration in Buffon's thinking in the course of his long scientific career, and it seems undeniable that the Buffon who composed the Dégénération des animaux and the Époques de la nature had moved considerably beyond certain conclusions advocated in the early 1750's.

However, the critical issue which we have not found adequately answered in the literature on Buffon, is the degree to which there can be discerned from the beginning of the Histoire naturelle a programmatic unity of theoretical principles and empirical practice that underlies Buffon's subsequent work, and renders the development of his thought a coherent and consistent articulation of this foundation.

It is our conclusion that a close reading of the following texts reveals such principles of intellectual unity. To support this interpretative claim requires, however, a discussion of the elusive roots and intellectual connections of Buffon's philosophical and scientific thought, a task rendered difficult by Buffon's persistent refusal to cite the works of contemporaries; his intentional destruction of his critical notes and early drafts; and his solitary work habits, which enabled his scientific and philosophical thought to retain a persistently idiosyncratic character that is difficult to unravel. Buffon's extant correspondence is also of only modest assistance. From it we can occasionally learn the titles of works Buffon has either purchased or loaned, but with the exception of mathematical discussions, almost nothing emerges of substantial intellectual content. We can, for example, determine from these sources that Buffon had read the Principia of Newton, and the works of major interpreters of Newton like Colin MacLaurin, John Keill, and Pieter van Musschenbroek. We also know that he had read the mathematical works of Fontenelle, Jakob and Johann Bernoulli, Abraham de Moivre, and Leibnitz. We also find evidence for a flurry of interest in chemistry in the late 1730's, and for an active interest in the major English deists--Wollaston, Toland and Tyndal--in the same period.

In their totality, however, the documentary sources do not supply a particularly revealing picture of Buffon's intellectual interests beyond a concern with mathematics, before the commencement of the Histoire naturelle. Buffon has often been characterized, perhaps too easily, as a French disciple of Newton and Locke.[17] Yet the name of Locke is never mentioned in any of Buffon's writings, as far as we can determine, and Buffon was a strong opponent of the great reformulator of Locke's epistemology for the Enlightenment, Etienne Condillac.[18] Buffon's Newtonianism is also, in some unusual ways, a rather unorthodox Newtonianism, and departs markedly from the Newtonian tradition on critical theoretical and epistemological issues, as we shall see shortly. In the revealing interview late in his life with Hérault de Séchelles, presented in the following texts, Buffon enumerated a somewhat unusual group of only four men whom (in addition to himself) he saw as worthy of serious study--Newton, Bacon, Leibniz and Montesquieu,[19] and a common intellectual theme in these men is difficult to locate.

The attempt to elucidate an underlying philosophical and methodological unity in Buffon's thought brings us always up against the enigmatic character of the early writings, and particularly the "Premier discours de la manière d'étudier et de traiter l'histoire naturelle" which prefaced the first volume of the <u>Histoire naturelle</u> in 1749. In presentation, this bears the character of a major program statement, but at the same time has led a long line of scholars to conclude that Buffon's starting point in natural history was grounded in radical nominalism and epistemological scepticism that apparently soon disappeared.[20] To assess this reading, it is first of all evident from these early writings that Buffon was deeply concerned not simply with empirical questions, but, even more profoundly, with the general problem of empirical knowledge, truth, and certitude, in the form these problems were beginning to take in the middle of the eighteenth century. Buffon's importance as a thinker whose impact extended beyond the narrow limits of eighteenth-century biology is undoubtedly tied to his willingness to confront the emerging "critical" problem as it bore on natural history, and it is our thesis that when viewed against the backdrop of the central philosophical problems in Enlightenment thought, a consistent unity in Buffon's approach is seen to emerge that is discernible clearly in this initial program statement.

The natural philosophy of Buffon's day had developed, particularly in France, out of a confrontation of two great alternatives. One, the comprehensive mechanism of Descartes, had reduced fundamental ontology to matter and a conserved quantity of motion, and, of course, mind. Opposing this was Newton's system of the world, admitting a concept of force that could act at distances and provide accelerations on matter. With this system, Newton had both been able to "save the phenomena," and also provide compelling mathematical rigor to his physics that was to give it a position of dominance in natural science by the mid eighteenth century. However, the root problem that necessarily accompanied any abandonment of Cartesian science and epistemology was that Descartes had, through his confrontation with pyrrhonist scepticism, inextricably linked the possibility of epistemological intelligibility with the acceptance of a strictly mechanical view of Nature. The brilliant Dutch physicist, Christiaan Huygens, expressed this point well in 1690:

> In the true Philosophy. . ., One conceives the cause of all natural effects in terms of mechanical motions. This, in my opinion, we must necessarily do, or else

> renounce all hope of every comprehending anything in Physics.[21]

By their break with the "mechanical" philosophy, and the admission of active forces and action at a distance, supported often by phenomenalist arguments, the early supporters of Newtonianism were, however, drawn into a close alliance with the mitigated scepticism of Gassendi and John Locke.[22] This wedding of empiricist epistemology and phenomenalist metaphysics, expressed for much of the eighteenth century by the Dutch Newtonians Willem s'Gravesande and Pieter van Musschenbroek, provided the needed philosophical justification for the Newtonian position in its struggle with the essentialism of Cartesianism.[23]

The result was an eclectic philosophical position which, at least if not pushed to the consequences to which Hume would ultimately force it,[24] had on one hand provided strong philosophical warrant for the rising _experimental_ sciences of the Enlightenment, by placing the foundation of knowledge on the gathering of simple ideas and their generalization into complex and abstract ideas.[25] By the same token, it rendered knowledge successively uncertain and probabilistic as one moved from the immediacy of sensation to more general claims about the natural world. When it came to making statements about origins, historical processes, and other dimensions of nature, one had, on these grounds, the most fragile of epistemological justification. While not directing his concerns specifically at this problem, the implications of an epistemology like Locke's can be discerned in the following passage:

> There remains that other sort [of probability], concerning which men entertain opinions with variety of assent, though _the things be such that falling not under the reach of our senses, they are not capable of testimony_. Such are, 1. the existence, nature and operations of finite material beings without us. . . . Or the existence of material beings which, either for their smallness in themselves or remoteness from us, our senses cannot take notice of--as, whether there be any plants, animals and intelligent inhabitants in the planets, and other mansions of the vast universe. 2. Concerning the manner of operation in most parts of the works of nature: wherein, though we see the sensible effects, yet their causes are unknown. . . . These and the like effects we see and know: but the causes that operate, and the manner they are produced in,

> we can only guess and probably conjecture. . . . <u>Analogy</u> in these matters is the only help we have, and it is from that alone we draw all our grounds of probability.²⁶

The mitigated scepticism imbedded in this epistemology has a significant consequence for the possibility of "historical" accounts of nature. While a host of figures--Thomas Burnet, Nicholas Steno, John Woodward, and William Whiston most prominently--had indeed attempted to elaborate historical cosmologies and geologies in the wake of Descartes' theories of 1644, the inherent tensions between the claims of such theories and the epistemological strictures placed on the possibility of historical knowledge by Locke and adopted by many Newtonians, served, it seems, to block the further elaboration of such speculations for much of the early Enlightenment. Newtonians, taking a clue from Newton's own strictures against "World Building" in the later editions of the <u>Opticks</u>, and anxious to distance themselves from the overweening rationalism and imperialistic epistemology of Cartesian mechanism, were in the forefront in attacking historical cosmology.²⁷ As the Newtonian John Keill argues in his attack against both Thomas Burnet and William Whiston in 1699:

> If [Burnet] had taken a right method and had made a considerable progress in those Sciences that are Introductory to the study of nature, I doubt not but that he would have made a very acute Philosopher.
> It was his unhappiness to begin at first with the <u>Cartesian</u> Philosophy, and not having a <u>sufficient</u> stock of Geometrical and Mechanical principles to examine it rightly. . . , in imitation of Mon. <u>Descartes</u> he would undertake to show how the World was made, a task too great even for a mathematician.²⁸

Furthermore, Newton's conclusion that the maintenance of the world order required the continued intervention and supervision of God implied, at least for some influential expositors of Newtonianism, a voluntarism that subverted the concept of natural necessity itself. The concept of fixed natural law, sufficient to supply the needed ontological grounding for strong inferences from the present to processes taking place at a previous time in the history of the earth or the solar system, are seen by the Dutch Newtonians, for example, as resting largely on God's arbitrary decree, and are accessible only through empirical induction with all of the uncertainties attendant on this. The Dutch Newtonian, Pieter van

Musschenbroek, states this clearly in his exposition of physics in 1734:

> All bodies are observed to move according to stated laws or rules, whatever may be the cause of their motions. By the name of <u>Laws</u> we call those constant appearances, which are always the same, whenever bodies are placed in like circumstances. . . .
> These laws are discoverable only by the use of our senses; for the wisest of mortals could not have discovered any of them by reason and meditation, nor can pretend to have any innate ideas of them in his mind. For they all result from the arbitrary appointment of the Creator, by which he has ordered, that the same constant motions shall always obtain in the same occasions. . . .[29]

With scientific law itself only discoverable through bare sensory induction, only the weakest support obtained for any claims to use a concept of natural law to reason to prior states and conditions of the earth unlike those immediately accessible in the present.

IV

The wedding of Newtonian natural philosophy, empiricist epistemology, and mitigated skepticism, with a radical theological voluntarism, was a common feature of the public expressions of the philosophy of science in the early Enlightenment, particularly on the Continent. It carried with it far-reaching consequences for the further development of historical accounts of the natural world, whether these were in cosmology, geology or biology.

<u>Experimental</u> and <u>observational</u> science, confined to the present time, presented no particular problems. No theoretical difficulty was placed in the way of a Baconian conception of "natural history" for the same reasons, since on this view natural history was solely concerned with describing the individual facts of nature, forming their systematic classification, and from this generalizing empirical laws. Microscopy, mineralogy, anatomy, taxonomy, experimental physics, chemistry, and geography could develop in many respects untroubled by the resultant positivism. However, by the same token, those sciences which attempted to reach beyond the sensible and synchronic were charged with "hypothesis" making, and with being victims of the Cartesian errors presumably put to rest by

Introduction

the more recent development of science since Newton. By a paradoxical dialectic, the seeming triumph of reason and speculative thought over sensory testimony and common experience that had marked the key development of rational mechanics from Galileo to Newton, was being undercut in the name of the new science itself. This scientific "pyrrhonism," whose full implications were only to be seen clearly by Hume, is well expressed by the Abbé Noël Pluche in the late 1730's:

> The universal incapacity men are in of going further than what is sensible and useful, naturally informs them of the limits within which they ought to confine themselves. In what escapes their senses it is, that the secret of the structure, and the mystery of the operation, lies hid. Their reason may and ought to exert itself on the effects and intensions which God shews us; but never on what he conceals . . . .He has not taught us what the nature of heaven and earth, of metals and fluids was, as he freed us from the care of producing them.[30]

Consequently, while it was fully possible to speculate on origins and developmental histories of natural phenomena in the style of Descartes and Burnet, the more difficult problem was that on the grounds of the reigning epistemology it was not at all clear that justification was available that could render these speculations, when challenged, anything more than vain chimeras.

A second level of difficulties specifically confronted any attempt to integrate biological phenomena into a comprehensive historical development of the world. These difficulties remained in force in the middle of the eighteenth century, even if some would claim to have overcome the epistemological problems we have outlined, and concerned the range of issues associated with the scientific account of organic origins.

The root problem at issue was that the architects of the mechanical philosophy, as well as their Newtonian successors, had simply failed, and failed miserably, to give a plausible mechanico-physical account, consistent with their accepted scientific ontologies, of the temporal origin of living beings that would accord with any of the proposed genetic theories of the development of the world. Descartes, in pointing the direction of his own desired solution to the question in 1644, indicated the kind of account which he considered paradigmatic, but which neither he, nor anyone else in the seventeenth and early eighteenth century, were able to supply satisfactorily:

> To comprehend the nature of plants and man, it is far preferable to consider the

> means by which they have arisen, little by little, from their seeds, rather than the means by which they have been created by God from the first origin of the world.³¹

Through a complex series of empirical, philosophical and theological developments, the only solution that eventually seemed admissible in terms of the available empirical evidence, and at the same time compatible with the basic principles of mechanistic philosophy, was the famous preexistence theory, which either in its panspermist or "encasement" versions, placed the true origin of organisms at the first foundation of the world. The Scotch anatomist George Garden expresses this basic thesis well in a review article written near the close of the seventeenth century:

> And Indeed, all the laws of Motion which are as yet discovered, can give but a very lame Account of the Forming of a <u>Plant</u> or <u>Animal</u>. We see how wretchedly Descartes came off, when he began to apply them to this Subject. They are form'd by Laws yet unknown to Mankind; and it seems most probable, the <u>Stamina</u> of all <u>Plants</u> and <u>Animals</u> that have been, or ever shall be in the World, have been form'd <u>ab Origine Mundi</u>, by the Almighty Creator, within the first of each respective Kind.³²

The elaborations, revisions and controversies that accompanied this doctrine in the late seventeenth and early eighteenth century have been well-described elsewhere, and need not detain us at this point.³³ The critical general issue that it presented was a reaffirmation of the basic incomprehensibility of underlying causes of natural phenomena, the assertion of miracle at the foundation of nature, and most importantly, a necessary removal of organisms from any essential participation in historical process. Thus, in a curious way, there was a congruence between the epistemological modesty of Newton and Locke, the reaffirmation of divine intervention in nature that Newton had made an integral part of his astronomical theory, the attack of Newtonians on "World-building," and the theory of preexistence of the embryo from the beginning of time. Pluche again summarizes this synthesis well, praising Newton on one hand for supporting Mosaic cosmology by his affirmation of divine action in nature, and on the other arguing that the origin of organisms precludes all plausible accounts by secondary causation, in the name of the same divine interventionism. Speaking as the Lord addressing the philosophers he writes:

> Now compare my work with yours, and see if it is possible to separate the formation of the minutest organ in the universe from the wisdom and express command of the Everlasting. I work differently from you. I have put in the ovarium of a mother the small egg which contains the [preformed] worm, whose formation you have missed. . . .I have known through the series of all ages, on what day and at what moment [the gnat] would break through all his tunicles, and become of the number of the living creatures. . . .Ye all of you think my majesty disgraced by these productions, and you chuse to ascribe it to some cause which you term a second cause. . . .[This] is transferring to a parcel of mud, or to a blind motion, a power and a glory which I have not granted to man. . . .No motion, no creature whatever, can form the skeleton and vessels which organize an animal. Much less can they give him life. This is the character of my handy work.[34]

This denial of genuine secondary causal relation in the apparently historical sequence of ancestor and descendant, and the "accidental" status this ultimately implied for the particular vicissitudes of time and environmental circumstance that might bear on any organism, rendered the consequences of a creationist metaphysic more radical than they had been for the western tradition prior to the seventeenth century. The fixity of species, for example, was not, in such a framework, simply due to a structuring of ontological reality in terms of an ordering of essences, but was grounded on God's direct creation of all organisms at one moment in time.

The development of new options on the question of generation in the Enlightenment required not simply new empirical observations, although certain of these were to play a critical role in the final outcome. More essential was the formulation of an ontology that could render tenable a genetic account of the origin of organic beings by secondary, natural causes. The stakes in this issue were very high, at least as many Enlightenment thinkers were to perceive the problem. At issue was the possibility of a naturalistic cosmology, extending from the more remote problem of the origin of the solar system, to the more immediately significant problem of the origin of life, and by extension, of man himself. If Darwin's work would bring these issues to full attention and significance, his reflections are but the culmination of concerns that arose in the middle of the eighteenth century.

The great transformation of "natural history" into a science concerned with grasping the "history of nature," was a conceptual development that required the resolution of the two proceeding issues we have outlined. On the one hand, it was necessary that foundations for empirical knowledge claims be worked out which could presumably circumvent the historical skepticism of the early eighteenth century, and provide some epistemological justification for claims about historical development. On the other hand, biology needed to be integrated into the natural order of the world in such a way that living phenomena could presumably be considered products of natural forces and activities, with the further consequence that they were, by this immanentizing, also subjected to the particularities of historical change and unique circumstance. Not accidentally, therefore, the issue of the generation of organic beings and that of the transformation of species became closely linked together for the biological tradition in the century after 1750. If the embryological development of organisms could be accounted for purely by inherent natural forces, then presumably the identity and historical permanence of species depended on similar forces, whose conservative characteristics possibly were themselves of only limited endurance.35

V

In determining Buffon's significance with reference to the complex of issues we have outlined, there are certain traditional interpretive options which, we conclude, must be revised. The first of these is the view of Buffon as making some kind of "Newtonian" revolution in natural history, a suggestion which is often made with good cause, in view of Buffon's background and his association with the Newtonian party at the Académie des Sciences in his early career.36 It is our suggestion, to the contrary, that it is precisely in his break with the underlying methodology and epistemology of Newtonian science, as it was understood in his period, that we achieve a key to the underlying philosophical and methodological unity of the Histoire naturelle. The second is the claim that Buffon's thought undergoes a radical change of position. If we are to localize a period of Buffon's basic intellectual transformation, it is to be placed between 1739 and 1745, in other words, prior to the beginning of the work on natural history. By the latter date, we suggest that Buffon had arrived at a coherent intellectual position and a scientific programme which was then elaborated and developed in the Histoire naturelle, but never significantly altered in its root problematic.

From the evidence that can be drawn from Buffon's earliest available writings, represented in the following selections by his preface to his translation of Stephen Hales' Vegetable Staticks, we see advocated an intellectual position congruent with the "positivistic" Newtonianism advocated by John Keill, Musschenbroek, and s'Gravesande, and even a veiled allusion to the sage pyrrhonisme advocated only two years previous to its publication by Bernard de Fontenelle as the proper epistemology of modern natural science.[37]

However, if we compare Buffon's statements in this selection to those then encountered in the "Premiere discours de la manière d'étudier et de traiter l'histoire naturelle," a work probably completed as early as 1744,[38] we see a subtle but profound shift in Buffon's basic philosophy of science. In the place of the advocacy of radical empiricism and the emphasis on the bare particularity of nature encountered in the Hales preface, Buffon has added a pronounced rationalistic underpinning. Observation and experiment, while being given due praise, are no longer the totality of valid scientific inquiry, but are to be subordinated to more comprehensive and unifying principles:

> . . . It is not necessary to imagine . . . that, in the study of natural history, one ought to limit oneself solely to the making of exact descriptions and the ascertaining of particular facts. This is, in truth, . . . the essential end which ought to be proposed at the outset. But we must try to raise ourselves to something greater and still more worthy of our efforts, namely: the combination of observations, the generalization of facts, linking them together by the power of analogies, and the effort to arrive at a high degree of knowledge. From this level we can judge that particular effects depend upon more general ones; we can compare nature with herself in her vast operations. . . .[39]

This new epistemological optimism about the possibility of truth and certitude in natural philosophy is no longer confined simply to the present, but presumably now gives even a warrant for making claims about the probable course of events involved in the formation of the earth and the solar system. General truths and the true system of the relations of causes and effects are now presumed accessible to man, enabling him to understand the system of nature, and more importantly, its historical formation, without reliance on scripture, a divine epistemological guarantee, or a naive intuitionism.

The problem, of course, is to discern how Buffon presumed he had resolved the "critical" problem he, and other perceptive Enlightenment thinkers realized was at stake. And it must be admitted that what we have identified to be his solution to these questions was never systematically developed, but only carried out in fragmentary reflections that must be pieced together to see a more coherent whole. Nevertheless, Buffon's claimed resolution of this issue, we suggest, is fundamental to understanding his thought. It is to be discerned in two central themes that can be followed through Buffon's mature thought: (1) his polemic against "abstract" knowledge and concepts, and (2) his conception of "physical" truth. These two recurrent themes provide both a reconciliation of Buffon's seemingly contradictory statements on a wide variety of issues, and also disclose the roots of a methodological and empirical research programme in natural history that inexorably leads him to the articulation of a truly <u>historical</u> conception of natural history.

To understand the significance of this requires that Buffon be understood not so much as a French disciple of Newton and Locke, but in many, if certainly not all, respects an adherent to the central theses of Leibnizianism, as they were accessible to him by the early 1740's.[40]

In 1736, Newton's unpublished work on the calculus, the <u>Fluxions</u>, had finally appeared in print, and served to reopen the long-standing dispute between the followers of Newton and of Leibniz over the priority of the discovery of the fundamentals of the infinitesimal calculus. In 1737, Buffon translated the work into French, adding to it in 1738 a long preface surveying the history of the controversy over the calculus.[41] In this preface, reprinted in part in the following selections, Buffon had argued strongly in favor of Newton's priority, but in so doing had become more directly involved in the Leibnizian literature, and also, it seems, in the more general issues in the Leibniz-Newton dispute, brought to a head in the Leibniz-Clarke controversy in 1717. Subsequent to 1738, one can detect references, locutions, and concepts in Buffon's writings that are distinctively Leibnizian. In this same period Buffon was in contact with an interrelated group of Enlightenment scientist-philosophers, including Pierre de Maupertuis, Samuel Koenig, Jean Jallabert, and Gabrielle du Châtelet, all of whom served as the primary vehicles for the introduction and spread of Leibnizianism-Wolffianism into France in the early 1740's. By 1740 Buffon had read and warmly praised du Châtelet's eclectic synthesis of Wolffian metaphysics and Newtonian mechanics, the <u>Institutions de physique</u>.[42]

Of direct interest is the light this "Leibnizian" connection sheds on Buffon's methodological and philosophical programme, and in particular on his distinction of "abstract" and "physical" concepts. The foundation for this distinction can be traced to the arguments Leibniz had formulated against the Newtonian concepts of absolute space and time as they were expressed publicly in the Leibniz-Clarke correspondence. Madame du Châtelet's summary of this position, essentially restating the arguments of Leibniz as they had subsequently been developed by Christian Wolff, is of particular relevance, inasmuch as it had been read by Buffon near the point at which he first articulated his plan for a great "history of nature."

Du Châtelet's argument, which is best read as a only paraphrase of arguments given by Christian Wolff in his Ontologia of 1729,[43] was that the Newtonian concepts of absolute space and time were only to be considered abstract ideas, meaning this in a Lockean sense, which accounted for the origin of the ideas of time, space and duration from an abstraction from empirical experience.[44] To this "Lockean" reading of Newton's arguments, Du Châtelet then posed in contrast the Leibniz-Wolff thesis of immanent and concrete time and space which gave, on her account, an ontological foundation for time independent of the mind. In this case, time and space were purely relations, but also real relations between concrete bodies. "Real" time resided in the material succession of bodies, and "real" space in their coexistence:

> The imaginary concepts [of absolute time and space], which infinitely assist one in the investigations of the truths which depend on their determination [and] constitute these entities formed by the imagination, become very dangerous, when they are taken for realities . . . .
>
> Those, therefore, who have wished to apply to actual space the demonstrations which they have deduced from imaginary space, cannot fail to be entangled in the labyrinths of error.[45]

Real time and space, by contrast, cannot exist "apart" from the bodies which comprise them:

> Time is therefore in actuality nothing but the order of successive beings, and one forms the idea of it insofar as one considers only the order of their succession. Thus there is no Time without beings truly and successively arranged in a continuous series.[46]

This Leibnizian argument, as it was refracted through Wolff and Du Châtelet, in effect made time and space immanent in the world, coexistent with its duration. "Abstract" time and space, by the same token, stood in an inferior ontological position to the concrete and "physical" meaning of time and space that was seen as imbedded in the material existence of things.

The Leibnizian tenet in this, we suggest, simply becomes generalized in Buffon's writings after 1740, and is applied to a whole array of scientific and methodological questions. As we read in the "Initial discourse," it is the ability to separate the "abstract" from the "concrete" dimensions of any scientific subject matter that is "the foundation of the true method of leading one's mind in the way of the sciences."[47]

The radical consequences of this methodological thesis are apparent in several ways in Buffon's subsequent scientific inquiry. In what would seem to be a surprisingly paradoxical position for a former Newtonian to take, for example, Buffon thus proceeds to exclude mathematics from any important role in natural history. Mathematics is, Buffon now claims, a purely "abstract" science, concerned only with the relations of ideas, whereas as Buffon now conceived it, natural history was to be a science dealing with the concrete and specific aspects of the natural world.[48]

As a further example, the same principle is at work in Buffon's somewhat curious argument on the status of Newton's inverse square law of universal attraction. This was to involve him in a significant theoretical debate in the late 1740's. In this controversy with Clairaut, Buffon focuses on the "abstract" character of Clairaut's attempt to fit Newton's law to the perturbations of the lunar orbit by the addition of a mathematical power series. Buffon's argument was that such mathematical manipulation was only a manipulation of arbitrary terms. To this he opposed the "concrete" simplicity of the law, and the need to seek real and physical forces and specifying initial conditions as the preferable solution to the moon anomaly.[49]

In the domain of biology and natural history, we can also observe the resolution and clarification of Buffon's seemingly paradoxical and contradictory statements on a variety of biological topics that have previously seemed explicable only by the assumption of radical change or inconsistencies in his thought. His attack on Linnean taxonomy in the "Premier discours," for example, has been read by his contemporaries and successors as a primary example of the epistemological skepticism and the advocacy of Lockean nominalism Buffon is alleged to have held initially.[50] Read in the light of the interpretive hypothesis we are advocating, however, it can be seen that his argument is more correctly articulating the claim that Linnean taxonomy is to be rejected because it is purely an <u>abstract</u> ordering of organisms that makes no attempt to connect its distinctions and categories with the "real and physical" relations of organisms as they concretely exist.

The opposition of the "abstract" and "concrete" orders is further displayed in Buffon's location of man in the order of nature. As he argues in the "Initial Discourse," man stands as the central starting point, the node from which the exploration of nature and its concrete natural history is to begin.[51]

Recalling Leibniz' thesis that all aspects of the universe are mirrored in each monad, the concrete beginning point in man advocated by Buffon does not commit him to a radical subjectivity since underneath his assertion seems to stand the belief that there is a set of <u>real</u> relations and connections of every being that are accesible to reason. As he writes in 1749:

> The animal unites all the powers of Nature; the forces which animate it are peculiar to it. It wills, acts, it determines itself, it operates and communicates by its senses with the most distant objects. Its self is a center where everything is connected, a point where the entire universe is mirrored, a world in miniature. . . .[52]

A significant consequence of this conception of a "physical" and "concrete" approach to natural history emerges in Buffon's general conception of an organic species and its relation to history. Just as he has separated other concepts into an "abstract" and a "real and physical" significance, the same applies to the concept of the organic species. Considered as a traditional universal in thought, denoting and grounded upon a class of similar individuals, a "species" is only an "abstract" entity of reason, whose "reality" is made problematic by all the potential problems entailed by empiricist epistemologies of the Enlightenment. But understood as a <u>physical</u> and <u>historical</u> connection of organisms united by the material bond of generation, Buffon's "species" are no longer abstract universals. This point is made clearly in the opening of his discussion of reproduction:

> We shall now examine more closely this property common to the animal and plant, this power of producing its likeness, this chain of successive existences of individuals which constitutes the real existence of the species.[53]

And as he elaborates on this point in his most extended and influential discussion of the species concept in his article on "The Ass" published in 1753:

> An individual is a being set apart, isolated, detached, and has nothing in common with other beings, except that he resembles or differs from them. All the similar individuals which exist on the surface of the earth are regarded as composing the species of these individuals. However, it is neither the number nor the collection of similar individuals which constitute the species, it is the constant succession and uninterrupted renewal of these individuals

> which forms it. For a creature which would
> last forever would not comprise a species,
> no more than would a million similar
> creatures which would last forever. The
> species is thus an abstract and general
> word, for which the thing exists only in
> considering Nature in the succession of
> time and in the constant destruction and
> just as constant renewal of creatures. It
> is in comparing Nature today to that of
> another time, and existing individuals to
> those past that we have drawn a clear idea
> of what is called a species, for the comparison of the number of the resemblances
> of these individuals is only an accessory
> idea, and often independent of the first.[54]

A further application of this principle is to be observed as operative in critical ways in Buffon's theory of generation, in which he attempted to reinstate an epigenetic account of development against the reigning preexistence theory. As we observe in the selection from his article De la reproduction en générale of 1749, Buffon relies on Newtonian microforces for his form-giving, organizing principles. But as we can observe when reading this in company with his Reflections sur la loi de l'attraction, such Newtonian forces are not to be regarded as "abstract" mathematical forces. Instead, they must be concrete, immanent causes able to serve as the structural foundation for the specificity and diversity of the different species, more Aristotelian substantial forms than simple analogues of attraction.

On one key issue, however, an important divergence from a Leibnizian position is illuminating of Buffon's intellectual position. For the main scientific traditions before the 1740's, in spite of the diversities of the Cartesian, Leibnizian and Newtonian traditions on numerous issues, a critical linkage still held between natural philosophy, theology and epistemology. For the Cartesian tradition, God's existence was at some point a necessary requirement if clear and distinct ideas were to be tied to a mechanically understood natural order.[55] For Leibnizians, God's existence and rationality underlay the all-important principle of sufficient reason, which gave an ontological foundation and a teleological necessity to the laws, forces and apparent contingencies in natural philosophy. For Newtonians, although God played a much more restricted epistemological role, His existence was needed as the ground for the concepts of absolute space and time, and as the periodic intervener in the repair of the world order.

The philosophes of the Enlightenment were increasingly to conclude that such theological premises were unnecessary.

Introduction

Finally Pierre Simon Laplace would argue in his famous conversation with Napoleon Bonaparte in 1802 that God had indeed become a superfluous entity in a scientific cosmos.[56]

On one level, this was a non-problematic assertion: the development of rational mechanics in the eighteenth century had increasingly eliminated the need for the periodic repair of the world-order by Newton's God. But on another level the problem was not so simple. At the same time Laplace was confidently making such assertions, a revolution in philosophy and science was taking place in the Germanies, which at its core was concerned with the possibility of finding in anthropology and the structures of consciousness some new foundation for those very ontological principles that the banished theological principle of order had supplied for science to that point. As the figure whose thought more than any other served as the watershed between the Enlightenment and the new movements of the nineteenth century, Immanuel Kant clearly saw that the easy solutions supplied by Enlightenment critics of religion also carried with them nagging epistemological problems over causality, natural necessity, and the ontological status of natural laws themselves that seemed to require for their solution a complete reversal in the traditional relationship between knower and known.[57] This problem of the necessity and status of natural laws was one that leads directly in to a third aspect of Buffon's thought.

Once the difficulty was perceived, solution to the problem was pressing if some justifiable distinction was to be made between merely empirical regularities, such as Galilean laws of accelerated motion, and the dynamic and explanatory natural laws which accounted for why such regularities held in a non-accidental way. In this context, probability theory provided for at least some natural philosophers a promise of a way out of the difficulty, and it is with reference to this issue that we can understand at least a part of Buffon's persistent concern with probability theory even within the context of discussions of natural history and geology.

On several points, Buffon's concerns with probability bear close similarities to arguments formulated by Jakob Bernoulli (1654-1705) in his important treatise <u>Ars conjectandi</u> of 1713. In the fourth part of this work entitled "The Use and Application of the Preceeding Doctrine [of Probability] in Politics, Ethics and Economics," Bernoulli had made a fundamental distinction between the real and determined order of events, which was <u>not</u> inherently probabilistic, and the assessment of human epistemological certitude concerning man's knowledge of that order.

On the first issue, Bernoulli had given a Leibnizian answer. God determines, through the principle of sufficient reason, one particular course of events as inherently necessary:

> Everything that exists or occurs in
> the past, the present and the future, has
> the highest certainty in itself. With
> regard to things in the present and past,
> this proposition is self-evident, since for
> everything that exists or has existed, the
> possibility that it does not exist or has
> not existed, is excluded. Also, with
> regard to the things in the future, it is
> not likewise to be doubted that they will
> occur, if not with the inevitable necessity
> of Fatalism, then indeed on the basis of
> divine foreknowledge and predestination.[58]

Those in the Leibnizian tradition commonly termed this theologically-grounded necessity beneath apparently contingent events "physical" necessity.[59] However, the problem recognized by Bernoulli and others concerned with the probabilistic calculus of decisional questions, was that of deciding on non-theological grounds the degree to which human knowledge could be concluded to conform to this metaphysically-necessary order.[60]

For orthodox Leibnizians, the problem had largely been resolved at the level of intuition. Intelligible truths, intuited through the principles of non-contradiction and sufficient reason, presumably gave access to this natural order. Bernoulli, and others interested in the early Enlightenment in the problem, sought to resolve the problem inductively through a calculus of probability. Through such a calculus, the degree of subjective certitude could be quantitatively assessed, with this running from complete certainty, to lesser degrees of certitude, such as "moral" certitude, that contained the possibility of error.

Discussion of decisional probability problems are an important feature in Buffon's early writings, and formed an important point of discussion between Buffon and Gabriel Cramer, the editor of Jakob Bernoulli's collected works, and an acquaintance of Buffon's after 1730. Buffon's main arguments on this point, as they can be determined from such works as the "Initial Discourse" to the <u>Natural History</u> and the <u>Essay on Moral Arithmetic</u>, both given in translation in the following texts, involved an interesting turning of the epistemological tables on a purely empiricist solution to the problem, while at the same time attempting to do without theological appeal.

Briefly put, this entailed linking together probability calculus and the concept of physical truth we have previously discussed. If "physical" truth rests in the repeated succession of events in time, this also, for Buffon, gives a possible means of solving the epistemological problem by probability calculus. An example of how this is conceived can be seen in his use of the example made famous by David Hume's employment of it in 1748--the rising of the sun. In section six of his Essay on Moral Arithmetic, Buffon uses this example, but draws from it exactly the opposite conclusions from Hume. Because this is a recurrent event, it satisfies the key condition for "physical" truth, and from the formula 2^{n-1}, where n is the number of recurrences, Buffon argues that our certitude increases to the point that the future non-recurrence of the event is almost inconceivable.

That solutions can be found to the problem of induction through such approaches is still a debated issue. At least for Buffon it seemed sufficient to skirt the fundamental questions his assumptions were leading him toward. In the eclectic blend of Leibnizianism, probability theory, Newtonian natural philosophy, and empirical scientific content, his scientific program seems to have attained a unified character, as difficult as this is to see from its fragmentary exposition. His repeated concern with the "physical" as opposed to the "abstract" levels of inquiry stands as an a priori in his methodology, reorienting his scientific inquiry into the exploration of the relations and connections of organic beings in their concrete spatial and temporal relationships. And it is this turn to the concrete which more than any other feature sets Buffon's program of research in natural history off from that of Linnaeus. In it we can see the transition from "natural history" as it had existed in the preceding tradition, to a full "history of nature," which would emerge in its influential form in his significant Époques de la nature of 1778, integrating historical cosmology, geology and biology into a single comprehensive scheme.

As the individual who would more rigorously systematize this in his distinction between Naturbeschreibung and Naturgeschichte, Immanuel Kant saw both the suggestiveness and also some of the problems in Buffon's claims:

> The only work in which the History of Nature is properly handled is Buffon's Époques de la Nature. However, Buffon gave free rein to his imagination and therefore has written more of a romance of nature than a true history of nature.[61]

But like Voltaire's judgment on that other great author of "romances" about nature, Descartes, such restrospective assessments are only made possible by the basic fertility of the initial insights of such pioneers, enabling a refinement and clarification beyond their original formulations. The degree of initial importance, and the potential consequences which Buffon's ideas had for his contemporaries can be determined from the accompanying reviews. For many, Buffon is an enigma. For others, he has vindicated a creative scientific method. For all his century he could not be ignored.

NOTES

[1] Immanuel Kant, "Von der Verschiedenheit der Racen uberhaupt," *Vorkritische Schriften: Kants Werke*, Vol. II (Berlin Akademie ed; Berlin: Reimer, 1912), p. 434. First published 1777.

[2] Francis Bacon, *Magna instauratio* (1620), "Aphorisms on the Composition of the Primary History," Aph. 2 in: F. Bacon, *Essays, Advancement of Learning, New Atlantis and other Pieces*, ed. R.F. Jones (New York: Odyssey Press, 1937), p. 352. See also *De dignitate et augmentis scientarum* in ibid., p. 384.

[3] John Harris, *Lexicon technicum* (London, 1710) II (no page number).

[4] I. Kant, *loc. cit.*

[5] J.E. Maguire, "Boyle's Conception of Nature," *Journal of the History of Ideas* 33 (1972), 523-42.

[6] J.J. Rousseau, *The First and Second Discourses*, ed. R.D. Masters, trans. R.D. and J.R. Masters (New York: St. Martin's, 1964), p. 178.

[7] This account is particularly indebted to the thorough study of Buffon's early years by Leslie Hanks, *Buffon avant l'histoire naturelle* (Paris: Presses Universitaires de France, 1966).

[8] Condorcet, "Éloge de Buffon," (1788) in *Oeuvres complètes de Buffon*, ed. G. Cuvier (Paris: Duméril, 1835), I, 1. First published in *Histoire de l'Académie royale des sciences*. Paris, 1791.

[9] Letter to Cramer of April 4, 1744, published in François Weil (ed.), "La Correspondance Buffon-Cramer," *Revue d'histoire des sciences et leur applications* 14 (1961), p. 123.

[10] The prospectus is given in *Journal des Sçavans*, (Octobre, 1748), pp. 639-40.

[11] This event seemed to be occasioned by Buffon's concern with the cost overruns of the original edition, but there is evident strain between Buffon and Daubenton as a result of this. See correspondence as published in G. Michaut, "Buffon administrateur et homme d'affaires: lettres inédits," *Annales de l'Université de Paris* (Jan.-Feb. 1931), 15-36.

[12] Important dimensions of the intellectual divergence between Buffon and Daubenton are discussed in Paul L. Farber,

"Buffon and Daubenton: Divergent Traditions within the *Histoire naturelle*," *Isis* 66 (1975): 63-74.

[13] See Daniel Mornet, *French Thought in the Eighteenth Century*, trans. L.M. Levin (New York: Prentice-Hall, 1929), p. 140. As an indication of the popularity of the *Histoire naturelle* in Enlightenment France, Daniel Mornet found in a survey of 500 catalogues of private libraries established between 1750-1780 that only Pierre Bayle's *Dictionnaire* was more popular than Buffon's work in the philosophical and scientific category, despite the relatively high cost of at least many of the editions of the latter. D. Mornet, "Les enseignements des Bibliothèques privées (1750-1780)," *Revue d'histoire littéraire de la France*, 17 (1910), esp. pp. 449-60.

[14] "Thus is stilled the tongue of nature" Obituary notice, *Magazin für das Neueste aus der Physik und Naturgeschichte*, ed. F.S. Voigt, 5 (1788), p. 193.

[15] Isidore Geoffroy St. Hilaire, *Histoire naturelle générale des règne organiques* (Paris: Masson, 1859) II, 385, 390.

[16] Jacques Roger, "Buffon," in *Dictionary of Scientific Biography*, ed. C.C. Gillispie (New York: Scribners, 1970) II, 577-78.

[17] Ibid., p. 577 and Hanks, *op. cit.*, p. 225.

[18] The relationship of Buffon and Condillac is somewhat unclear. Buffon's philosophical thought seems to have taken its mature shape before the appearance of Condillac's writings, and there are evident personal tensions between them. See comments in Hérault de Séchelles, *Voyage à Montbard*, trans. J. Lyon below, p. 372.

[19] Ibid.

[20] This interpretation was immediately put on Buffon's remarks by the review of the *Histoire naturelle* in the *Nouvelles ecclésiastiques* of February, 1750. See translation below in following selections, p. 241. This is a common reading of Buffon's initial beginning point, and has been repeated for much of recent Buffon scholarship in A.O. Lovejoy's classic "Buffon and the Problem of Species," first published in 1909 and revised in *Forerunners of Darwin*, ed. B. Glass, O. Temkin and W. Strauss (Baltimore: Johns Hopkins Press, 1959), pp. 90-93. See also J. Roger, *Les sciences de la vie dans la pensée française du xviiie siècle*, 2nd ed. Paris: Colin, 1971), pp. 528-42.

[21] C. Huygens, *Treatise on Light*, trans. S. P. Thompson (New York: Dover, 1962), p. 3. (First published 1912).

[22] On this see especially H.G. Van Leeuwen, _The Problem of Certainty in English Thought, 1630-1690_ (The Hague: Martinus Nijhoff, 1963), chps. iv-v, and the suggestive remarks by Keith M. Baker, _Condorcet_, (Chicago: Univ. of Chicago Press, 1975), chp. iii.

[23] Namely, by giving epistemological support to the phenomenalist arguments Newton employs to justify his "mathematical" conception of forces.

[24] Hume's conclusion was that Newtonian science had served with finality to disclose the _unknowability_ of nature. See his _History of England_, 9th ed. (New York, 1851), VI, 374.

[25] Inasmuch as Locke gave greatest epistemological certitude to simple ideas of sensation. See especially _Essay Concerning Human Understanding_, ed. A.C. Fraser (New York: Dover, 1959), II, 347-50.

[26] Ibid., p. 379-80.

[27] We are indebted here to the important work of David C. Kubrin, "Providence and the Mechanical Philosophy," (unpublished Doctoral dissertation, Department of History, Cornell University, 1968), chp. xi.

[28] John Keill, _Examination of the Reflection on the Theory of the Earth_, (Oxford, 1699), quoted in Kubrin, _op. cit._, p. 325.

[29] Petrus van Musschenbroek, _The Elements of Natural Philosophy_, trans. John Colson (London, 1734), I, 5.

[30] Noël-Antoine Pluche, _The History of the Heavens Considered according to the Notions of the Poets and Philosophers compared with the Doctrines of Moses_ (London, 1740: French edition 1738-39) II, 223-24.

[31] Descartes, _Principes de philosophie_, pt. 3, art. 45 in _Oeuvres de Descartes_, ed. Ch. Adam and P. Tannery (Paris: Cerf, 1897-1913), VIII, 100.

[32] George Garden, "A Discourse concerning the Modern Theory of Generation," _Philosophical Transactions of the Royal Society of London_ 17, No. 192 (1691), 476-77.

[33] See Roger, _Les sciences de la vie_.

[34] Pluche, _op. cit._, II, 138-39.

[35] It should be noted that Darwin's own initial reflections on the problem of transformism are in the context of developing

Introduction   31

a strong analogy between the embryological development of individuals and a parallel development of species. See his "B" notebook, published as "Charles Darwin's First Notebook on the Transmutation of Species 1837-1838," ed. Gavin De Beer, <u>Bulletin of the British Museum of Natural History (Historical Series)</u>, 2 (1960), esp. pp. 41, 49-50.

[36] Hanks, <u>op. cit</u>., p. 90.

[37] See quote and reference below, p. 35.

[38] See letter from Buffon to Jean Jallabert, 2 August, 1745 quoted in <u>Oeuvres philosophiques de Buffon</u>, ed. Jean Piveteau (Paris: Presses Universitaires de France, 1954), p. viii. In this Buffon speaks of readying the prefactory discussion for press.

[39] Buffon, "Initial Discourse to the Natural History," trans. by John Lyon, in following texts, p. 121.

[40] The one study on French Leibnizianism remains that of W.H. Barber, <u>Leibniz in France</u> (Oxford: Clarendon, 1955). In addition see <u>idem</u>., "Mme. du Châtelet and Leibnizianism: the Genesis of the <u>Institutions de physique</u>," in <u>The Age of Enlightenment: Studies Presented to Theodore Besterman</u>, ed. W.H. Barber, <u>et. al</u>. (Edinburgh: Oliver and Boyd, 1967), pp. 200-22.

[41] Hanks, <u>op. cit</u>., p. 108. See also the partial translation in Part I, below.

[42] Letter of Helvétius to Mme du Châtelet in <u>Les Lettres de la Marquise du Châtelet</u>, ed. T. Besterman (Gèneve: Institut et musée Voltaire, 1958) II, 36n.

[43] The opening philosophical discourse by du Châtelet is more properly seen as an exposition of Wolff's rather than Leibniz' philosophy. French manuscripts of Wolff's metaphysics and logic had been sent by Frederick the Great to du Châtelet in 1736, and Wolff's disciple, Samuel Koenig, served as du Châtelet's private tutor in 1739. See Barber, "Madame du Châtelet . . .," p. 215 and I.O. Wade, <u>Voltaire and Candide</u> (Princeton: Princeton Univ. Press, 1959), p. 35.

[44] Du Châtelet sees Newton's concept of absolute and independent time and space as defended by ". . . tous les Disciples du livre de l'Entendement Humain," (du Châtelet, <u>Institutions de physique</u> [Paris, 1740], p. 97.)

[45] <u>Ibid</u>., pp. 105-7.

[46] <u>Ibid</u>., p. 120.

[47] Buffon, "Initial Discourse", below, p. 127.

[48] Ibid., p. 123.

[49] See below, p. 83.

[50] See, for example, letter of James Burnet (Lord Monboddo) to Linnaeus in J.E. Smith, A Selection of the Correspondence of Linnaeus and Other Naturalists, From the Original Manuscripts (London, 1821), II, 555. See also similar claims in references cited above, note 20.

[51] Buffon, "Initial Discourse. . .," below, p. 112.

[52] Buffon, "Comparison des animaux & des vegetaux," Histoire naturelle (Paris, 1749) II, as reprinted in Oeuvres philosophiques de Buffon . . ., p. 234.

[53] Ibid., p. 233.

[54] Buffon, "De l'asne," Histoire naturelle (Paris, 1753) IV, in ibid., pp. 355-6.

[55] On this connection see P.R. Sloan, "Descartes, the Sceptics and the Rejection of Vitalism in Seventeenth Century Physiology," Studies in History and Philosophy of Science 8 (1977), 1-28.

[56] See R. Hahn, "Laplace and the Vanishing Role of God in the Physical Universe," in: The Analytic Spirit, ed. H. Woolf (Ithaca: Cornell University Press, 1981), 85-95.

[57] See valuable historical remarks on this problem in G. Tonelli, "La nécessité des lois de la nature au XVIIIe siècle," Revue d'histoire des sciences 12 (1959), 225-41.

[58] J. Bernoulli, Ars conjectandi as translated from German translation Wahrscheinlichkeitsrechnung, trans. R. Haussner (Ostwald's Klassiker der exakten Wissenschaften, Vol. 108; Leipzig: Englemann, 1899), pp. 71-72.

[59] See Jean Des Champs, Cours abrégé de la philosophie wolfienne en forme de lettres (Amsterdam and Leipzig, 1763) I, pp. 254-55.

[60] See the useful discussion of this context in I. Hacking, The Emergence of Probability (Cambridge: Cambridge University Press, 1975), Chp. 11.

[61] Kant's marginal note to his Lectures on Physical Geography, Barth ed. as given in J.A. May, Kant's Concept of Geography (Toronto: Univ. of Toronto, 1970), p. 63.

PART I:
The Formative Years

2. Buffon's Preface to the *Vegetable Staticks* of Stephen Hales (1735)
Phillip R. Sloan

The short preface added to his translation of Stephen Hales' <u>Vegetable Staticks</u>, coupled with the notes and minor revisions made to the text,[1] provide the main available insight into Buffon's methodological and philosophical thought in the period prior to his migration into natural history in 1739.

In drawing attention to this work, Buffon's translation served as the main route of transmission to Continental circles of Hales' plant physiology, and more importantly, his pneumatic chemistry.[2] The presentation of material by Hales primarily in the form of simple reports of experiments, united by only scant theoretical discussion, could not have been better suited to the stated philosophy of the French Royal Academy of Sciences. As Bernard de Fontenelle, its powerful secretary, had written two years before Buffon's translation in a spirit of commendation:

> To the present, the Academy of Sciences takes nature only in small parcels. No general system [is embraced], for fear of being drawn in to the inconvenience of rash systems, to which the human mind is only too easily accustomed, and which once established, oppose themselves to the truths which arise unexpectedly. Today one is convinced of one fact, tomorrow of another which is unrelated to it. . . . Thus the collections that the Academy presents each year to the public are only comprised of detached contributions, independent of each other.[3]

In this selection, we see Buffon echoing this claim, and presenting Hales as a paradigm case of this method, one seen as also embraced by all the great names of modern science. The "positivism" extolled in this preface will, however, rather shortly disappear from Buffon's writings, and alongside a concern with particular facts, Buffon will increasingly speak of the need for a more comprehensive "general view" of the natural world.

NOTES

[1] For a detailed discussion of the textual changes and other emendations to Hales' text, see L. Hanks, *Buffon avant l'histoire naturelle* (Paris: Presses Universitaires de France, 1966), pp. 73-100.

[2] H. Guerlac, "The Continental Reputation of Stephen Hales," *Archives internationales d'histoire des sciences* 4 (1951): 393-404.

[3] H. Potenz (ed.) *Pages choisies des grands écrivains: Fontenelle* (Paris: Colin, 1909), p. 143.

PREFACE OF THE TRANSLATOR*

Translated by Phillip R. Sloan

The first time I read the works of Mr. Hales, I perceived that they would be well worth the effort to reread them. Inasmuch as I wanted to do this with all the attention that they merit, I thought that it would scarcely be more trouble to translate them, and my desire to please the public has confirmed my determination here. My translation is literal, especially of those parts where the author gives the detail of his experiments. I have taken a little more liberty in those parts which are less important, but in general, I have striven to render the sense well, and to clarify that which has appeared obscure to me. I have even added to the plates, in order to make some interesting sections better understood, which have not appeared to me developed enough in the original.

The novelty of the discoveries and of most of the ideas which make up this work, will, without doubt, surprise natural philosophers. I know of nothing better of its kind, and this form of writing is excellent in itself, because it is only experiment and observation. However, this is not the place for me to give an oration about this work. The merit of an author must not be measured by the praises of the translator. The public is distrustful of this, and not without reason. Thus I ask Mr. Hales not to be offended if I do not extend myself on the merits of his book. The care I have taken in its translation, testify enough to the esteem in which I hold it. But it seems to me that one must never define the taste of the public by his own, and when a work is submitted to public judgment, it is presumptuous to dictate the mood in which it is to be read. In return for my suppression of lengthy praises, I only ask one favor, namely that this book be read with some real confidence. Works based on experiment merit this more than others. I can even say, that with regard to natural philosophy, experiments must be sought and systems feared. I admit that nothing would be so desirable, as to establish initially a single principle, by which the universe could then be explained. And I agree that if one were so fortunate as to be able to divine this, all the effort that is given to making experiments would be quite unnecessary. But men of judgment see sufficiently the extent to which this idea is vain and chimerical. The system of nature perhaps is dependent on several principles. These principles are unknown to us, and their combination is no less unknown. How can one dare to fancy himself able to unveil these mysteries, with no other guide than his imagination? And how can one forget that the cause is only known by the effect? It is by precise, rational, and coherent experiments that

Nature is forced to disclose its secrets. All the other methods have never succeeded, and true natural philosophers cannot prevent themselves from viewing the ancient systems as so many antique dreams, and are compelled to read most of the read most of the new ones, as one reads novels. The collections of experiments and observations are thus the only books which can augment our knowledge. It is not necessary in order to be a natural philosopher to know what would happen according to this or that hypothesis, one postulating, for example, a subtle matter, vortices, an attraction, etc. It is a question of sufficiently knowing what does happen, and of being well acquainted with what is evident to the senses. Knowledge of the effects will conduct us insensibly to that of the causes, and no longer will one be drawn into the absurdities which seem to characterize all systems. In fact, hasn't experiment successively destroyed all of these? Has it not shown us that the elements, which were formerly believed to be so simple, are also composite, like other bodies? Has it not taught us what we must think about heat, cold, dryness, and moisture; weight and absolute levity; the abhorrence of the vacuum; the laws of motion formerly agreed upon; the unity of colors; the rest and the sphericity of the earth; and, if I dare say, the vortices?

Therefore, let us constantly collect experiments, and flee, if it is possible, from the system-building mentality, at least until we are well informed. Someday we will be able to "...situate these materials with confidence, and even if we should not be so fortunate as to be able to build the edifice in its entirety, the materials will certainly serve us as foundation for it, and perhaps even for extending it beyond our hopes." This is the method my author has followed. It is that of the great Newton; it is that which Verulam, Galileo, Boyle and Stahl have recommended and embraced. It is the one which the Academy of Sciences has chosen as its rule, and which its illustrious members--Huygens, Réaumur, Boerhaave, etc.--have followed so well, and have always placed so much value upon. In short, it is the view which has led the way at all times, and which still today leads the great men. Example alone must be sufficient to guide us, and must render the public favorable to the work that is today presented to it. I even dare to say that from the little acquaintance of it one has, it will easily be seen that England itself rarely produces such excellent things, and that in spite of so many brilliant discoveries that we owe to the superior geniuses of this wise nation, it does not cease to distinguish itself by intellects perhaps more brilliant than the majority of those who have preceeded them. But to put it briefly, these discoveries would have been still move brilliant if Mr. Hales had presented them differently. His book is not made to be read, but to be studied. It is a collection of an infinitude of useful and unusual facts, whose connection is not seen at first glance. Certain relationships, necessary for some minds, are neglected. Details need not be entered in to here.

Finally, he has written his book only for those who are seekers for the barest truth, and he presupposes a great deal of knowledge in his readers, and still more penetration. The beginning of the "Analysis of Air" is the finest part of his book, and one which he has least developed. I have tried to supplement it here by adding to the plate. Everything is new in this part of his work. It contains one fertile idea which he submits to a new kind of proof, from which flows an infinity of discoveries on the nature of different bodies.

These are some of the surprising facts which he scarcely condescends to announce: Would one have imagined that air can become a solid body? Could one have believed that its elasticity could be removed and restored? Could we have thought that certain bodies, like bladder stones and tartar, are more than 2/3 solid and metamorphosed air? Mr. Hales knows how to return [air] to its original state. He informs us up to what point flame, the respiration of animals, and lightning destroy the elasticity of the air. He measures the force of respiration, and simulates its motion, to the point that a dog was made to respire and live more than eight hours after the trachea was cut. He finds the means to purify the air, and render it suitable for respiration for a longer period. He demonstrates its effects on fire, and on plants and animals.

These are some samples of his discoveries. I will say nothing about all those which he has made on plants, on the quantity of their nutriment, and on their transpiration, their growth, their respiration, diseases, the force, quantity, motion, rarefaction, and quality of the sap, etc. I will be content to assure myself that the fanciers of agriculture will find here something to amuse themselves, and the natural philosophers something to be instructed by.

The author has given a second work to the public, entitled _Animal Statics_.[1] As he is presently working on these matters, and must combine his new discoveries with his old ones to form a single work, it has not been deemed opportune to translate this work, and I have been content to give the translation of one appendix he has added to that work, in which will be found some excellent observations that relate to plant statics or the analysis of the air.

NOTES

*Buffon, "Préface du traducteur" _La Statique des végétaux, et l'analyse de l'air_ (Paris: Jacques Vincent, 1735), pp. iii-viii.

[1][_Haemastaticks; or an Account of Some Hydraulic and Hydrostatical Experiments made on the Blood and Blood-Vessels of Animals_, first published as Volume II of _Statical Essays_ (London, 1833). Volume I of this edition contained the third edition of the _Vegetable Staticks_. The _Haemastaticks_ was only

translated into French by François Boissier de Sauvages in 1744. Throughout our presentation of primary sources, notes added by editors will be given in brackets. Those appearing in original texts will be unbracketed.]

3. Buffon's Preface to Isaac Newton's *Fluxions* (1740) (selected)
John Lyon

Isaac Newton had worked on his "fluxions" ("velocities," "velocities of motion," "velocities of increase") as early as 1664, but the first general knowledge of this was gleaned by the public from the Principia Mathematica of 1687.[1] The De analysi per aequationes numero terminorum infinitas (1669; publ. 1711), and the Methodus fluxionum et serierum infinitorum (1671, publ. 1736),[2] formed the basis of his work. The latter was translated into English by J. Colson in 1736, nine years after Newton's death, under the title The Method of Fluxions and Infinite Series.[3] It was this English edition which Buffon says he used as the basis for his translation into French. The text was translated in 1737, and Buffon added the lengthy preface the following year. The volume appeared in print in 1740.[4]

The section of Buffon's preface translated here contains a brief outline of the Newton-Leibniz controversy, and a partial history of the development of the calculus. He continues the history in a final section of the preface, which has been omitted from this translation. But what is most interesting for our purposes is the section on the conception of the infinite and the metaphysical errors to which it leads. This material is repeated verbatim in Buffon's "Essay on Moral Arithmetic" [Section XXIV], and has been omitted in our translation of the latter. The substance of the matter also forms part of the basis of Buffon's lengthy disquisition on truth, metaphysics, and the methods of the sciences and mathematics near the close of the "Initial Discourse" (pp. 122-27, below). This section is in its turn provides the occasion for Lamoignon-Malesherbes' extended critique of Buffon's method in Part IV (see pp. 331-45, below).

Just as we know the "phenomena which offer themselves daily to our eyes," that is, the general effects, but not the causes (p. 124, below), so too do we have "distinct ideas of magnitude," of augmentation and diminution (p. 45, below), but not of infinity. In each case our experience is of the finite, the limited, the sensed. On the basis of such repeated experience, however, we extrapolate to the idea of the infinite, or to the idea of the cause of numerous effects. But the infinite is simply a privation, Buffon argues, and the idea of cause, or of laws of nature, is simply an abstraction. And we cannot "come back," as it were, from such privations or abstractions and work with absolute and necessary certitude on the concrete realities entailed in our daily experience of life. Approximation is the best we can hope for, even though it be at times an approximation which for all practical purposes we can consider as an identity.

NOTES

1. Carl Boyer, The Concepts of the Calculus (Wakefield, Mass: Hafner Publ. Co; 1949), pp. 196-97.

2. Ibid, pp. 190, 193.

3. J. Itard in "De l'algebre symbolique au calcul infinistesimal," (in Rene Taton, ed; Histoire generale des sciences, Vol. II: La Science moderne [Paris: Presses Universitaires de France; 1958], p. 232), erroneously calls the translator of the Fluxions from Latin (1736) "John Collins." Florian Cajori, A History of Mathematics (New York: Macmillan, 1919), p. 193, notes that J. Colson did the translation in 1736. (See also Cajori's A History of the Conceptions of Limits and Fluxions in Great Britain (Chicago and London: Open Court; 1919), p. 149; and Dirk Struik, A Source Book in Mathematics (Cambridge, Mass: Harvard Univ. Press; 1969), p. 284. Buffon refers to Colson as the translator.

4. See Leslie Hanks, Buffon avant l'histoire naturelle (Paris: Presses Universitaires de France; 1966), p. 108.

"PREFACE" TO ISAAC NEWTON'S
METHOD OF FLUXIONS AND INFINITE SERIES*

Translated by John Lyon

The work translated here was begun in 1664 and completed in 1671.[1] Newton, still little known at this time, wanted to have it printed as a continuation of an introduction to the algebra of a certain Kinckhuysen, which he had corrected and augmented. It is not obvious why this work was not printed; it is only obvious that in the same year Newton changed his mind, and took up the plan of publishing it with his Optics, of which he had already composed the greatest part. But the objections which had been made to his principles and to his optical experiments, and the quibbling concerning them, angered him and prevented him from giving these two works to the public. Here is what he himself said of the matter: "Because of letters from various people containing many objections to my theory, repeated interruptions to answer such objections arose suddenly and distracted me completely from deeper deliberation. The result of this, for the imprudence of which I blame myself, was that in chasing shadows I lost both the main chain of my reasoning and the contemplative quiet of mind, which for me is something absolutely essential."[2] It even appears that he had entirely forgotten his work until 1704, when he extracted his "Traite des Quadratures" from it. Many years after, M. Pemberton[3] obtained his consent to have the complete work printed, but for no apparent reason the project was not carried off. Ultimately, Newton died before the work appeared, and, furthermore, even then it only appeared in translation. Newton composed it in Latin, and M. Colson, into whose hands the manuscript had been given, did not want to publish it in the original. He translated it and in 1736 had it printed in English, so that, he said, his English compatriots would be able to enjoy the works of the great Newton before other nations. He added another reason which, to me, appears better and more natural, namely, he wanted to affix a "Commentary" and "Notes" of his own. These were in English, and apparently he wished to avoid the trouble of putting them into Latin.

Whatever the case may be, it is upon this English version that my translation is based. It is none the worse for that, for I have always followed the spirit of the author rather than the literal meaning of his words. In matters of this sort it suffices to understand things in order to translate them well; and, besides, Geometry, above all the Geometry of Newton, only has one mode. I have not translated the commentary of M. Colson, though I value it and admit that it contains much that is worthwhile. But it must also be admitted that these valuable items are found buried in a morass of calculations which discourage their pursuit. Furthermore, this long commentary is only the beginning of a commentary, for the author has promised us a more complete sequel should this one be well-received. Add to all that that these long glosses are followed by two large chapters which have no connection with the work or the commentary, and my reluctance to translate it will appear justified.

Newton alone appears here. But Newton clearer, more manageable, and more accessible to the generality of geometers than he is in any other of his works. In 1671, at the time that this work was composed, it might have needed a commentary. But geometry has made great progress in the intervening 70 years, and I do not believe that geometers will be inhibited in the reading of this work, which is clear enough and comprehensible enough to be easily understood, and whose main articles have already been commented on.[4] Besides, it contains nothing entirely novel, or anything of which the results, at least, are unknown. This has come about as much from the short pieces which Newton himself published in 1704 and 1711, etc., as from the various pieces and treatises which other geometers have published on these matters.

It is quite easy to survey differential calculus and integral calculus with all their applications in one single, small, volume. To do so one would explore the manner in which these subjects are treated by the hand of their master and the genius of their inventor. Throughout this process one remains convinced that Newton is the sole author of these marvellous calculi, as he is also of many other extremely remarkable works.

Everyone knows that Leibnitz claims part of the glory of the invention of the calculus, and many people give him at least the title of "second inventor." He published in 1684 the rules of differential calculus, and praise has been heaped upon him by extremely accomplished geometers, who, not content to offer these brilliant compliments to him, work further to enhance his reputation, attributing their own discoveries to him. On the other side, Newton's claims were sustained by the mass of his works, and appeared to repose in the superiority of which he was conscious. He passed many years without any complaint, without claiming this discovery, but finally he launched proceedings, proceedings in which entire nations interested themselves and which are not yet concluded. These are filled right to this day with chicanery, which perhaps is the source of most quarrels which have occured concerning the infinitesimal calculus. Perhaps it will be useful to present here an abbreviated account of this epoch in the republic of letters,[5] while occasionally noticing the main events in the history of the geometry and calculus of the infinite.[6]

From the very first steps one takes in geometry, one comes across the concept of infinity; and since the most ancient

Buffon's Preface To Isaac Newton's _Fluxions_

times geometers have glimpsed it. The quadrature of the parabola and the treatise de Numero Arenoe of Archimedes show that this great man had some most appropriate ideas about infinity. These ideas were extended, arranged in various ways, and, finally, the art of applying the calculus to the concept of infinity was developed. But the basis of the metaphysics of infinity has not changed; and it is only in these latter days that several geometers have given us views on infinity which differ from those of the ancients, views so far removed from the nature of things that they have escaped notice as such even in the works of these great men. From this source have come all the opposition and contradiction which have been made to the infinitesimal calculus and which it still undergoes. From thence have come the disputes between geometers over the manner in which calculus should be done, and over the principles from which it derives. It is astonishing what prodigies this calculus works; but astonishment is followed by confusion. It appeared that the concept of infinity was producing all these marvels, and one deluded oneself by thinking that the knowledge of this "infinity" had been refused to all previous centuries and reserved for ours. Finally, we have constructed on that concept systems which have only served to confuse facts and obscure ideas. Before proceeding further, let us say two words on the nature of this "infinity", which, in enlightening men, seems to have bedazzled them.

We have distinct ideas of magnitude. We see that, in general, things are capable of being augmented and diminished, and the idea of a thing becoming greater or lesser is an idea which is as familiar to us as the idea of the thing itself. Anything whatsoever, then, being presented to us, or even imagined by us, we see that it is possible to augment or diminish it. Nothing prevents, nothing destroys, this possibility; one can always conceive of the smallest thing imaginable halved, and the largest thing doubled. It is even conceivable that the thing in question could be increased or decreased a hundred times, a thousand times, or a hundred-thousand times. And it is this possibility of augmentation or diminution without limit which constitutes the true idea that one ought to have of infinity. This idea arises out of the idea of the finite. A finite thing is a thing which has bounds or limits. An infinite thing is simply this same finite thing from which we have removed the bounds and limits. Thus the idea of infinity is only the idea of privation, and has no real object corresponding to it. This is not the place to show that space, time, and duration are not real "infinites." It suffices for our purposes to show that there is no number actually infinite, or infinitely small, or greater or lesser than an "infinite," etc.

Number is only an assemblage of units of the same kind. Unity is not a number, designating rather a single thing in general. But the prime number 2 designates not only two things, but, further, two things of the same kind. The same is true of all the other numbers. But these numbers are only representations, and do not ever exist independently of the things they represent. The characters which designate them do not make them real. It is necessary that they represent a subject, or rather an assemblage of subjects, in order that their existence might be possible. I mean their intelligible existence, for they can have no real existence. Now an assemblage of units or of subjects can ever and only be finite, that is to say, one can always designate the parts of which it is composed. Consequently, number cannot be infinite, no matter what augmentation is given to it.

But may we not say that the last term in the natural series 1, 2, 3, 4, etc., is infinite? And are there not last terms of other series even more infinite than the last term of the natural series? It seems that numbers ought, finally, to become numberless, since they are always susceptible of augmentation. To that I reply that this augmentation to which they are susceptible proves conclusively that they are not capable of being infinite. I say, further, that in these series there are no last terms, and even to imagine a final term is to destroy the essence of the sequence which consists in the succession of terms which can be followed by other terms, and these other terms in turn by others, all of which are of the same nature as their precedent terms, that is to say, all finite, all composed of units. Thus, when we suppose that a series has a final term and that this final term is an infinite number, we go against the definition of number and against the general law of series.

Most of our errors in metaphysics come from the reality with which we endow ideas of privation. We are familiar with the finite and see in it real properties; we deprive it of these, and, considering it after this deprivation, we no longer recognize it, and think that we have created a new thing, whereas we have only destroyed some part of that which formerly was known to us.

Thus, we should consider the infinite, whether it be the infinitely large or the infinitely small, only as a privation, an excision of the idea of the finite, which is useful in some cases as a supposition to assist us in simplifying our ideas and generalizing their results in the practice of science. Thus, all art reduces to turning this supposition to account, and trying to apply it to subjects which one is considering. All the merit is thus in the application; in a word, in the way that one uses the supposition.[7]

Before Descartes had applied algebra to geometry, the principles and the metaphysics of geometry were well known and quite definite. However, this application has greatly augmented our geometrical knowledge, and it has been extended to all the operations of this science. Likewise, the concept of infinity was known by, and the metaphysics of infinity was familiar to the Ancients. But the application of calculus to this "infinite" which has been made in our days has raised us

Buffon's Preface To Isaac Newton's Fluxions

above the ancients and has won for us all the new discoveries. Archimedes, Appolonius, Viviani, Gregory of St. Vincent, were familiar with "infinity." Their method of approximation and exhaustion is drawn from it, and served them to square and rectify some curves. But these dealings with infinity in the absence of calculus only produced special methods, often inconvenient and always confined to some simple cases. The generality of cases was reserved for calculus; it embraces all, it resolves all. Consequently, the geometry which preceded the invention of calculus has become less necessary; and perhaps it has also been a bit neglected.

The ancient geometers considered curves as polygons composed of infinitely small sides. They inscribed and circumscribed around curves figures composed of finite and continuous segments, the number of which they augmented and diminished to infinity. They thus came in the end to measure some curves. Cavalieri, and 20 years later Fermat and Wallis, were the first who may have applied some ideas of calculus to this geometry of the infinite. Their methods of "summing up" are the forerunners of calculus, and the first germs of that kind which may actually have developed.

Cavalieri however had not taken the true route. He had ideas which, reduced to actual calculus, were fruitful;[8] but he was only able to draw out of this previously known conclusions. He considered the line as an indivisible part of the surface, the surface as an indivisible part of the solid, and he attempted to measure surfaces and solids by the infinite sums of lines and surfaces. The results of his method are positive, his method quite generally applicable. Yet despite this advantage he did not go beyond the Ancients, offering nothing new. He himself appeared to limit the merit of his work to the perfect accord between the results of his method and the truths of the geometry of the Ancients.

Fermat accomplished far more than Cavalieri. He found a means of calculating the infinite, and gave an excellent method for the resolution of the largest and the smallest quantities. This method is quite close to the notation which is used today. Indeed, this method was the differential calculus, had its author but applied it generally.

But Wallis took another route. He actually applied arithmetic to ideas of the infinite. He reduced compound fractions to infinite series. He used the same method quite felicitously for his arithmetic series for the quadrature and rectification of curves. However, he proceeded cautiously, and despite a quite powerful and general calculus, he continued to use combinations, particular and individual attachments of numbers, etc. Brownker and Mercator profited from the views of Wallis. They expanded his method, and it could be said that they were the first who dared to advance along this route and to trace out this fruitful path. Brownker squared the hyperbola by means of an infinite series composed entirely of finite and familiar terms, while Mercator gave a demonstration of the same by infinite division, after the fashion of Wallis.

Almost at the same time as Mercator, Jacques Gregori[9] gave a demonstration of this same quadrature of the hyperbola, and the epoch which specifically marked the birth of the new calculi had arrived. It is quite astonishing that these Geometers did not become students of the general method of series after having found the particular series of the hyperbola. It would appear that a moment's reflection would have given them at least the quadrature of the ellipse and the circle by the same method. However, they did not find it, and one cannot see that they made any other use of this theory of infinite series, than that of squaring the hyperbola.

But it is true that Newton did not pay any attention to them. In the month of June, 1669, all these methods had been sent to Barrow as remarkable novelties, and he wrote of them to Newton, who did not think much of them. For he returned to Barrow's hands some papers which contained: 1.) the general method of series which he had discovered some years previous, a method by which he did for all curves that which others had only done for the hyperbola; 2.) the numerical and factual [litterale] solution of the affected equations; 3.) the method of fluxions; 4.) the inverse method of tangents, the quadrature and rectification of curves, and a word on the measurement of solids and the discovery of centers of gravity, etc., that is to say, that as these measures amount to those of surfaces, it would not be necessary that he show that his method would yield all these. Thus as early as 1669 Newton had discovered the infinite series, the differential calculus, and the integral calculus. . . .

NOTES

*Buffon, "Preface," La methode des fluxions et des suites infinies (Paris: Bure l'âine, 1740), pp. iii-xiii.

[1]See the Com[mentariolis] Epistolicum, pp. 101, 102, etc. and Newton, Principia, 3rd ed. p. 246.

[2][Et Subortae statim (per diversorum Epistolas objectionibus refertas) crebrae interpellationes me prorsus a goncilio deterruerunt & effecerunt ut me arquerem imprudentiae quod umbram captando, catenus perdideram meam rem prorsus substantialem." Latin text of Newton as given in French original. I owe a debt of gratitude to my colleague Dr. William Hunt, of the Department of Modern and Classical Languages at the University of Notre Dame, for the basic framework of this translation.]

[3]See A View of Sir Isaac Newton's Philosophy.

[4]See the works of Sterling and Maclaurin.

Buffon's Preface To Isaac Newton's _Fluxions_

⁵["De cette époque litteraire."]

⁶[At this point begins section XXIV of the "Arithmétique morale."]

⁷[Section XXIV of the "Arithmétique morale" ends here.]

⁸_Geom. Indivisibil._ Bonon, 1635.

⁹[The reference is apparently to the Scotch mathematician, James Gregorie (1638-75), who published his _Vera circuli et hyperbolae quadratura_ in 1667. The Mercator referred to is Nicolaus, who published in London in 1668 a work on the quadrature of the hyperbola.]

4. Buffon's Moral Arithmetic (1730, 1777) (selected)
John Lyon

Although not published until 1777 in Volume IV of the Supplément to the Histoire naturelle, the "Essai d'arithmétique morale" gives internal evidence of containing a congeries of issues, some of which Buffon worked on as early as 1730.[1] Yet the "Essai" possesses a certain thematic unity, which, as Jacques Roger suggests, can be described as a concern for the relation between mathematics and reality, or between mathematics and the daily problems of human life.[2]

Buffon was not a great mathematician. He was a philosophe, broadly interested in the arithmetical and geometrical calculus of probability and sharply fascinated by the metaphysical problems brought to the forefront by the development of the infinitesimal calculus, problems centering around the conceptual propriety of various understandings of "infinity" and the appropriate use of the term.[3]

The functioning of the adjective "moral" in the title of this work may seem puzzling to contemporary syntactical sensibilities. Yet its double meaning here is rather patent. First of all, it signifies right conduct or right use, and Buffon is directly concerned with the proper application in life of certain forms of mathematics, that is, with how certain forms of abstract computation ought to be led to assist, to conduce to, proper human behavior. Secondly, "moral" signifies probable or generally convincing, as opposed to necessarily convincing or apodictic, and Buffon is concerned in this work with a form of computational reasoning which falls somewhere between what he calls "mathematical" and "physical" certainties.

The sections of the "Essai" contained in the present translation (sections 1-8, 12-20 complete; and parts of sections 9 and 10) are those in which Buffon concerns himself with the questions of the role of mathematics in scientific method, the nature of "infinity," the calculus of probability and its role as a guide in life, and the morality or immorality of gambling. Constraints of space have not allowed us to include other sections of the "Essai," e.g., those in which geometric probability is worked out (containing the only source for the "Memoir sur le jeu de franc-carreau," presented to the Academie des Sciences in 1733, and the sections on "Buffon's needle"), or the final section on the nature and presuppositions of measurement.[4]

For other places in this volume at which Buffon takes up the issues extensively dealt with here, see the references in the last paragraph of our introduction to Buffon's "Preface" to Newton's Fluxions, pp. 41-42, above.

NOTES

[1]See the note containing Buffon's letter to Cramer, Oct. 3, 1730, in section XV of the present translation. For further dating of sections of the text (e.g., parts of the section on the "Jeu de franc-carreau," 1733), see Jacques-Louis Binet's "Preface" and Jacques Roger's "Introduction" and annotations to their recent volume, Un autre Buffon (Paris: Hermann; 1977). See also Lesley Hanks' Buffon avant l'histoire naturelle (Paris: Presses Universitaires de France; 1966), pp. 30-33, 42-44. Maurice Frechet's introduction, "Buffon comme philosophe des mathématiques," in Jean Piveteau, ed., Oeuvres philosophiques de Buffon (Paris: Presses Universitaires de France; 1954), pp. 436, 438, 446, contains an analysis of the several mathematical problems of the text.

[2]Binet & Roger, op.cit., pp. 26-27.

[3]See Julian Lowell Coolidge, The Mathematics of Great Amateurs (Oxford: Clarendon Press; 1949), pp. 171-177.

[4]For an outline of the issues raised in the text and notes as to other places where Buffon discussed them, see Jacques Roger's introduction to the "Essai," in Binet and Roger, op.cit., pp. 25-31. For a detailed discussion of Buffon's conception of infinity, see Pierre Brunet, "La notion d'infini mathématique chez Buffon," Archeion, 13 (1931), pp. 24-39.

AN ESSAY ON MORAL ARITHMETIC*

Translated by John Lyon

I

It is not my intention here to offer essays on morality in general, for that would require more insight than I can muster, and more skill than I possess. The foremost and soundest part of morality is more an application of the maxims of our divine religion than a human science; and I shall be careful to avoid rashness in treating matters wherein the law of God establishes the principles for us, and faith provides our means of reckoning. The respectful recognition, or, rather, the adoration which man owes to his Creator; the brotherly concern, or rather, the love which he owes his neighbor, are natural sentiments and virtues deeply inscribed in the properly constituted soul. Everything that emanates from that pure source bears the hallmark of truth. The light from that source is so strong that the glamour of error is not able to obscure it. Its evidence is so overwhelming that it allows of neither reasoning, nor deliberation, nor doubt, and it allows of no other degree of certainty than conviction.

The measure of things uncertain is my object here, and I shall try to give some rules for gauging the relations of verisimilitude, the degrees of probability, the weight of evidence, the effects of chance, the disadvantage of risk; and to judge at the same time of the real value of our fears and our hopes.

II

There are truths of different sorts, certitude of different orders, and probabilities of different degrees. Truths which are purely mental, such as those of geometry, all amount to truths of definition. In order to solve the most difficult problem, it is only necessary to understand it properly. For there is nothing in mathematics and other purely speculative sciences other than the difficulty of disentangling what we ourselves have put there, and of untying the knots which the human mind has made a practice of tying and tightening [serrer], according to the definitions and suppositions which serve as the foundation and the structure of these sciences. All their propositions can always be shown to be obvious, for one can always move back from each of these propositions to

53

other antecedent propositions which are identical to them, and from these latter in turn one can move back to others which are matters of definition. It is for this reason that evidence, properly speaking, belongs to the mathematical sciences, and only to these sciences. For one ought to distinguish the evidence given by reasoning from the evidence given us by the senses, that is to say, mental evidence [l'évidence intellectuelle] from intuition of the senses [l'intuition corporelle]. For this latter is only a clear apprehension of objects or images, while the former is a comparison of like or identical ideas, or, above all, it is the immediate perception of their identity.

III

In the physical sciences [sciences physiques], evidence is replaced by certitude. Evidence does not admit of degrees [n'est pas susceptible de mésure], for it has only one absolute characteristic, which is the clear negation or affirmation of the thing which it demonstrates. But certitude, never being absolute has relations which admit of comparison and can be measured. Physical certitude, that is to say, the most certain certitude of all, is still for all that only near-infinite probability that a result, an event, which has never failed to happen, will happen again. For example: since the sun has always risen, it is consequently physically certain that it will rise again tomorrow. A reason for existing is to have existed but a reason for ceasing to be is to have begun to be; and, consequently, one is not able to say that it is equally certain that the sun will always rise unless one supposes an eternal precedent equal to the eternal sequel. For otherwise it shall cease to rise at some time for it began to rise at some time. For we ought to judge of the future only on the basis of what we know of the past. Since a thing has always been, or has always acted in the same fashion, we ought to be assured that it always will be or will act in the same fashion. By always, I mean a very long time, and not by any means an absolute eternity, the "always" of the future only being equal to the "always" of the past. The absolute, of whatsoever kind it may be, lies neither within the province of nature nor within that of the human spirit. Men have regarded as ordinary and natural effects all those events which possess that sort of physical certitude. Something that always happens ceases to astonish us. Contrariwise, a phenomenon which might never have appeared or which, always happening in a given manner, should cease to occur, or occur in a different manner, would give us good reason for astonishment. It would be an event which appeared to us to be so extraordinary that we might regard it as supernatural.

IV

These natural effects which do not surprise us have nevertheless everything that makes for astonishment. What a concourse of causes, what an assemblage of principles would not be required in order to produce a single insect, or a single plant! What a prodigious combination of elements, movements, and springs in the animal machine! The tiniest works of Nature are matters of the greatest admiration. That which causes us not to be astonished by all these marvels is that we are born into a world of wonders which are always in sight, and that our understanding and sight are equally accustomed to them. Finally, we know that things have been this way before our observation of them, and will remain so after. If we were born in another world with another form of body and other senses, we would have had other relations with external objects, would have viewed other marvelous things and not been any more surprised by them. Our situation in one case would parallel our situation in the other, both resting on ignorance of causes, and upon the impossibility of knowing the nature [réalité] of things of which we are only allowed to perceive the relations in which they stand to ourselves.

There are then two ways of considering natural operations. The first way is to view them in the fashion that they present themselves to us, without attention to causes, or rather without searching for their causes. The second way is to view these operations in their connection with principles and causes. These two points of view are quite different and produce different reasons for wonder, one way leading to the sensation of surprise, the other giving birth to the sentiment of admiration.

V

We shall only speak here of the first manner of looking at the operations of nature noted above. However incomprehensible, however complicated they might appear to us, we shall judge them as most evident and most simple, and singularly by their results. For example, we are unable to conceive of or even imagine why particles of matter mutually attract, and we shall content ourselves with being sure that in fact they do so. We shall judge consequently that matter is always attracted to matter and that it will continue to be so attracted. It is the same with other phenomena of all kinds. However unbelievable they might appear to us, we shall in fact believe them so long as we are assured that they have happened quite often, and we shall doubt of them if they have failed to take place with an appropriate regularity. Finally, we shall deny them if we come to be sure that they have never happened. In a word, we shall grant them credence according to the measure in which we shall have seen them and known them to happen, or seen and known the contrary.

But, if experience is the foundation of our physical and moral knowledge, analogy is the prime instrument of such knowledge. When we observe that a certain event invariably takes place after a certain fashion, we are assured by our experience that it will take place, subsequently, in the same fashion. And when it is related to us that something takes place in such and such a fashion, if these events are analogous to other facts which we already know, then we believe them. On the contrary, if the reported event has no analogy with ordinary events, that is to say, with things with which we have been familiar, then we ought to doubt of the reported event. And if the related event is directly opposed to what we know, we ought not hesitate to deny it.

VI

Experience and analogy are capable of giving us various forms of certitude which are nearly the same and sometimes of the same kind. For example, I am almost as certain of the existence of the city of Constantinople, on which I have never laid eyes, as I am of the existence of the moon, which I have seen quite often. This is so because the testimony of a great number of persons is capable of producing a certitude almost on a par with physical certitude, when it is brought to bear on things which have an obvious analogy with things with which we are familiar. Physical certitude ought to be measured by an immense number of probabilities, since that certitude is established through a constant set of observations, which constitute what one calls the experience of all times. Moral certitude ought to be measured by a lesser number of probabilities, since it only supposes a certain number of analogies with that which is known to us.

Let us suppose a man who had never seen nor heard anything: let us then search out how belief and doubt would be produced in his mind. Let us suppose him struck for the first time by the appearance of the sun. He sees it shining high up in the heavens, next sinking, and finally he sees it disappear. What can he possibly conclude? Nothing, except that he has seen the sun, that it has traced a certain path in the sky, and that he sees it no more.

But then this star appears once more the following day, and likewise disappears. This second viewing of it is an initial experience which ought to produce in him the hope of

Buffon's *Moral* Arithmetic 57

seeing the sun again; and he begins to believe that it might appear again, although he also doubts it much. The sun reappears again; and this third sighting of it gives him a second experience which diminishes his doubt to the degree that it increases the likelihood of a third reappearance. A third experience increases his faith in its regularity so much that he can scarcely doubt of its fourth reappearance. And, finally, after he shall have seen this shining star appear and disappear regularly ten, twenty, or a hundred times in a row, he will come to believe it certain that he will always see it appear, disappear, and move about in the same fashion. His certainty of seeing the sunrise tomorrow will be in proportion to the number of times he has seen it rise in the past. Each observation, that is to say each day, produces a probability, and the sum of these probabilities taken together, as soon as these are quite numerous, gives him physical certitude.

Consequently, we can always express this certitude numerically, dating from the beginning of the time of our experience; and our certitude will be the same as it is for all other natural operations. For example, if one should wish to reduce here the age of the earth and our experience of it to 6,000 years, the sun has only risen for us[1] 2,190,000 times; and to date from the second day that it has risen, the probability of its rising tomorrow increases as the series 1, 2, 4, 8, 16, 32, 64.... or 2^{n-1}. When, in the natural series of numbers n=2,190,000, one has then I say, $2^{n-1}=2^{2,189,999}$. This is a number so prodigious that we can form no idea of it. And it is for this reason that one ought to regard physical certitude as built up out of an immense number of probabilities. For, if we assume the date of creation to be only 2,000 years earlier than above the immense likelihood of the recurrence becomes 2^{2000} times more than $2^{2,189,999}$.

VII

But it is not as simple to estimate the value of analogy, nor, consequently, to find the measure of moral certainty. In truth, it is the degree of probability which gives force to analogical reasoning. And of itself analogy is only the sum of the relations with things known. Nevertheless, according as that sum or relation in general shall be more or less great, the consequence of analogical reasoning will be more or less sure, without, however, ever being absolutely certain. For example: a witness whom I suppose to be in his right mind, says to me that an infant has been born in this town. I believe the account without hesitation, the birth of an infant being nothing out of the ordinary, but rather being connected with an infinite number of other similar events, that is to say, with the birth of all other children. Thus, I believe this event, without, however, being absolutely certain of it. If this same witness should tell me that this infant was born with two heads, I should still believe it, though more weakly, an infant with two heads being less consonant with things known. And if he were to add that this new-born child not only had two heads, but also six arms and eight legs, I would with reason find it hard to believe the fact. However, no matter what difficulty this might give to my belief, I would not be able to disbelieve his account entirely. For this monster, no matter how extraordinary, was supposedly only composed of parts each of which have some relation with things known, only their arrangement and number being extraordinary. The force of analogical reasoning, then, will always be proportionate to the analogy in question, that is to say, proportionate to the number of relations with other observed events. And, in order to reason properly analogically, it will only be a question of: placing the event in question properly in all its circumstances; comparing these latter with analogous circumstances; adding up the number of these; next, taking a model for the sake of the comparison to which one shall relate this probability-value. In this way one will have the precise probability, that is to say, the degree of force of analogical reasoning.

VIII

There is thus a prodigious distance between physical certitude and the kind of certitude which can be deduced from analogies. The former works by amassing an immense sum of probabilities which force us to believe. The latter only produces a probability more or less great, often, indeed, so little as to leave us in perplexity. Doubt is always inversely proportionate to probability, that is to say, the doubt is greatest when the probability is weakest.

Moral certainty ought to be placed in the order of certitudes produced by analogy. Indeed, it appears to occupy a middle ground between doubt and physical certainty. And this middle ground is not a point but a greatly extended and indefinitely circumscribed line. One feels that, given a certain number of probabilities, moral certainty follows. But what number of probabilities suffices here? And dare we hope to determine this number as precisely as we do the number of probabilities necessary for physical certitude?

After having reflected on this matter, I concluded that, of all the possible moral probabilities, that which most affects man in general is the fear of death. And I felt that total fear or total expectation, a fear or expectation which would equal that produced by the fear of death, might perhaps be taken in the moral realm to be the standard [unite] against which one ought to compare other fears. Furthermore I thought that the same could be said of expectations for there is no

Buffon's *Moral Arithmetic*

difference between expectation and fear except that between positive and negative. The probabilities of all cases of these two ought to be measured in the same manner.

I look, thus, for the actual probability that a man who appears in good health, and who, consequently, has no fear of death, might die nevertheless within 24 hours. Going to the mortality tables, I find that one may conclude that there is only one chance in 10,189 that a man of 56 years will not live out the day. Now since all men of that age, when reason has matured and experience has taught all it can teach, have nevertheless no fear of death within 24 hours, despite the fact that there is one chance out of 10,189 that they will die in this brief interval of time, I therefore conclude on the matter that all probabilities equal to or less than this number ought to be disregarded, and that all fear or all hope that finds its index below the number of 10,000 ought not to affect us, or even bother us for an instant, either in heart or head.[2]

In order to better understand what I mean, let us suppose a lottery wherein there is but one prize, and one has but one ticket, while 10,000 tickets have been issued. I say that the probability of obtaining the prize being only one out of 10,000, one's hope is practically nil, since there is practically no probability that this will come about. In other words, one's chances of winning the prize are the same as one's chances of dying within 24 hours. And since this fear does not affect one in any fashion whatsoever, the hope of winning the prize in the lottery ought not affect one any more; indeed, it should affect one much less, since the intensity of the fear of death is much greater than the intensity of all other fears or all other hopes. If, despite the evidence of such a demonstration as we have been through, a man should obstinately continue to hope, and continued to enter a lottery with similar odds every day, taking each day one new chance, counting each day on winning the prize, one might be able to undeceive him by betting with him at even odds that he will die before winning the prize.

And so, in all games, wagers, risks, and chances, and in every case, in a word, where the probability is less than 1/10,000, it ought to be, and in effect is for us, absolutely negligible. And for the same reason, in all cases wherein this probability is greater than 10,000 [sic], it gives us more complete moral certitude.

IX

From this we may conclude that physical certitude is to moral certitude as $2^{2,189,999}$ is to 10,000; and that every time that an effect, of which we are absolutely ignorant of the cause, happens in the same manner 13 or 14 times in a row, we are morally certain that it will take place once more in the same manner for a 15th time: for 2^{13} = 8192, and 2^{14} = 16384, and, consequently, when a given effect has taken place 13 times, the odds are 8,192 to 1 that it will happen a 14th time; and when it has taken place 14 times, the odds are 16,384 to 1 that it will occur in a similar fashion a fifteenth time, which is a probability greater than that of 10,000 to 1, that is to say, greater than the probability which gives moral certitude.

. . . .

Before ending this article, I ought to observe that it is necessary to take care not to be mistaken about what I just said concerning effects whose causes are unknown. For I mean only those effects whose causes, howsoever unknown, ought to be supposed to be constant, such as those of natural events. Every new discovery in physics established by 13 or 14 experiments, all of which confirm the discovery, has already a degree of certitude equal to moral certitude, and that certitude doubles with each new experiment, so that as the instances increase, physical certitude is gradually approached. But it is not necessary to conclude from this process of reasoning that the effects of chance follow the same law. It is true that in one sense these effects are one example of those events of whose immediate causes we are ignorant. But we know that in general those causes, far from being able to be considered constant, are on the contrary necessarily as variable and inconstant as is possible. Thus, by the very notion of chance, it is evident that there is no connection, no dependence, among its results, and that consequently, the past can have no influence whatsoever on the future. One may be much, or even totally deceived if one should wish to draw inferences from anterior events to the occurrence on non-occurrence of future events. That one card, for example, might have been thrown 3 times in a row is no reason why it should not be thrown a fourth time, and the odds are even that it will or will not be thrown this time, whatever number of times it might have been thrown or not, since the laws of the game are such that the chances here are equal. To presume or believe the contrary, as certain gamesters do, is to go against the very principle of chance, or never to remind oneself that by the rules of the game the lots are always equally divided.

X

Concerning effects for which we see the causes, one single proof suffices in order to bring about physical certitude. For example, I see that in a clock the weights cause the wheels to turn, and the wheels make the pendulum [balancier] move. Thereafter, I am certain, without needing repeated experience, that the pendulum will move always in the same fashion, so long as the weights shall turn the wheels. This is a necessary consequence of the arrangement which we ourselves have contrived in constructing the machine. But when we observe a novel phenomenon, an effect in Nature still unknown (since we do not know the cause of it, or whether it might be regular or variable, permanent or intermittent, natural, or accidental), we have no other means of acquiring certitude than expèrience repeated as often as necessary. Here nothing depends on us, and we can know only to the degree that we experiment. We are only assured by the result itself, and the repetition of this result. As soon as it shall have occurred 13 or 14 times in the same fashion, we already have a degree of probability equal to moral certitude that it will occur a 15th time in like fashion. At this point we readily make a broad assumption, and conclude from analogy that this effect depends on the general laws of nature, and that it is consequently as old as all the other effects, and that there is a physical certitude that it will always take place in the fashion that it has always taken place, and that it simply has not been [regularly] observed.

In games of chance which we have contrived, balanced, and calculated ourselves, we ought not say that we are ignorant of the causes of effects. To be sure, we are ignorant of the immediate cause of each effect in particular. But we see clearly the prime and general cause of all effects. I do not know, for example, and I cannot even imagine in any fashion whatsoever, what the difference is in the movements of the hands, in exceeding or not exceeding ten with a throw of three dice. Yet such movements are nevertheless the immediate cause of the event. But I do see evidently by the number and the markings of the dice, which here are the prime and general causes, that the chances are absolutely equal, that it is indifferent to bet whether one's throw will exceed ten or not. Further, I see that these same events, when happening successively, have no connection, since at each throw of the dice the chance is always the same, yet the situation unprecedented; that the previous throw has no influence upon the throw that comes next; that one is always able to bet evenly for or against exceeding ten; and that, finally, the longer one plays the more the number of throws above 10 and below 10 will approach equality. Thus each experience here gives a result completely opposed to the experience we have of natural effects; I mean to say, we experience the inconstancy rather than the constancy of causes. In the case of natural effects, each event doubles the probability that the same effect will recur - that is to say, the certitude of the constancy of the cause. On the contrary, each try in games of chance increases the certitude of the inconstancy of the cause, thus always showing us that, all in all, it is absolutely inconstant and totally indifferent to the production of one or another set of results.

When a game of chance is by its construction perfectly fair [égal], the player has no reason to make up his mind on this or that course. For, in fact, the odds being even in this case, there are no good reasons for choosing either course. And consequently, if one should consider the matter deliberately, one could choose a given course only for poor reasons. Thus, the logic of gamblers appears to me to be completely vicious, and even those carefree spirits [bons esprits] who indulge thus degrade themselves to the level of gamblers, becoming involved in absurdities which before long embarrass them as reasonable men.

. . . .

XII

It is generally known that gambling is an avid passion, and that the habit is ruinous. But this truth has been able to be demonstrated only by sad experience, upon which one has not sufficiently reflected in order to correct oneself by conviction. A player, whose fortune is exposed each day to the blows of chance, wastes away little by little and necessarily finds himself ultimately destroyed. He attributes his losses to this same chance, which he accuses of injustice. He equally regrets what he lost and what he did not win. Avidity and false hope make him study the luck of others. Consequently, humiliated by finding himself necessarily without the means of further satisfying his cupidity, in his despair he takes it as his evil star which did this to him, and does not imagine that this blind force, the fortune of gaming, marches to the truth with an indifferent and uncertain step, but that with each step she tends nevertheless toward a goal and holds for a definite limit, which is the ruin of those who tempt her. He only sees the apparent indifference she has for good or ill, producing with time the necessity of ill, for a long chain of chances is a fatal chain, the prolongation of which leads to evil. He does not feel, independently of the harsh tax the cards levy on him and the even more harsh tribute he pays to the knavery of several adversaries, that he passes his life making ruinous agreements; that gambling of its very nature is a vicious contract by its very principle, a contract harmful to each contractor in particular, and contrary to the good of society at large.

This is not some vague discourse on morality. These are exact truths of metaphysics which I submit to the calculation or rather to the force of reason; truths which I claim to be able to demonstrate mathematically to all those who have a mind sufficiently clear and an imagination sufficiently strong to

Buffon's Moral Arithmetic

bring things together without geometry and to calculate without algebra.

I shall not speak of those games invented by artifice and designed by avarice, wherein chance loses a part of its force [droits], where fortune is never able to balance things, since it is invincibly drawn and always constrained to favor one side: I mean to say all games where the hazards are inequally divided, offering winnings as assured as dishonest to one, and leaving to the other only a sure and detestable loss – games such as Faro [Pharaon], where the banker is nothing more than an avowed knave and the punter simply a dupe who has agreed to act the role of clown.

It is in gaming in general, gaming the most above-board and honest, that I find a vicious essence; I include under the name of gaming even all the agreements, all the bets, wherein one risks one thing with its advantage in order to obtain an equal choice advantage from others. And I say that in general gaming is an evil pact, a contract disadvantageous to two parties, a pact the effect of which is to render losses greater than gains, and to take away from good to add to evil. The demonstration of this is as easy as it is evident.

XIII

Let us take two men of equal fortune who, for example, each might have 100,000 livres in assets, and let us suppose that these two men gamble in one or many throws of the dice 50,000 livres, that is to say, the half of their wealth. It is certain that he who wins will only augment his wealth by a third, and that he who loses diminishes his by a half. For each of them had 100,000 livres before playing, but after the event one will have 150,000 livres, that is to say, a third more than he had, and the other will only have 50,000 livres, that is to say, half less than he had. Thus the loss is 1/6 greater than the gain, for there is this difference between 1/3 and 1/2. Thus the agreement is harmful to both, and consequently essentially vicious.

This reasoning is not captious; it is true and exact, for whatever one of the players would have lost, that the other has won. This numerical equality of the sum does not obscure the true inequality of the loss and the gain. The equality is only apparent, the inequality quite real. The pact which these two men make in gambling the half of their assets is in fact the same as another pact which no one would be advised to make, namely, to each agree to throw into the sea one-twelfth of their wealth. For one can demonstrate to them before they risked the halves of their fortunes, that the loss being necessarily a sixth greater than the gain, this sixth ought to be regarded as a real loss, which, since it could fall indifferently either on one or the other, ought consequently to be equally divided.³

If two men should bethink themselves to gamble all of their fortune, what would be the result of this convention? One might double his fortune, and the other would reduce his to zero. Now, what proportion is there here between the loss and the gain? That between everything and nothing. The gain of one is only equal to a sum rather moderate, and the loss of the other is numerically infinite, and morally so great that the work of all his life would perhaps not suffice to regain his fortune.

The loss is thus infinitely greater than the gain when one bets his all; it is greater by a sixth when one bets half of his wealth; it is greater by a twentieth when one wagers a quarter of his fortune. In a word, whatever small portions of one's fortune are hazarded at play, there is always more to lose than to gain. And thus an agreement to gamble is a vicious contract, and one which tends to ruin both contractors. This new but useful truth I wish all might know who, through cupidity or idleness, spend their lives tempted by gambling.

It has often been asked why one should feel the loss more than the gain. It is possible to make a plainly satisfactory response to this question, a response that can remove all doubt about the truth of that which I am going to present. Now the response is easy: one is more sensible of the loss than the gain because, in effect, supposing them numerically equal, the loss is nevertheless always and necessarily greater than the gain. Feeling is in general only an implicit reason less clear, but often more subtle, and always more certain than the direct product of reason. One indeed feels that the gain does not make us as happy as the loss pains us. This feeling is only the implicit result of the reasoning which I am going to present.

XIV

Money ought not be estimated by its numerical quantity. If the metal, which is only the sign of riches, were riches themselves, that is to say, if happiness or the advantages which result from riches were proportional to the quantity of money, men would have reason to estimate it numerically and by its quantity. But it is not the case that the advantages to be drawn from money are in proper proportion to its quantity. A man with an income of 100,000 ecus in rent is not 10 times as happy as a man who has only 10,000 ecus. Further, money, when it goes beyond certain limits, ceases to have real value, and is not capable of augmenting the good of him who possesses it. A man who should discover a mountain of gold would not be any

Buffon's _Moral Arithmetic_

richer than he who should only find a cubic fathom [toise cube] of it.

Money has two sets of values, both of them arbitrary, both conventional, one of which is the measure of particular advantages, and the other makes up the tariff of society. The first of these values has only been estimated in a most vague manner. The second is capable to being properly evaluated by means of a comparison of the quantity of money with the produce of the land and the work of men.

In order to succeed in giving some precise rules about the value of money, I shall examine several particular cases of which the mind can easily recognize the structure, and which, as examples, will lead us by induction to a general estimation of the value of money for the poor, for the rich, and even for the man more or less wise.

For the man who, in his station in life, whatever it may be, only has the necessities, money has an infinite value. For the man who in his station abounds in superfluities, money has hardly any value. But what things are necessities, and what superfluous? I mean by the necessary the expenses which one is obliged to make in order to live as one always has lived. With this "necessity" one thus is able to have his comforts and even pleasures. But before long habit has made necessities of them. Thus in the definition of superfluities I count for nothing the pleasures to which we are accustomed, and I say that the superfluous is the expense which can procure for us any new pleasures. The loss of the necessary is a loss which we resent infinitely, and when one risks a considerable part of this necessity, the risk cannot be compensated for by any expectation, however great it might be. On the contrary, the loss of the superfluous has limited effects. And if even in the case of superfluities one is still more conscious of loss than gain it is because, in fact, losses always being greater than gains, this sentiment [the awareness of losses of superfluities more than the gain therein] turns out to be grounded on this principle [that the loss of the necessary cannot be compensated for, though the loss of the superfluous can]. For the ordinary sentiments are founded upon common notions or on easy inductions. But the refined sentiments depend on refined and exalted ideas, and are indeed only the product of many combinations [of ideas], which are often too subtle to be distinctly perceived, and almost always too complicated to be reduced to a process of demonstrative reasoning.

XV

The mathematicians who have calculated the play of chance, and whose researches in this category deserve praise, have only considered money as a quantity susceptible of augmentation and diminution, without other value than that of number. They have estimated the proportions of gain and loss by the numerical quantity of money. They have calculated risk and expectation relative to that same numerical quantity. We shall consider here the value of money from another point of view, and by our principles we shall give the solution of several cases which are embarrassing for the ordinary calculator. Take this question, for example: in a game of heads or tails, where one assumes two men, (Peter and Paul), playing against each other, the conditions being that Peter will throw in the air a piece of money as many times as it shall be necessary for it to come up heads. If that happens on the first throw, Paul gives him one _ecu_. If that only takes place on the second throw, Paul will give him two _ecus_; if on the third throw, he gives him four _ecus_; if on the fourth throw, Paul gives him eight _ecus_; if on the fifth try, he gives him 16 _ecus_, and so on, doubling the number of _ecus_. It is obvious that under these conditions Peter can only win, and that his gain will be at least one _ecu_, perhaps 2, 4, 8, 16, 32, etc., _ecus_, even 512 _ecus_, or 16,384 _ecus_, 580,448 _ecus_, - perhaps even 10,000,000, 100,000,000, 1,000,000,000 _ecus_: perhaps, finally, an infinite number of _ecus_. For it is not impossible to throw the coin 5 times, 10 times, 15 times, 20 times, a thousand times, 100,000 times without having heads come up. One asks how much Peter ought to give to Paul as indemnity or, what comes to the same thing, what is the sum equivalent to the hope which Peter has, who can only win.

This question was first presented to me by the late M. Cramer, celebrated Professor of Mathematics at Geneva, on a journey which I made to that city in the year 1730. He said to me that it had been proposed previously by M. Nicholas Bernoulli to M. de Montmort, as in fact one finds it on pages 402 and 407 of the Analysis of games of chance of that author. I dreamt several times of that question without being able to resolve it. I did not see how it would be possible to mesh mathematical calculation with common sense without having to bring in moral considerations. And, having expressed part of my ideas to M. Cramer,[4] he said to me that I was right to think thus, and that he had also resolved this question in a similar manner. He then showed me his solution approximately as it has since been printed in the Memoirs of the Academy of Petersburg in 1738, following an excellent memoir by M. Daniel Bernoulli On the Measure of Chance, where I saw that most of the ideas of M. Dan Bernoulli accorded with mine, which gave me great pleasure. For, independently of his great talents in geometry, I have always regarded and recognized M. Dan Bernoulli as one of the best minds of this century. I also found M. Cramer's idea quite fitting and proper from a man who has given us proofs of his ability in all the mathematical sciences. To the memory of this man I return his due recognition, with as much pleasure as business, for it is to the friendship of this savant that I owe a part of the initial knowledge which I acquired in this area. M. de Montmort gives the solution of

Buffon's Moral Arithmetic   67

the problem by the regular rules, and he says that the sum equal to the expectation of him who cannot fail to win is equal to the sum of the series 1/2, 1/2, 1/2, 1/2, 1/2, 1/2, 1/2, ecu, etc., continued to infinity, and that consequently this equivalent sum is an infinite sum of money. The ratio upon which this calculation is founded is that there is a probability of 1 in 2 that Peter, who cannot but win, will have one ecu; a probability of 1 in 4 that he will have two of them; a probability of 1 in 8 that he will have four of them; of 1 in 16 that he will have 8; of 1 in 32 that he will have 16, and so on to infinity. And thus his expectation in the first instance is a half-ecu, for the expectation is measured by the probability multiplied by the sum which is obtained. Now the probability is one-half, and the sum obtained by the first throw is one ecu. Likewise, his expectation for the second throw is again a half-ecu, for the probability is one-fourth, and the sum obtained is two ecus. And one-fourth times two ecus gives again one-half ecu. One will find his expectation to be likewise on a third, a fourth, and in all subsequent cases to infinity, since the number of ecus augments in the same proportion as the number of probabilities decreases. Thus the sum of all his expectations is an infinite sum of money, and consequently it is necessary that Peter give to Paul as equivalent, half of an infinite [number] of ecus.

This is mathematically true, and one cannot contest the calculation. Also, M. de Montmort and the other geometers have regarded that question as being well resolved. However, that solution is so far from being the truth that instead of giving an infinite sum, or even a very great sum, which is already quite different, there is no man of common sense who would give 5 ecus let alone 10 in order to purchase that hope in putting himself in the place of him who can only win.

XVI

The reason for this extraordinary contrareity between calculation and common sense is attributable to two causes. The first is that probability ought to be regarded as nothing when it is very small, that is to say below 1/10,000. The second cause is the slight proportion there is between the quantity of money and the advantages which result from it. The mathematician, in his calculation, estimates money by its quantity. But the moral man ought to estimate it otherwise. For example, if one should propose to a man of middling fortune to put 100,000 livres in a lottery because the odds are only 100,000 to 1 that he will win 100,000 times 100,000 livres, it is certain that the probability of winning 100,000 times 100,000 livres, being one out of 100,000 — it is certain, I say, mathematically speaking that his expectation will be worth his betting 100,000 livres. However, it would be a great error to hazard this sum, and an even greater error in that the probability of winning was so small, although the money to be gained augments proportionally. This is so because with 100,000 times 100,000 livres he will not have twice the advantages that he would have with 50,000 times 100,000 livres, nor 10 times as much advantage should he have 10,000 times 100,000 livres. This is so since the value of money is not proportionate to its quantity to the moral man but rather proportionate to the advantages which money brings. It is obvious that this man ought to hazard only in proportion to the expectation of his advantages, that he ought not calculate the numerical quantities of the sums that he might obtain, since the quantity of money, beyond certain limits, cannot augment his happiness, and that he will not be happier with one-hundred-thousand millions in rent than with a thousand millions.

XVII

In order to feel the appropriateness and the truth of all that I have proposed, let us examine more closely than the geometers have done the question that comes to be proposed. Since ordinary calculation is unable to resolve the issue on account of moral considerations which complicate mathematical matters, let us see if we are able by other rules to arrive at a solution which does not offend common sense and which might at the same time conform to experience. This research will not be useless, and we shall furnish a certain manner for estimating properly the price of money and the value of expectation in all cases. The first thing that I note again is that mathematical calculation which sets the equivalent to Peter's expectation to be an infinite sum of money. This infinite sum of money is the sum of a series composed of an infinite number of terms which always are worth a half-ecu. And I see that this series which mathematically ought to have an infinity of terms, is not morally able to have more than 30 of them, since if play were to continue only to the 30th term, that is to say, if heads appeared only after 29 throws, Peter would have owed to him 520,870,912 ecus, that is to say, more money than there is perhaps in the entire realm of France. An infinite sum of money has only a notional existence, and does not exist in reality; and all the expectations founded upon those terms in an infinite series above 30 simply do not exist either. There is here a moral impossibility which destroys the mathematical possibility. For it is possible mathematically and even physically to throw thirty times, 50 times, 100 times in a row, etc., the piece of money without having heads turn up. But it is impossible at this point to satisfy the condition of the problem,[5] that is to say, to pay the number of ecus which should be due, in case that should take place. For all the

Buffon's _Moral_ _Arithmetic_   69

money on earth could not suffice to make the sum thus owed only to the 40th throw, since that would suppose 1,024 times more money than exists in the realm of France, for there are not in the whole earth 1,024 realms as rich as France.

Now the mathematician has only found that infinite sum of money equal to Peter's expectation because the first case gives him half an _ecu_, the second case a half an _ecu_, and each case to infinity always a half an _ecu_. Therefore the moral man, on first reckoning the same, will determine 20 _ecus_ to be sufficient instead of an infinite sum, since all the terms which are beyond 40 give a sum of money so great that it does not exist. So that it is only necessary to count a half an _ecu_ for the first case, the same for a second, third, etc., as far as 40, which all comes to 20 _ecus_ as the equivalent of Peter's expectation, a sum already quite a bit smaller than, and a good deal different from, an infinite sum. This sum of 20 _ecus_ will be reduced further by considering that the 31st term will give more than 1,000,000,000 _ecus_, that is to say, it supposes that Peter would have much more money than there is in the most wealthy realm in Europe, a thing impossible to suppose. And consequently the terms from 30 to 40 are also imaginary, and the hopes founded on them ought likewise to be considered imaginary. Thus the equivalent of Peter's expectation has already been reduced to 15 _ecus_.

The sum will be further reduced by taking into consideration that since the value of money ought not be estimated by its quantity, Peter ought not count 1,000,000,000 _ecus_ as twice as useful to him as 500,000,000 _ecus_, nor 4 times as useful as 250,000,000 _ecus_, etc., and that consequently the expectation of the thirtieth term is not a half an _ecu_, no more than the expectation of the 29th, nor the 28th, etc. The value of this expectation which, mathematically, is found to be a half an _ecu_ for each turn, ought to be diminished as early as the second term, and always further diminished even to the last term of the series: for the value of money cannot be estimated by its numerical quantity.

XVIII

But how then estimate it, how find the proportion of this value to different quantities? What then is 2,000,000 pieces of silver if not double 1,000,000 of the same metal? Can we give the general and precise rules for such estimation? It appears that each ought to judge his own estate, and next estimate his fortune and the quantity of money proportionate to that estate and the usage which he might make of it. But this manner is still vague and too particular to be of service as a principle, and I believe that more general and more exact means are to be found to make this estimation. The first means which presents itself is to compare the mathematical calculation with experience, for in most cases we are able by repeated experiences to arrive, as I said, at a knowledge of the effect of chance as surely as if we deduced it immediately from causes.

I have thus made 2,048 trials of this question, that is to say, I have played this game 2,048 times by having a child throw the piece in the air. The 2,048 separate throws produced 10,057 _ecus_ in all, and so the sum equivalent to the expectation of him who cannot fail to win is approximately 5 _ecus_ for each throw. In this trial there were 1,061 throws which only produced one _ecu_; 494 which produced 2 _ecus_; 232 throws that produced 4; 137 that produced 8; 56 that gave 16; 29 that gave 32 _ecus_; 25 that gave 64; 8 that gave 128, and finally 6 throws which produced 256. I hold this result generally valid, because it is founded on a great number of trials, and that moreover it agrees with another incontestable and mathematical line of argument, by which one finds after a bit this same equivalent of 5 _ecus_. Here is the reasoning. If one plays 2,048 times, there ought naturally to be 1,024 times which will only produce one _ecu_ each; 512 times which will only produce two, 256 times which produce eight, 64 times which produce 16, 32 times that gave 32, 16 times which yield 64, 8 times that give 128, 4 times which yield 256, 2 times which gave 512, and 1 time which will produce 1,024. And, finally, there is one throw of which one cannot calculate the value, but which can be neglected without appreciable error, because I can suppose, offending only quite slightly the equality of chance, that there might be 1,025 rather than 1,024 throws which only produce one _ecu_; besides, the equivalent of this throw cannot be more than 15 _ecus_ at the outside, since we have seen that for one throw of this game, all the terms beyond the 30th of the series give sums of money so great that they do not exist, and that consequently, the greatest equivalent one can suppose is 15 _ecus_. Adding together all these _ecus_ that I ought naturally to expect from the indifference of chance, I have 11,265 _ecus_ for 2,048 throws. Thus this line of reasoning gives very close to 5 1/2 _ecus_ for the equivalent, which nearly accords with the experience of 1/11. I am quite aware that it may be objected that that kind of calculation which gives 5 1/2 _ecus_ as equivalent when playing 2,048 throws, will give a greater equivalent if one added a much greater number of throws. Because, for example, it appears that if instead of playing 2,048 times, one only played 1,024, the equivalent is a little less than 5 _ecus_; and if one played only 512 throws, the equivalent is very close to 4 1/2 _ecus_; and if one played 256 times, it is not more than 4 _ecus_, and thus it always diminishes. But the reason for this is that the throw which one is not able to calculate the value of makes then a considerable part of the whole, and becomes more considerable as one plays fewer throws. Consequently, it is necessary that in a large number of throws, such as 1,024 or 2,048, this throw ought to be considered as of slight value, or even of no value. Following the same line of thought, one

Buffon's <u>Moral Arithmetic</u>   71

finds that if one plays 1,048,576 throws the equivalent by this line of reasoning will prove to be close to 10 <u>ecus</u>. But one ought to consider all "morally," and thus one will see that it is not possible to play 1,048,576 in possible to play 1,048,576 in this game, for supposing that it only takes two minutes for each throw, paying, etc., one would find that it would be necessary to play 2,097,152 minutes, or more than 13 years in a row, playing 5 hours per day, which is a supposition morally impossible. And if one pays attention to it, one will find that between one throw and the greatest number of throws morally possible, the reasoning which gives various equivalents for all the different number of throws, gives 5 <u>ecus</u> as a median [moyen]. Thus I persist in saying that the sum equivalent to Peter's expectation is 5 <u>ecus</u>, instead of half of an infinite sum of <u>ecus</u>, as the mathematicians would have it, and as their calculations appear to prove.

XIX

Let us see now if after this determination it will not be possible to draw out the proportion of the value of money to the advantages which result from it.

The progression of probabilities is 1/2, 1/4, 1/8, 1/16, 1/32, 1/64, 1/128, 1/256, 1/512...$1/2^\infty$.

The progression of sums of money to be obtained is 1, 2, 4, 8, 16, 32, 64, 128, 256...$2^{\infty-1}$.

The sum of all the probabilities multiplied by that of all the sums of money to be obtained is $\infty/2$, which is the equivalent given by mathematical calcuation as the expectation of him who can only win. But we have seen that this sum $\infty/2$ is not able, in reality, to be more than 5 <u>ecus</u>. It is thus necessary 2 to look for a series such that the sum multiplied by the series of probabilities would be equal to 5 <u>ecus</u>; and, that series being geometrical, as that of the probabilities, one will find that it is 1, 9/5, 81/25, 729/125, 6561/625, 59049/3125, instead of 1, 2, 4, 8, 16, 32. Now, this series 1, 2, 4, 8, 16, 32, etc., represents the quantity of money, and, consequently, its numerical and mathematical value.

And the other series, 1, 9/5, 81/25, 729/125, 6561/625, 59049/3125, represents the geometrical quantity of money given by experience, and consequently its moral and actual value.

There, consequently, is a general estimation of the value of money in all cases, as just as possible, and independent of any supposition. For example, one sees, in comparing the two series, that 2,000 <u>livres</u> does not produce double the advantage of 1,000 <u>livres</u>, that it lacks 1/5 of this, and that 2,000 <u>livres</u> is only in moral and real experience equal to 9/5 of 2,000 <u>livres</u> [sic: should be 1,000 <u>livres</u>] that is to say, 1800 <u>livres</u>. A man who has 20,000 <u>livres</u> of wealth ought not consider it as double the wealth of another who has 10,000

<u>livres</u>, for it is really only equivalent to 18,000 livres of that same kind of money whose value is computed by the advantages which result from it. And by the same token, a man who has 40,000 <u>livres</u> is not 4 times as rich as he who has 10,000 <u>livres</u>, for in reality he is only worth 32,400 <u>livres</u> by comparison. A man who has 80,000 <u>livres</u>, has, by the same rule, only 58,300 <u>livres</u>; he who has 160,000 <u>livres</u>; ought only to count 104,900 <u>livres</u>, that is to say, although he may have 16 times as much wealth as the first man, he hardly has 10 times as much of our true money. Again, a man who has 32 times as much money as another, for example, 320,000 <u>livres</u> in comparison to a man who has 10,000 <u>livres</u>, is only really rich to the extent of 188,000 <u>livres</u>, that is to say, 18 or 19 times richer, instead of 32 times richer, etc.

The miser is like the mathematician: both estimate money by its numerical quantity. The man of sense considers neither the mass nor the number of money, he only sees the advantages which he is able to get from it. He reasons better than the miser, and feels better than the mathematician. The <u>ecu</u> which the poor man has set aside for payment of a necessary tax, and the <u>ecu</u> which tops off the sacks of a financier are both the same value to the miser and the mathematician; the latter counts them as two equal units, while the other takes them with equal pleasure. The man of sense, however, counts the <u>ecu</u> of the poor as a <u>louis</u>, and the <u>ecu</u> of the financier as a <u>liard</u>.

XX

Another consideration which comes to the support of this estimation of the moral value of money is that a probability ought to be regarded as nul when it is less than 1 in 10,000, that is to say, when it is as small as the insensible fear of death within 24 hours. It could even be said that, considering the intensity of all the other sentiments of fear or hope, one ought to regard as nul a fear or hope which would have only 1/1000 probability. The most credulous of men would be capable of drawing lots without any emotion if the death ticket were mixed with 10,000 life tickets; and the steady man could draw without fear if this ticket were mixed with a thousand. Thus, in every case wherein the probability is less than a thousandth, one ought to regard it for all intents and purposes as nil. Now since in our question, the probability is found to be 1/1024, from the 10th terms of the series 1/2, 1/4, 1/8, 1/16, 1/32, 1/64, 1/128, 1/156, [sic: i.e. 256], 1/512, 1/1024, it follows that from a moral point of view we ought to disregard all the following terms and limit all our hopes to this tenth term, which gives 5 <u>ecus</u> again as the equivalent for which we are looking, and consequently confirms the justice of our determination.

Buffon's <u>Moral Arithmetic</u>

In thus reforming and abridging all calculations where the probability becomes less than one is a thousand, there remains no more a contradiction between mathematical calculation and common sense. All difficulties of this sort thus disappear. The man informed by this truth will no longer give himself over to vain hopes or unfounded fears. He will not voluntarily risk his <u>ecu</u> in order to get a thousand, unless he sees clearly that the probability is greater than one in a thousand. Finally, he will cure himself of the delusion of acquiring a large fortune with small means.

I acknowledge to M. Bernoulli that since the ten-thousandth is taken from the Tables of Mortality which always represent only the average man, that is to say, men in general, well or sick, healthy or infirm, vigorous or weak, the odds are perhaps a little more than 10,000 to one that a well, healthy, and vigorous man will not die within twenty-four hours. But still much is lacking to show that this probability ought to be increased to 100,000. For the rest, this difference, however great, does not change any of the main consequences which I elicit from my principle.

NOTES

*George Louis Le Clerc, Comte de Buffon, "Essai d'Arithmetique morale," in <u>Histoire naturelle, générale et particulière. servant de suite a l'Histoire naturelle de l'homme. Supplement.</u> Vol. IV. pp. 46-148 (Paris, 1777.) In Sir Harold Hartley and Duane H.D. Roller, eds., <u>Landmarks of Science.</u> New York: Readex Microprints; 1969.

[1] I say for us, or rather, for our latitude [<u>climat</u>], for that would not be exactly true for the polar latitude.

[2] I communicated this idea to M. Daniel Bernoulli, one of the greatest geometers of our century, and most versed of all in the science of probabilities. Here is the response which he made by his letter dated from Basle, March 19, 1762:

> I strongly approve, Monsieur, your manner of estimating the limits of moral probabilities. You consult the nature of man, in his actions, and you suppose in fact that no one disturbs himself in the morning with the thought that he might die that day. The chances of dying being, according to you, one in 10,000, you conclude that one ten-thousandth of a probability ought not to make any impression on man's mind, and consequently this ten-thousandth ought to be regarded as

> absolutely nothing. This is without doubt
> to reason like a mathematical philosopher
> [en mathematicien philosophe]. But this
> ingenius principle appears to lead to a
> smaller quantity, for the exemption from
> fear is not assuredly for those who are
> already ill. I do not argue with your
> principle, but it appears rather to lead to
> 1/100,000 rather than to 1/10,000.

[3][This seems to be an inconsistent analysis, and appears to have been seen as such by later commentators. See, e.g., Julian Lowell Coolidge, The Mathematics of Great Amateurs (Oxford: Clarendon Press; 1949), p. 173: "I can see no justification for estimating the value of the sum in one case in comparison with what he had at the end and in the other case on what he had at the beginning; it would be more logical to say that the winner had gained what was one-third of what he then had while the loser had lost 100 percent of what he still owned." Also, Isaac Todhunter, A History of the Mathematical Theory of Probability (Cambridge & London: Macmillan; 1865), p. 345: "Buffon does not seem to do justice to his own argument such as it is ...."]

[4]Here is what I said then in a note to M. Cramer, for I have preserved a copy of the original.

> M. de Montmort is content to reply to M.
> Nic. Bernoulli that the equivalent is equal
> to the sum of the series 1/2, 1/2, 1/2,
> 1/2, etc., ecus continued to infinity, that
> is to say, $= \infty$, and I do not believe that,
> as a matter of fact, one can challenge his
> mathematical calculation. However, far
> from giving an infinite equivalent, there
> is no man of common sense who would give 20
> ecus, or even 10.
> The reason for this contradiction
> between mathematical calculation and common
> sense appears to me to consist in the
> slight proportion which there is between
> money and the advantage which results from
> it. A mathematician in his calculation
> only thinks of money as a quantity, that is
> to say, according to its numerical value.
> But the moral man ought to think of it
> otherwise, and uniquely according to the
> advantages or the pleasure which it can
> procure. It is certain that the man of
> common sense ought to conduct himself
> according to this view, and only think of
> money in proportion to the advantages which

result from it - not relative to its quantity which, beyond certain limits can in no way increase his happiness. For example, he would hardly be happier with 1,000,000,000 than with 100,000,000, nor with 100,000,000,000 than with 1,000,000,000. Thus, past certain limits, he will make a very great mistake should he risk his money. If, for example, 10,000 ecus should be his total wealth, it would be an infinite wrong to chance it; and the more these 10,000 ecus should be central to him, the more wrong would he be to do so. I say then that his error [wrong] would be infinite, in so far as these 10,000 ecus should be a part of the necessities of his life, that is to say, insofar as these 10,000 ecus should be absolutely necessary for living in the fashion after which he was raised and as he has always lived. If these 10,000 ecus are part of the superfluities of his life, his error diminishes, and the more this amount is a small part of his superfluities, the more his error will diminish. But it will never cease being an error, unless he is able to regard this part of his superfluous income as indifferent. Otherwise he may regard the sum hoped for as necessary for success in a scheme which will give him proportionate returns, that is, returns which will bring him as much pleasure as the sum he hopes for is greater than that which he risks. And one cannot give rules for this fashion of envisaging a bounty to come. There are people for whom the hope itself is a pleasure greater than those which they might be able to procure by the enjoyment of their stake. In order to reason then more certainly about all these things, it would be necessary to establish some principles. I might say, for example that the "necessary" is equal to the sum which one is obliged to spend in order to continue to live as one has always lived. The "necessary" sum for a king will be, for example, 10,000,000 in rent (for a king who should have less would be a poor king). The "necessary" sum for a man of station will be 10,000 livres in rent (for a man of position who should have less would make a poor lord). The "necessary" sum for a peasant will be 500 livres, because unless he would live in misery, he needs at least this amount in order to live and feed his family. I would suppose that the "necessary" would not be able to procure new pleasures for us, or, to speak more exactly, I would count for nothing the pleasures or advantages which we have always had, and after that, define the superfluous as that which would secure for us other pleasures or new advantages. I would further say that the loss of the necessities is something which makes for infinite pain, and that thus it cannot be compensated for by any hope. But on the contrary, the sense of the loss of the superfluous is limited, and consequently it is possible to be compensated for it. I believe that one feels this himself when one plays, for the loss, be it small or large, always pains us more than an equal gain pleases us, and that without going into consideration of mortified <u>amour propre</u>, since I am supposing play of complete and sheer chance. I should also say that there is a direct proportionately between the quantity of money in "the necessities" and the amount of these necessities which it secures to us; but in the superfluous, the proportionately decreases as the superfluities grow. You, Monsieur, may be judge of these ideas, etc. Geneva, this 3rd day of October, 1730. (Signed) Le Clerc de Buffon.

[5] It is for this reason that one of our most capable geometers, the late M. Fontaine, inserted into the solution which he offered to this problem the declaration of Peter's estate, for, in effect, Peter could only offer as an equivalent his total assets. See this solution in the Mathematical Memoirs of M. Fontaine (Paris, 1764).

5. Buffon on Newton's Law of Attraction (1749) (selected)

Phillip R. Sloan

The short treatise which follows provides a critical insight into Buffon's theoretical reflections at the point when he was moving from work in the physical sciences to his mature work in geology, biology and natural history.

Occasioned by Alexis Clairaut's claim that the perturbations of the lunar apse could not be accounted for by Newton's original formulation of the inverse square law of universal attraction, Clairaut had generated a significant controversy in the French Academy of Sciences that drew Buffon to the defense of Newton.

In his reply to Clairaut, we observe first Buffon's enunciation of a claim that was to be critical to his subsequent cosmological theory--namely, that attraction and the inertial force of rectilinear motion could be considered as <u>two separate</u> and <u>opposed</u> forces, which produced the planetary motions as their resultant dynamic equilibrium, and could even serve to generate the solar system from an initial mass of matter, knocked off the sun and given its initial impulsion by the force of a colliding comet.[1] Although this would seem implicit in Newton's own formulations, Buffon's concretizing of inertia as an opposing force is opening the way to a concept of a polarity of natural forces that would be developed particularly by the German natural philosophers subsequently.

We also observe in this discussion a principle that would subsequently prove important for his theory of form and organization in biology. In his discussion, Buffon argues that the action of the law of attraction in the physical world requires that it be manifest by means of immanent, particularizing forces, whose action accounts for the gap between theory and observation, rather than allowing this discrepancy to be accounted for by purely "abstract" mathematical manipulation.[2] In this Buffon has laid a theoretical foundation for arguing that specifying forces can so instantiate the general force of attraction that one can even account for the structuring and permanence of natural species, and also the organization of the embryo, purely by the action of Newtonian microforces.

Clairaut, it should be noted, came eventually to grant the adequacy of the inverse square law, but the underlying point of dispute between Buffon and him was never to be fully resolved.[3]

NOTES

[1] See selection from Buffon's "On the Formation of the Planets," below, p. 156.

[2] See discussion in "General Introduction," above, pp. 21-22.

[3] On this controversy, see Philip Chandler, "Clairaut's Critique of Newtonian Attraction: Some insights into his Philosophy of Science," Annals of Science 32 (1975): 369-78. Chandler adopts a generally favorable attitude to Clairaut's position on this matter and attributes Buffon's position to metaphysics. The reasons why Buffon would argue such a point are not, however, explored.

REFLECTIONS ON THE LAW OF ATTRACTION*

Translated by Phillip R. Sloan

The motion of planets in their orbits is a motion composed of two forces. The first is a projective force, which would be exerted along the tangent of the orbit if the continuous effect of the second [force] were to cease instantaneously. The second force tends toward the sun, and, by its effect, would drive the planets toward the sun if the first force happened, on its part, to cease for a single instant.

The first of these forces can be regarded as an impulsion whose effect is uniform and constant, and which has been communicated to the planets since the formation of the planetary system. The second may be considered as an attraction toward the sun, and must be measured, like all qualities originating from a center, by the inverse ratio of the square of the distance, just as, in fact, one measures the qualities of light, smell, etc. and all other quantities or qualities which propagate themselves in a right line, and are related to one center. Now it is certain that attraction is propagated in a right line, since there is nothing straighter than a plumb line, which hangs perpendicular to the surface of the earth, and tends directly to the center of this force, and deviates only very little from the direction of a centrally-directed ray. Thus, it can be said that the law of attraction must be the inverse ratio of the square of the distance, solely because it extends from or, what is the same thing, tends toward a center.

But as this preliminary account, as well founded as I believe it to be, might be contradicted by men who give little weight to the force of analogies, and who are accustomed only to be convinced by mathematical demonstrations, Newton believed that it would be much more valuable to establish the law of attraction by the phenomena themselves than by any other means. He has, in fact, demonstrated geometrically that if several bodies are moved in concentric circles, and if the squares of the times of their revolutions are as the cubes of their distances from their common center, the centripetal forces of these bodies are reciprocally as the squares of the distances, and that if the bodies are moved in slightly non-circular orbits, these forces are also reciprocally as the squares of the distances, provided that the apsides of these orbits are immobile. Thus the forces by which the planets tend toward the centers or to the foci of their orbits, obey, in fact, the inverse square law. And since gravitation is general and universal, the law of gravitation is uniformly that of the inverse ratio of the square of the distance. I do not believe that anyone can doubt Kepler's law, or deny that it holds for

79

Mercury, for Venus, for the Earth, for Mars, for Jupiter, and for Saturn, especially considering them separately, and as not disturbing one another, paying attention only to their motion around the sun.

Thus, every time that we shall consider only one planet or one satellite in motion in its orbit around the sun or another planet, or simply two bodies, either both in motion, or with one at rest and one in motion, we may be assured that the law of attraction exactly follows the inverse-square ratio, since, by all observations, Kepler's law is found to be true, as much for the principal planets as for the satellites of Jupiter and Saturn. However, an objection could be drawn here from the motions of the moon (which are irregular to the point that Mr. Halley called it "an obstinate star,")[1] and principally from the motion of its apsides, which are not immobile, as is demanded by the geometrical supposition, upon which is based the result drawn from the inverse ratio of the square of the distance as the measure of the force of attraction between the planets.

To that [objection] there are several means of responding. First, it could be said that as the law is observed to be generally exact for all the other planets, a single phenomenon in which this same exactitude is not found to hold cannot destroy this law. It can be regarded as an exception for which a special reason must be sought. In the second place, one can respond, like Mr. Cotes, that even if one were to grant that the law of attraction is not, in this case, exactly in the ratio of the inverse square of the distance, but is a little greater, this difference can be estimated by the calculation, and it will be found almost insignificant, since the ratio of the centripetal force of the moon, which, is the most troubling of all, approaches sixty times closer to the ratio of the square of the distance than to its cube:

> ...One may give this answer, that, though we should grant that this very slow motion arises from a slight deviation of the centripetal force from the law of the square of the distance, yet we are able to compute mathematically the quantity of that aberration, and find it perfectly insensible. For even the ratio of the lunar centripetal force itself, which is the most irregular of them all, will vary inversely as a power a little greater than the square of the distance, but will be well-nigh sixty times nearer to the square than to the cube of the distance. But we may give a truer answer,...[2]

In the third place, we must more positively respond that this motion of the apsides does not follow from the fact that the law of attraction slightly exceeds the inverse ratio of the square of the distance, but from the fact that the sun acts on the moon by a force of attraction which must disturb its motion and produce that of the apsides. As a consequence, it alone could properly be the cause which prevents the moon from following the Keplerian rule exactly. Newton has calculated, in this light, the effects of this perturbing force, and he has drawn from his theory the equation and the other motions of the moon with such precision, that they correspond very exactly, and almost to a few seconds, to the observations made by the best astronomers. But, to speak only of the motion of apsides, he makes evident from the XVth proposition of the first book [of the _Principia_], that the progression of the apogee of the moon comes from the action of the sun, so that up to this point, all is in accord, and his theory is found to be as true and also as exact in the most complicated case, as in the least.

However, one of our great geometers,[3] has claimed that the absolute quantity of motion of the apogee cannot be derived from the theory of gravitation, as it has been established by Newton, because in employing the laws of this theory, it is found that this motion can only be completed in eighteen years, whereas it is completed in nine years. In spite of the authority of this able mathematician, and the reasons he has given to support his opinion, I have always been convinced, as I still am today, that Newton's theory agrees with the observations. I will not here undertake the necessary examination in order to prove that it has not fallen into the error with which it is reproached. I find that it is quicker to affirm the law of attraction as it is, and make one see that the law which Mr. Clairaut has wished to substitute for that of Newton, is only a supposition which implies contradiction.

Let us allow for a moment that which Mr. Clairaut claims to have demonstrated -- that by the theory of mutual attraction, the motion of the apsides must be made in eighteen years, rather than nine. And let us remember at the same time that with the exception of this phenomenon, all others, however complicated they might be, accord, on this theory, very exactly with the observations. To judge first by probabilities, this theory must stand, since there are a very considerable number of cases[4] where it corresponds perfectly with nature, but only a single case where it differs from it, and it is easy to be deceived in the enumeration of the causes of a single particular phenomenon. Thus it would appear to me that the first notion which must occur to us is the necessity to seek the particular explanation of this singular phenomenon. And it seems to me that such a one can be imagined. For example, if the magnetic force of the earth could be, as Newton said, entered into the calculation, perhaps it would be found that it influences the motion of the moon, and that it could produce this acceleration in the motion of the apogee. And in this case it would, in fact, be necessary to employ two terms in order to express the measure of the forces which produced the motion of the moon. The first term of the formula will always be that of the law of universal attraction, that is to say, the exact inverse ratio of the square of the distance, and the second term will represent the measure of the magnetic force.

Without doubt, this supposition is better founded than that of Mr. Clairaut, [whose thesis] would seem to be much more hypothetical and subject, moreover, to some insurmountable difficulties. To express the law of attraction by two or more terms, and to add to the inverse ratio of the square of the distance a double squared fraction, replacing $1/x^2$ with $1/x^2 + 1/mx^4$, would seem to me to be nothing other than adjusting a formula in such a way that it corresponds to every case. This formula no longer expresses a physical law. Because, by once allowing the addition of a second, a third, or a fourth term etc., a formula could be found which would, in all laws of attraction, be able to represent any given case, adjusting it at the same time to the motions of the apogee of the moon and to the other phenomena. As a consequence, this supposition, if it were allowed, would not only destroy the law of attraction as an inverse-square of the distance, but would even grant admission to all the possible laws imaginable. A law in physics is only a law because it is simply measured, and the scale[5] which represents it is not only constant, but is also unique and cannot be represented by some other. Each time the expression of a law is not represented by a single term, the simplicity and unity of the expression, which forms the essence of the law, no longer remains, and as a consequence, there is no longer a physical law.

As this last reflection might appear to be only a metaphysical one, and because there are few people who would know how to judge this, I will try to make it evident by explaining myself further. Thus I say that every time one would wish to establish a law on the augmentation of dimunition of a physical quality or quantity, one is strictly compelled to employ only one term to express this law. This term is the representation of the measurement which must vary as the quantity being measured in fact varies. So that if the quantity, at first being only an inch, becomes in turn a foot, an ell, a toise, a league etc., the term which expresses this becomes successively all of these things, or rather represents them in the same order of magnitude. And it is the same for all other ratios in which one quantity is able to vary.

By whatever means we thus would suppose a physical quality could vary, since this quality is unitary, its variations will be simple and always expressible by a single term, which will be its measurement. But as soon as one would wish to employ two terms, the unity of the physical quality will be destroyed, because these two terms will represent two different variations in the same quality, i.e. two qualities in place of one. Two terms are, in fact, two measurements, two separate and unequal variables. Thenceforth, they cannot be applied to a simple subject, or to a simple quality. If two terms are allowed to represent the effect of the central force of a star, it is necessary to admit that in place of one force there are two, in which one will be relative to the first term, and the other relative to the second. From which it is clearly seen in the present case, that Mr. Clairaut must necessarily acknowledge another force, different from attraction, if he utilizes two terms in order to represent the total effect of the central force of one planet.

I do not know how it can be imagined that a physical law, such as that of attraction, could be expressed by two terms related to the distances. Because, if there were, for example, a mass M in which the attractive force[6] was expressed by $a^2/x^2 + b/x^4$, would there not result the same effect as if this mass was composed of two different materials, as, for example, by $1/2\,M$, for which the law of attraction would be expressed by $2a^2/x^2$, and $1/2M$, for which the attractive force would be $2b/x^4$? This seems absurd to me.

But independent of these impossibilities that the supposition of Mr. Clairaut implies, which also destroy the unity of the law on which the truth and beautiful simplicity of the system of the world is founded, this supposition suffers from many other difficulties that Mr. Clairaut must, it seems to me, resolve before we grant it. He should at least begin by first examining all the particular causes which could produce the same effect. I feel that if like Mr. Clairaut I had solved the three-body problem, and if I had found that the theory of gravitation in fact gave only half the motion of the apogee, I would not have drawn the conclusion that he draws against the law of attraction. Furthermore it is this conclusion that I oppose, and I do not believe that one is forced to support it, when even Mr. Clairaut could have shown the inadequacy of all other particular causes.

Newton says:

> In these computations I do not consider the magnetic attraction of the earth, whose quantity is very small and unknown: If this quantity should ever be found out, and the measures of degrees upon the meridian, the lengths of isochronous pendulums in different parallels, the laws of the motions of the sea, and the moon's parallax, with the apparent diameters of the sun and moon, should be more exactly determined from phenomena: we should then be enabled to bring this calculation to a greater accuracy.[7]

Does this passage not prove very clearly that Newton had not claimed to make the enumeration of all the particular causes, and does it not in fact indicate that if some differences are found between his theory and the observations, this could come from the magnetic force of the earth, or from some other secondary cause? As a consequence, if the motion of the apsides does not also accord exactly with his theory as a whole, does this necessitate subverting the theory at its foundation by changing the general law of gravitation? Or rather, would it not be necessary to attribute this deviation, which can only be found in this single case, to other causes? Mr. Clairaut has proposed a difficulty for the Newtonian system, but this is only, at most, a difficulty which must not be allowed to become a principle. It is necessary to search for its resolution, and not make it a theory for which all the consequences are only supported by calculation. Because, as I have said, all things cannot be represented by calculation, and nothing is made real by it. And if the addition of one or several terms to the formula[8] of a physical law, like that of attraction, is allowed, we are given no more than something arbitrary, in place of having reality represented to us.

For the remainder, it is sufficient for me to have established the reasons which make me reject the supposition of Mr. Clairaut. These lead me to believe that far from having struck a blow at the law of attraction and overthrowing physical astronomy, they seem to me, to the contrary, to remain within it in all its forcefulness, and have the potential to extend it much further. And that, without claiming that I have said a great deal, is much closer to a complete account of this matter, which I would hope would be given unprejudice and full attention in order to form a better opinion on it. . . .

NOTES

*"Reflexions sur la loi de l'attraction," Histoire naturelle, générale et particulière, servant de suite a la theorie de la terre et d'introduction a l'histoire des mineraux (Supplément à l'histoire naturelle, tome I) (Paris: Imprimerie royale, 1774), pp. 126-142. First published in Memoires de l'Académie royale des Sciences année 1745 (Paris, 1749). In register of the Académie this is listed as having been delivered in 20 and 24 January, 1748. See L. Hanks, Buffon avant l'histoire naturelle (Paris: Presses Universitaires de France, 1966), p. 280. This text has been translated once previously in the highly-flawed translation of Buffon's Natural History by William Kenrick (London, 1775) V, 43-51. I have consulted this on some issues of translation.

[1][Sidus contumax.]

²[Editor's preface to the second edition of _Principia Mathematica_ edited by Roger Cotes (London, 1713) as given in: _Sir Isaac Newton's Mathematical Principles of Natural Philosophy_, trans. Andrew Motte, rev. F. Cajori (Berkeley: University of California Press, 1971), I, p. xxiii. Buffon quotes the Latin of Cotes' original.]

³See the _Memoires de l'académie des Sciences_, année 1745.

⁴[_choses._]

⁵[_échelle._]

⁶[_vertu attractive._]

⁷[_Newton's Mathematical Principles_, Motte-Cajori trans., op. cit., II, 484. Buffon's quote is from the Latin second edition, vol. III, p. 547.]

⁸[_la suite de l'expression._]

PART II:

The Emergence of the Natural History

Figure 2 - Stylized depiction of the *Cabinet du Roi*

6. The "Initial Discourse" to Buffon's *Histoire naturelle* (1749)
John Lyon

The "Premier discours: De la manière d'étudier et de traiter l'histoire naturelle" is the locus classicus of Buffon's methodological reflections, though similar considerations may be found at other places in the body of his works.[1]

Buffon's Histoire naturelle has been called the De rerum natura of the eighteenth century.[2] And Buffon obviously intended the "Initial Discourse" to be the proper introduction to his massive work. It is also the best introduction to the mind of Buffon himself. Though the Histoire naturelle has been published, or republished, in English at least four times,[3] in each case the "Initial Discourse" was eliminated from the English text by the translator.

"Style is the man," Buffon declared in his "Discours sur le Style" (1753). Of no one would this assertion be more characteristic than of Buffon himself. One finds Buffon the naturalist emphasizing either as principles of procedure or as hallmarks of nature precisely those qualities that characterize his prose style: "continuity," "plenitude," and Cartesian precision.[4] This grand view of nature in the "Discours" (a view Buffon later modified significantly), introduces the reader to a delicately nuanced and profoundly ambiguous approach to nature that was to become explicitly evolutionary in the work of Buffon's disciple Lamarck, and to be brought to maturity in the work of Charles Darwin.

For the Buffon of the "Initial Discourse" all nature was a matter of gradations or nuances. Species were made by the mind of man rather than by the hand of God. Although there is no suggestion of the transforming power of time in the "Initial Discourse," Buffon proposes there a scheme of the nature of things in which each individual is part of an infinitely nuanced chain of being. It is a "metaphysical error," he insists, to search for a basis for systematization in nature. The more divisions a would-be systematizer makes in nature, the closer he approximates "truth."

This brings us to a philosophic problem of which Buffon seems to be unaware, for he touches upon it several times without comment. Essentially, it is a problem of epistemology, and it involves Buffon's use of certain unexamined presuppositions that are prejudicial to much of his argument, turning it into a tautology at critical points. Two examples of the problem will be given here.

The problem is perhaps most strikingly presented at the end of the essay, when Buffon is prescribing the "natural order" of procedure in natural history. Working from the general to the particular, or from the particular to the general, are both allowable, he says: "Both ways are good, and the choice of one or the other depends more on the bent of the author than on the nature of things, which always allows of being treated equally well by either method" (emphasis mine). A second example of this difficulty occurs about a quarter of the way through the essay when Buffon, commenting on "artificial" methods of systematization notes that the division of the plant kingdom according to size is inadequate, for, as others have noted, "there are within the same species, such as that of the oaks, great variation in sizes."

Lurking behind both these examples seems to be a commonsense view of nature in which, for example, it is just as indubitable that a real species of oaks exists as it is that man is intuitively aware of the nature of nature. Otherwise one could not beg one of the questions at issue by appealing to the real objective existence of the species "oak" to disprove a system of taxonomy based on size, in a general argument in which what is at issue is the natural manner of classifying nature.

How does Buffon know that nature always allows of being treated equally well by induction or by deduction? Intuitively? Pragmatically? For one certainly gets the sense that Buffon, or the investigator of nature, is in fact in possession of real, objective knowledge of the truth about the ultimate nature of things. Yet at the same time he is affecting to investigate the ultimate nature of things with an eye to finding out what he does not yet know. If the plant kingdom were in fact articulated according to size, there would be no place for a category or species such as "oaks." Yet Buffon writes as if all men of good faith know indubitably that there is indeed an objective species "out there" in nature, namely, the oaks, large and small, majestic and shrubby. However, Buffon appeals to this knowledge that supposedly all men of good faith possess in the course of an inquiry into the nature of nature, an inquiry in which the presumption or pretension is that the inquirer is not initially in possession of the knowledge he seeks.

This problem, under whatever rubric one wishes to subsume it, runs throughout the essay. To conceive of it in other terms than those previously used, one might consider what Buffon calls the "metaphysical error" of judging a whole by a part. In order to show the inadequacy and artificiality of methods that organize "wholes" in nature according to the relations of a single part, Buffon has to presuppose that nature is a chain of beings infinitely nuanced, <u>and that this presupposition is incontestable</u>. This indeed he does, though it does not seem that he does it consciously. For if his position that only the individual is real while species are the creation of the mind of man[5] is to be held, it can only be held as an unexamined presupposition. How could it be verified? Among the problems of verification would be the logical problem of how the "mind" of "man" --both terms being collective abstractions -- could possibly discover that everything in nature is individual without thereby extruding itself so completely from nature as to make any comprehension of nature impossible to man. The substitution of "minds" for "mind" and "men" (or "persons") for "man" would accomplish little here either, for the problem of communication would then replace the problem of knowing.

The operating synthesis out of which Buffon works seems to rest on two diametrically opposed postulates: (1) that it is man's mind which categorizes nature into species, while nature knows only the individual; and (2) that truth is a correspondence between man's understanding and the objective nature of things. But the ability not to be troubled by what one trusts will be a <u>fruitful</u> inconsistency almost seems to be a pre-requisite for dealing with the problem of species.[6]

To move from the logical to the psychological and morphological, Buffon's criticism of Linnaeus is manifestly unfair and, it appears, ignobly motivated. His treatment of Linnaeus's system of nature reflects a confusion about the general and the particular, and the nature of "truth." In the text, for example, Buffon criticizes Linnaeus for dividing the animal kingdom into only six classes. The more particulate divisions the merrier, Buffon implies, for so do we come closer to the nature of reality. But he also suggests that combination and generalization are a higher degree of knowledge than mere observation and description; and he says that the more general are the divisions we make, the less danger do we run of making inaccurate descriptions. Though serious qualifications follow this statement, and though it is obvious that what Buffon is proposing here is that the true method in the sciences treads between the obvious and the incomparable, it also appears obvious that he is trimming his logic in order to attack Linnaeus (who, of course, admitted that his system was not "natural").[7] The attack then takes the form of a <u>reductio ad absurdum</u>; for example, who in his right mind would classify a mole as one of the <u>Ferae</u>, ferocious beasts of prey?

It is at this point that Buffon makes the most perplexing comment of the entire essay, and a slight digression must be made to accommodate this anomaly. In criticizing Linnaeus's divisions of nature, he points to one of the general characters that Linnaeus uses for this purpose, namely, the mammary glands, and says that if this character is to be useful as an identifying mark of the quadrupeds, then all quadrupeds should have it. "However," he continues, "it has been known since Aristotle that the horse has no mammary glands whatsoever" (p. 115).[8]

There are aspects of Buffon's "Discourse" that <u>appear</u> to be quaint. For instance, he suggests an arrangement or systematization of the things of nature according to their relation to man. This method he calls "simple" and "natural," while all other methods are artificial and contrived. And even on this issue Buffon seems somewhat inconsistent, for later in the essay, he denigrates the "Ancients" utilitarian approach to botany, an approach in which only those plants somehow useful to man and thus having some practical relation to him were catalogued and described.

Buffon may, however, be more consistent and less anthropocentric than he at first appears. The relativism and subjectivism implied in his utilitarian systematization may only be instrumental to an ultimately objectivist view of nature. Buffon may simply be suggesting that, whether we would prefer to do so or not, we must start with subjective perceptions, though we need not end with them. This interpretation would also undo the charge of inconsistency leveled at Buffon in this instance. For his charge against the ancient botanists would not then be that they started from human convenience in their arrangement of plants, but that they never progressed beyond such a short-sighted view.[9]

Another apparently paradoxical aspect of Buffon's approach in the "Discourse" is his rejection of the use of the microscope. This may, however, be partially a ploy designed to make Linnaeus look ludicrous. For example, one of the apparent reasons why he opposes Linnaeus system is that the microscope is necessary to such classification as this system proposes. Yet Buffon had no apparent objection to the use of the magnifying glass in such pursuits; and with the assistance of John Turberville Needham, Buffon had done notable work with the microscope.[10] Later on in the essay the gentlemanly dilettante shows up once more, though again in the somewhat strained context of making debater's points against Linnaeus. In the identification of living beings, Buffon relegates such things as internal or minute organs to the rank of mere "minor details," while calling easily recognizable external features such as size, form, and color the "main and essential things."[11]

The core of the "Discourse" lies in Buffon's suggestions about the "manner of properly conducting one's mind in the sciences." It is best to let Buffon speak for himself here, and it is to be hoped that the translation allows him to do just that.

Buffon himself was obviously trying to tread the line between the two extremes he warns of: the absence of a system, which results from too diffuse observation, and too restricted a system, which results from the willingness to apply mental categories on the basis of too narrow an observation. He is attempting to fit the precise observation of many things within a framework of human usefulness tempered by an awareness of the independently real. Though he does not seem to be as reflectively aware of the limitations and structure of the mind as we would like to think we are, Buffon is presenting us with an ideal of scientific humanism that is both admirable and useful. The "manner of properly conducting one's mind in the sciences," which is the burden of the "Discourse," depends more on the "esprit de finesse" than on the "esprit geometrique," and lies in the ability to distinguish what we put in an object from what we find there.

Perhaps, as Buffon suggested about the philosopher's stone early in the essay, men need an "imaginary goal" to sustain them in their work.

NOTES

[1]See, e.g., Buffon's earlier translation (1735) of Stephen Hales' Vegetable Staticks (1727); his contemporaneous work Histoire naturelle de l'homme (vol. III of the Histoire naturelle, 1749); his later Époques de la nature (supplement to vol. V, of the Histoire naturelle, 1778); and his "Essai d'arithmétique morale" (dated 1777, but concerned with issues Buffon had dealt with since 1730).

[2]John Herman Randall, The Career of Philosophy (New York: Columbia University Press, 1962), I, 902.

[3]W. Kendrick, London, 1775; W. Smellie, Edinburgh, 1781-1785; J. S. Barr, London, 1792; and W. Wood, London, 1812 (Based on Smellie). We have also discovered an anonymous partial translation of the "Initial Discourse" which appeared in Vol. 9 of The Universal Magazine of Knowledge and Pleasure (London: John Hinton), July-December issue, 1751. The translation is not announced as being a translation, and no credit is given to Buffon as author of the work. It is noted at the end of the article that the subject will be continued. But we

have been unable to locate the sequel in any subsequent volume of the Magazine.

⁴See Charles Augustin Sainte-Beuve, Portraits of the Eighteenth Century, trans. Katherine P. Wormerley (New York: Ungar, 1964), II, 260-261. This essay on Buffon is Wormerley's conflation of the material in several essays in Sainte-Beuve's Causeries du lundi.

⁵Buffon of course soon dropped this radical assertion about the sole reality of the individual and turned to a concept in which species alone were real. But he did not rest there either. See Lovejoy, "Buffon and the Problem of Species," in B. Glass, O. Temkin, W.L. Strauss, Jr., eds: Forerunners of Darwin: 1745-1859. (Baltimore: John Hopkins Press; 1968); and Paul L. Farber, "Buffon and the Concept of Species," J. Hist. Biol., 5 no. 2 (Fall 1972), esp. pp. 260-261. See also on this point the assertion of David Hume: "'tis a principle generally receiv'd in philosophy, that everything in nature is individual"; A Treatise of Human Nature, bk. I, pt. I, sec. 7, in T. H. Greene and T. H. Grose, eds, David Hume: The Philosophical Works (Darmstadt: Verlag Aalen, 1964 reprint), I, 327.

⁶See Etienne Gilson, D'Aristote à Darwin et Retour (Paris: Librarie Philosophique J. Vrin; 1971). On the general issue of this inconsistency, fruitful or otherwise, and the contrary postulates it implies. cf. Arnold Toynbee: "We cannot think about the universe without assuming that it is articulated; and, at the same time, we cannot defend the articulations that we find, or make, in it against the charge that these are artificial and arbitrary, that they do not correspond to anything in the structure of reality, or that, even if they do, they are irrelevant to the particular mental purpose for which we have resorted to them. It can always be shown that they break up something that is indivisible and let slip something that is essential. . . . yet, without mentally articulating the universe, we ourselves cannot be articulate - cannot, that is, either think or will." A Study of History, Vol. XII: Reconsiderations (New York: Oxford University Press; 1964), pp. 9-10.

⁷See James L. Larson, "Linnaeus and the Natural Method," Isis, 58, no. 193 (Fall 1967) 304-320, esp. pp. 312ff.

⁸Neither causal observation nor a cursory search through Aristotle would seem to bear this out. In bk. 3 of the Historia animalium Aristotle notes: "All animals have breasts that are internally and externally viviparous, as for instance all animals that have hair, as man and the horse" (The Works of Aristotle, trans. D'Arcy Wentworth Thompson, [Oxford: Clarendon Press, 1910], IV, 521b). However, earlier in this

The "Initial Discourse" 95

same work (bk. 2, sec. 1, 500a) Aristotle writes: "Of solid-hooved animals the males have no dugs, excepting in the case of males that take after the mother, which phenomenon is observable in horses."
   The source of the difficulty here might possibly be thought to lie in the polite conventions of Buffon's age (or, more precisely, of the age of Louis XIV, in which his first years were spent). For Buffon simply says: "cependant depuis Aristote on fait que le cheval n'a point de mamelles." And "cheval" here might seem to be a "preciosite" or euphemism for "étalon" or "cheval entier," that is, "stallion." For it appears that, though stallions and geldings usually have inconspicuous rudimentary mammary glands, even these are not always present. See Charles W. Turner, The Mammary Gland. I. The Anatomy of the Udder of Cattle and Domestic Animals (Columbia, Mo.: Lucas Bros., 1952), p. 356. I am thankful to John C. Greene for providing me with this reference. It might be noted here also that in the essay on the horse in vol. IV (1753) of the Histoire naturelle Buffon sometimes uses "horse" to refer to the male of the species, and sometimes "stallion." But when he comes to describe the act of copulation, the male is an "étalon," the female a "jument"! There is no "preciosité" here. And so this statement of Buffon's remains enigmatic.

   [9]Cf. Aristotle, Historia animalium I, 6, 491a, 20, for the suggestion that because man is the animal we are most familiar with, all classification will of necessity start from him. Buffon's exact meaning in endorsing a similar starting point is not fully clear. In the introduction (p. 22 above) the possibility that it could have a Leibnizian interpretation is discussed. Another possibility is the claim that Buffon began from a position of radical sujectivism. Thus Jacques Roger insists that by the 1760's "Buffon a renonce à classer les êtres d'après les rapport qu'ils ont avec l'homme, comme il le faisait en 1749, pour les classer desormais d'après les rapports réels qu'ils ont entre eux. C'est-a'dire qu'il croit maintenant que l'homme peut saisir ces rapports réels et voir la nature telle qu'elle est: l'ordre qu'il y decouvre n'est plus relatif a l'observateur, il existe dans les choses mêmes" (Jacques Roger, Les sciences de la vie dans la pensée française du XVIIIe siècle [Paris: Colin, 1963], p. 567). Thus, Roger's interpretation leaves open the question of Buffon's quaintness and inconsistency in 1749, suggesting only that by the 1760's he had come to greater maturity of thought.

   [10]Sainte-Beuve claimed that Buffon avoided the microscope because of his "shortsightedness"; Wormeley, Portraits of the Eighteenth Century, II, 246, 265. See, however, the introduction to the selection on the generation of animals below, pp. 166-67.

[11] See text below, p. 105. This statement does not seem consistent with Buffon's simultaneous anatomical work with Daubenton, which gave an important role to internal anatomy. See "Discours sur la nature des animaux," *Histoire naturelle generale et particuliere: histoire naturelle des quadrupedes* (Paris: Imprimerie royale, 1753), IV, pp. 11-13.

Figure 3 - Original woodcut prefacing *Premier discours*

INITIAL DISCOURSE:

ON THE MANNER OF STUDYING AND EXPOUNDING NATURAL HISTORY

Translated by John Lyon

Natural history, taken in its fullest extent, is an immense subject. It embraces all objects which the universe displays to us. This prodigious multitude of quadrupeds, birds, fishes, insects, plants, minerals, etc., offers to the curiosity of the human mind a vast spectacle, the totality of which is so grand that it appears, and indeed is, inexhaustible in its details. A single division of natural history, such as the history of insects, or the history of plants, is vast enough to occupy the attention of many men. The objects which these particular branches of natural history present are so multitudinous that the most capable observers, after many years' work, have given only very imperfect rough outlines of those branches to which they have been singularly devoted. However, they have done all that they were capable of doing. And, far from blaming these observers for the trifling advancement of the science to which their work has been devoted, one could not give them too much praise for their assiduity and patience. It is impossible to deny that they possess the very highest qualities, for it takes a peculiar force of genius and courage of mind to be able to envisage nature in the innumerable multitude of its productions without losing one's orientation, and to believe oneself capable of understanding and comparing such productions. It takes a particular predilection to love these things, a predilection beyond that which has as its goal only particular objects. For it can be said that the love of the study of nature supposes two qualities of mind which are apparently in opposition to each other: the grand view of an intense intellectual power which takes in everything at a glance, and the detailed attention of an instinct which concentrates laboriously on a single minute detail.

The first obstacle encountered in the study of natural history comes from this great multiplicity of objects. But the variety of these same objects, and the difficulty of bringing together the various productions of different regions is another apparently insurmountable obstacle to the advancement of our understanding, an obstacle which in fact work alone is unable to surmount. It is only by dint of time, care, expenditure of money, and often by lucky accidents, that one is able to obtain well-preserved specimens of each species of animal, plant, or mineral, and thus form a well-ordered collection of all the works of nature.

But when specimens of everything that inhabits the earth have been collected; when, after much difficulty, examples of all things that are found scattered so profusely on the earth have been brought together in one location; and when for the first time this storehouse filled with things diverse, new, and strange is viewed, the first sensation that results is bewilderment, mixed with admiration. And the initial thought that follows is a humbling self-reflection. It seems unimaginable that, even with time, one could come to the point of distinguishing all these different objects, or that one could succeed not only in distinguishing them by their form, but further by knowing all that pertains to the birth, the generation, the organization, the habits--in a word, all that pertains to the history of each thing in particular. However, as these objects become familiar, after they have been seen often and, so to speak, without any plan, they slowly create lasting impressions, which are soon bound together in our mind by fixed and invariable relationships. Furthermore, despite ourselves, we construct more general views by which we are able to embrace at one and the same time many different objects. And it is thus that we find ourselves in a position to undertake disciplined study, to reflect fruitfully, and to open up for ourselves routes by which we may arrive at useful discoveries.

Thus, a beginning should be made by observing things often and by frequently reexamining them. However necessary attention to the whole may be, here, at the beginning, one may dispense with this responsibility: I mean that scrupulous attention which is always useful when a great number of things are undertaken, and often detrimental to those who are beginning to learn natural history. The essential thing is to fill the heads of such beginners with ideas and facts, and thus prevent them, if possible, from prematurely establishing schemata. For it always happens that through ignorance of certain facts and through a limited stock of ideas, such neophytes use up their energy in false combinations, and load their memories with vague consequences and results contrary to truth, which form in the sequel preconceptions that are difficult to erase.

In order to avoid such shortsightedness, I have said that it is necessary to begin the study of nature by very broad observation. And it is also necessary that this observation be almost at random. For if you have resolved to consider things only from a certain point of view, or in a certain order, or in a certain system, although you may have taken the best road, you will never arrive at the same breadth of knowledge to which you might lay claim if, at the outset, you allowed your mind to follow its own lead, to get to know itself, to acquire a degree of certainty without extraneous assistance, and to fashion by itself the first chain of connections which depicts the order of its ideas.

This is true without exception for all persons of mature mind and disciplined intellect. Young people, on the contrary, ought to be guided and advised in these matters. It is even necessary to encourage them by means of that which is most stimulating in science, by calling their attention to the most remarkable things without giving to such things any precise explication. For the mysterious, at that age, excites curiosity, whereas at a mature age it would only inspire aversion. The young easily lose interest in things which they have already seen. They review things with indifference unless they are presented with these same things from other points of view. And instead of simply repeating to them what has already been said, it is better to add other details, even strange or useless ones. Less is lost by deceiving them than by disgusting them.

After having seen and reviewed things many times, the young will begin to describe such things in a comprehensive way, and by themselves make divisions and perceive general distinctions. At this point the taste for science may be born, and one should step in and assist the birth. This enthusiasm, so necessary for all things, yet so hard to come by, cannot be supplied by precepts. In vain would education wish to provide it; in vain do parents compel their children to learn it. Such efforts will lead only to that end common to all men, that is, to that degree of intelligence and memory which suffices for social life or ordinary affairs. But it is to nature that one ought to ascribe that initial spark of genius, that first hint of interest of which we speak, which subsequently develops in various directions contingent upon various circumstances and purposes. In addition, the minds of young people ought to be presented with things of all kinds, with all manner of studies, and with objects of all sorts, so that they might be able to recognize the type toward which their mind tends with greater inclination, or to which they would devote themselves with greater pleasure. For its part, natural history ought to be presented to them precisely at that time when their reason is beginning to develop, or at that age when they might begin to think that they already know quite a bit. Nothing is more apt to lessen their conceit and make them feel how much there is that they are ignorant of. And, independently of this initial result, which cannot but be useful, even a slight study of natural history will elevate their ideas and give them a knowledge of an infinity of things of which the common man is ignorant and which are often encountered in the course of life.

But let us return to the man who would apply himself seriously to the study of nature, and take up again a consideration of the subject at the point at which we let it drop, namely, at the point at which the adept begins to generalize ideas and to form for himself a method of arrangement and systems of explication. It is at this point that he should consult those who are proficient in the field, read solid authors, examine their various methods, and borrow insights wherever he comes upon them. But since it ordinarily happens that one is easily carried away at this point by his affection and taste for certain authors, or for a certain method, and that often, without a sufficiently mature examination, it is easy to adopt a system which is sometimes ill-conceived, it is proper that we give here several preliminary notions about the methods that have been devised in order to facilitate a knowledge of natural history. The methods are very useful, when applied with appropriate restrictions. They shorten the work, assist the memory, and offer to the mind a series of ideas composed indeed of objects which differ among themselves but which nevertheless have certain common relations. These common relations then form stronger impressions than would be the case with discrete objects which have no connection among themselves. Therein lies the utility of the various methods. But the disadvantage here is the tendency to overextend or to unduly constrict the chain of connections, to wish to subject the laws of nature to arbitrary laws, to wish to divide this chain where it is not divisible, and to wish to measure its strength by means of our weak imagination. Another drawback which is no less serious, and which is the contrary of the one just described, is the temptation to restrict oneself to a regime of overly-detailed methods, and thus to wish to judge of the whole by a single instance, to reduce nature to the status of petty systems which are foreign to her, and, from her immense works, to fashion arbitrarily just as many unconnected assemblages of data as there are petty systems. The final disadvantage of such methods is that, in multiplying names and systems, they make the language of science more difficult than science itself.

We are naturally led to imagine that there is a kind of order and uniformity throughout nature. And when the works of nature are only cursorily examined, it appears at first that she has always worked upon the same plan. Since we ourselves know only one way of arriving at a conclusion, we persuade ourselves that nature creates and carries out everything by the same means and by similar operations. This manner of thinking causes us to invent an infinity of false connections between the things nature produces. Plants have been compared with animals, and minerals have been supposedly observed to vegetate. Their quite different organization and their quite distinct means of operation have often been reduced to the same form. The common matrix of these things so unlike each other lies less in nature than in the narrow mind of those who have poorly conceived her, and who know as little about appraising the strength of a truth as they do about the proper limits of comparative analogy. For example, since blood circulates, must it be asserted that the sap of plants circulates also? Or should it be concluded that there is a growth in minerals like that known in plants? Is it proper to proceed from the movement of the blood to that of the sap, and from that to the movement of the petrifying juice? Isn't what we are doing in these cases only bringing the abstractions or our limited mind to bear upon the reality of the works of the Creator, and granting to him, so to speak, only such ideas as we possess on the matter? Nevertheless, such poorly founded statements have been made and are repeated every day. Systems are constructed upon uncertain facts which have never been examined, and which only go to show the penchant men have for wishing to find resemblances between most disparate objects, regularity where variety reigns, and order among those things which they perceive only in a confused manner.

For, when not stopping at superficial knowledge—which only gives us incomplete ideas of the productions and methods of nature—we wish to penetrate further and examine more meticulously the form and behavior of nature's works, it is surprising what variety of design, and what a multiplicity of means we see. The number of the productions of nature, however prodigious, is only the least part of our astonishment. Nature's mechanism, art, resources, even its confusion, fill us with admiration. Dwarfed before that immensity, overwhelmed by the number of wonders, the human mind staggers. It appears that all that might be, actually is. The hand of the Creator does not appear to be opened in order to give existence to a certain limited number of species. Rather, it appears as if it might have cast into existence all at once a world of beings some of whom are related to each other, and some not; a world of infinite combinations, some harmonious and some opposed; a world of perpetual destruction and renewal. What an impression of power this spectacle offers us! What sentiments of respect this view of the universe inspires in us for its Author! And what would be the case in this regard if the weak light which guides us became sufficiently keen to allow us to perceive the general order of causes and of the dependence of effects? But the greatest mind, the most powerful genius, will never lift itself to such a pinnacle of knowledge. The first causes of things will remain ever hidden from us, and the general results of these causes will remain as difficult for us to know as the causes themselves. All that is given to us is to perceive certain particular effects, to compare these with each other, to combine them, and, finally, to recognize therein more of an order appropriate to our own nature than one pertaining to the existence of the things which we are considering.

But seeing that this is the only route open to us, and since we have no other means of arriving at a knowledge of the things of nature, it is necessary to follow that route as far as it can lead us. We must gather together all the objects, compare them, study them, and extract from the totality of their connections all the insights which may be able to assist us to see them clearly and to know them better.

The first truth which issues from this serious examination of nature is a truth which perhaps humbles man. This truth is that he ought to classify himself with the animals, to whom his whole material being connects him. The instinct of animals will perhaps appear to man even more certain than his own reason, and their industry more admirable than his arts. Then, examining successively and by order the various objects which compose the universe, and placing himself at the head of all created beings, man will see with astonishment that it is possible to descend by almost imperceptible degrees from the most perfect of creatures to the most formless matter, from the most perfectly formed animal to the most amorphous mineral. He will recognize that these imperceptible nuances are the great work of nature, and will find them not only in the size and shape of things, but in changes, productions, and successions of every sort.

In thoroughly studying this idea, one sees clearly that it is impossible to establish one general system, one perfect method, not only for the whole of natural history, but even for one of its branches. For in order to make a system, an arrangement--in a word, a general method--it is necessary that everything be taken in by it. It is necessary to divide the whole under consideration into different classes, apportion these classes into genera, subdivide these genera into species, and to do all this following a principle of arrangement in which there is of necessity an element of arbitrariness. But nature proceeds by unknown gradations, and, consequently, it is impossible to describe her with full accuracy by such divisions, since she passes from one species to another, and often from one genus to another, by imperceptible nuances. As a result, one finds a great number of intermediate species and mixed objects which it is impossible to categorize and which necessarily upset the project of a general system. This truth is too important for me not to insist on whatever might make it clear and evident.

Take botany, for example, that admirable part of natural history which, by virtue of its utility, has deserved at all times to be the most cultivated. Let us call to mind the principles of all methods which botanists have given us. We shall see with some surprise that they have always had in view the aim of comprehending in their methods generally all species of plants, and that none of them have been completely successful. It always turns out that, in each of these methods, a certain number of plants must be considered anomalous, their species falling between two genera; and it has been impossible to categorize them, because there is no more reason to ascribe them to the one genus than to the other. Indeed, to propose to devise a perfect system is to propose an impossibility. It would necessitate a work which would represent exactly all the works of nature. But, contrary to such hopes, it always happens that, despite all known methods and despite any assistance which can be had from the most enlightened system of botany, species are constantly being discovered which it is not possible to assimilate to any of the genera posited by such systems. Experience accords with reason on this point, and one ought to be convinced that it is not possible to design a general and perfect system in botany. However, it appears that the search for such a general system may be the search for a kind of "philosopher's stone" for botanists, a search which they have pursued with infinite pains and infinite labor. Some have taken forty years, and some fifty, in the creation of their systems, and what has happened in botany is the same as what has happened in chemistry, namely, that in the pursuit of the philosopher's stone—which has not been found—an infinite number of useful things have been discovered. Thus, from wishing to design a general and perfect system in botany, plants and their usages have been studied in more detail and have come to be better known. In this respect it is true that men need an imaginary goal in order to sustain them in their work. For if they had been persuaded that they would do only what in effect they are capable of doing, they would do nothing at all.

This proclivity which botanists have for establishing general systems with pretensions of perfection and methodological rigor is thus poorly founded. Consequently, their labors deliver to us only defective systems which have been successively destroyed, the one by the other, and have undergone the common fate of all systems founded on arbitrary principles. And what has contributed the most to this process of successive destruction is the freedom which botanists have allowed themselves of choosing arbitrarily a single feature of plants as a distinguishing characteristic. Some have established their method on the basis of the configuration of leaves; others on their position; others on the form of the flowers; some on the number of flower petals; other, finally, on the basis of the number of stamens. I would never finish if I wished to report in detail all the systems which have been imagined. But at the present time I wish to speak only of those systems which have had a good reception and have been followed, each in its turn, without sufficient attention having been given to that erroneous principle which all these systems share, namely, the desire to judge a whole or a combination of many wholes on the basis of a single part, and by comparing the differences of such single parts. For to desire to discern the differences of plants using solely the configurations of their leaves or their flowers as criteria is as if one set out to discern the differences between animals by means of the variations in their skins or generative organs. For who does not see that whatever proceeds in such a manner cannot be considered a science? It is at the very most only a convention, an arbitrary language, a means of mutual understanding. But no real cognizance [<u>connaissance réele</u>] of things can result from it.

Might I be permitted to speak my mind upon the origin of these various systems and upon the causes which have multiplied them to the point that botany itself is actually easier to apprehend than the nomenclature which is merely its language? May I be permitted to say that it would be preferable for a man to have engraved in his memory the forms of all plants and have clear ideas of them, which is what botany really is, than to memorize all the names which the various systems give to these plants, as a result of which scientific terminology has become more difficult than science itself? Here, then, is how it appears to me that this state of affairs has arisen. In the first place, the members of the plant kingdom were divided according to their various sizes. There are, after this fashion, large trees, small trees, shrubby trees, bushes, large plants, small plants, and herbs. This is the foundation of a classification which itself has subsequently been divided and subdivided according to other relations of size and form in order to give each species a particular character. After classification according to this plan, some people came along who have examined such a distribution, and who said that this method has been on the relative size of plants cannot be maintained, for there are within the same species, such as that of the oaks, great variations in sizes. There are some kinds of oak which rise a hundred feet in height, and others which never grow more than two feet tall. The same is true, allowance being made, of chestnuts, pines, aloes, and of an infinity of other kinds of plants. Thus it is said that the genus of plants ought not be determined by their size, since this distinction is equivocal and uncertain. And, with reason, this method has been abandoned. Next, others have appeared on the scene who, believing they can do better, have said that in order to know plants it is necessary to stick to the most obvious part of them. And, since the leaves are the most obvious feature, one should arrange plants according to the form, size, and position of the leaves. Thus one becomes familiar with another scheme or method, and follows it for awhile. But then it become evident that the leaves of almost all plants vary prodigiously with age and terrain, that their form is no more constant than their size, and that their position is still more uncertain. Thus this method has proven more satisfactory then the preceding one. Finally, someone

The "Initial Discourse"

imagined--Gesner[1],--I believe--that the Creator had put in the reproductive structures of plants a certain number of different and unvariable characters, and that it was on this assumption that one ought to try to create a system. And, since this idea turns out to be true up to a point, in that the organs of generation of plants are found to have some unique features more constantly than all the other organs of the plant taken separately, there have suddenly arisen many systems of botany, each founded nearly upon the same principle. Among these methods that of M. de Tournefort[2] is the most remarkable, the most ingenious, and the most complete. This illustrious botanist was aware of the shortcomings of a system which could be purely arbitrary. As a man of intellect, he avoided the absurdities which are found in most of the other systems of his contemporaries, and he made his allocations and his exceptions with boundless knowledge and skill. In a word, he had put botany in a position to do without other methods, and he had made it capable of a certain degree of perfection. But there arose another methodologist who, after having praised de Tournefort's system, tried to destroy it in order to establish his own. This same person, having adopted with M. de Tournefort the distinguishing characteristics drawn from fructification, then employed all the organs of generation of plants, and above all the stamens, for the purpose of dividing his genera. Holding in contempt the wise concern of M. de Tournefort not to push nature to the point of confusing, for the sake of his system, the most various objects--like trees and herbs--he put together in the same class the mulberry and the nettle, the tulip and the barberry, the elm and the carrot, the rose and the strawberry, the oak and the bloodwort. Now, isn't this to make sport of nature and of those who study her? And if all that classification were not presented with a certain appearance of mysterious order and wrapped up in Greek and botanical erudition, would one be long in perceiving the ridiculousness of such a system, or rather in pointing out the confusion which results from such a bizarre assemblage? But that is not all; and I am going to persist in this assertion, because it is proper to preserve for M. de Tournefort the glory which he merited by his sensitive and persistent labor, and because it is not necessary that those who have learned botany according to de Tournefort's system should waste their time studying that new system wherein everything is changed, even to the names and surnames of plants. I say, then, that this recent method which brings together in the same class genera of plants which are entirely dissimilar, has, furthermore, independently of its incongruities, essential shortcomings and drawbacks greater than all the methods which have preceded it. As the characters of the genera come to be set by distinctions almost infinitely small, it becomes necessary to proceed to the identification of a tree or plant with a microscope in one's hand. The size, the form, the external appearance, the leaves, all the obvious features are useless for purposes of identification. Nothing is important except the stamens, and if one is unable to see the stamens, one can do nothing, one has seen nothing of significance. This large tree which you see is perhaps only a bloodwort. It is necessary to count its stamens in order to know what it is, and, since its stamens are often so small that they escape the naked eye or the magnifying glass, one must have a microscope. But unfortunately for this system there are plants which do not have stamens. There are also plants in which the number of stamens varies, and therein lies the shortcoming of this method of classification, just as in the others, in spite of the magnifying glass and microscope.[3]

After this frank exposition of the bases upon which the various systems of botany have been constructed, it is easy to see that the great shortcoming here is a metaphysical error in the very principle of such systems. This error involves disregarding the progression of nature, which is always a matter of nuances, and wishing to judge the whole by a single part. This is a manifest error, and one that it is astonishing to find so widespread. For almost all who have systematically named things have employed only a single feature, such as the teeth, the claws, or the spurs, as a means of classifying animals, and the leaves or the flowers in classifying plants, rather than making use of all parts of the organism, and searching out the differences and similarities of complete individual specimens. To refuse to make use of all the features of objects which we are considering is voluntarily to renounce the greatest number of advantages which nature offers us as a means of knowing her. And even if one were assured of finding constant and invariable characters in the several parts taken by themselves, it would not be necessary to restrict thus the knowledge of the productions of nature to a knowledge of these constant characters, which only give particular and very imperfect ideas of the whole organism. And it appears to me that the sole means of constructing an instructive and natural system is to put together whatever is similar and to separate those things which differ. If the individual entities resemble each other exactly, or if the differences between them are so small that they can be perceived only with difficulty, such individuals will be of the same species. If the differences begin to be perceptible, while at the same time there are always many more similarities than differences, such individuals will be of different species, but of the same genus. And if the differences are even more marked, without however exceeding the resemblances, then such individuals will be not only of another species, but even of another genus than the first and the second instances, but of the same class, for they resemble each other more than they differ. But if, on the contrary, the differences exceed the similarities, such individuals are not even of the same class. This is the systematic order which ought to be followed in arranging the productions

The "Initial Discourse"

of nature. Certainly the similarities and differences will be taken not only from one feature but from the whole organism. And likewise this method of inspection will be brought to bear on form, size, external bearing, upon the various parts, upon their number and position, upon the very substance of the thing. Similarly, these elements will be used in large or small number as the occasion necessitates. And these principles will be applied in such a way that if an individual specimen, whatever its nature may be, is so singular as to be always recognizable at first sight, it will be given but one name. But if this specimen has a form in common with another, and differs constantly from it in size, color, substance, or by any other obviously sensible quality, then it is given the same name as the other, to which is added an adjective to mark the difference. And thus one proceeds, putting in as many adjectives as there are differences. By this means one will be certain to express the various attributes of each species, and there need be no fear of falling into the inconveniences of the two restricted methods of which we have spoken, and about which we have discoursed at length. This is so because of a common shortcoming of all systems of botany and natural history and because the systems which have been devised for animals are even more defective than the systems of botany. For, as we have already hinted, there has been a desire to pronounce on the resemblance and difference of animals by employing in such proceedings only the number of fingers or claws, of teeth and breasts--a project which greatly resembles that of recognition by stamens, and which is in effect of the same author.

It follows from all that we have just set forth that, in the study of natural history, there are two equally dangerous positions: the first is to have no system at all, and the second is to try to relate everything to a restricted system. In the great number of persons who currently apply themselves to this science there could be found striking examples of these two approaches, so opposed to each other and yet both equally vicious. Most of those who, without any prior study of natural history, wish to have collections of this sort are people of leisure with little to occupy their time otherwise, who are looking for amusement, and regard being placed in the ranks of the curious as an achievement. Such persons start out by purchasing indiscriminately everything that catches their eye. They appear to desire passionately whatever they have been told is rare and extraordinary. They esteem the things they have acquired in terms of the price which they have paid for them. They arrange their collection with smugness, or stack things up confusedly; and they soon end by being sick of the whole thing. Those who take the other approach, however (and these are more erudite), after having filled their heads with names, phrases, and restricted systems, come finally to adopt one of these methods, or else busy themselves with creating a new one. They thus labor all their life upon one particular approach and in a false direction, and, desiring to bring everything to their particular point of view, they restrict their minds, cease to see objects as they really are, and end by embarrassing science and loading it with the burden of ideas which have nothing to do with science.

Thus, one ought not regard as fundamental to science the methods that these authors have given us concerning natural history in general or those designed for one of its parts. Such methods should be used only as systems of artificial signs which are agreed upon for purposes of mutual understanding. Actually, they are only arbitrary connections and differing points of view under which the objects of nature have been considered. Only by making use of such methods in this spirit is it possible to draw from them some utility. For although it may not appear very necessary, it might be good if one knew all species of plants whose leaves resemble each other, all those whose flowers look alike, all those which may nourish certain kinds of insects, all those which have a given number of stamens, and all those which have certain excretory glands. The same is true of animals: there might be a point in knowing all animals which have a given number of digits. To speak precisely, each of these methods is only a dictionary in which one may find names arranged according to an order derived from a certain idea, and, consequently, arranged as arbitrarily as the alphabetical. But the advantage which may be had from such arrangements is that, in comparing all the results, one may finally come across the true method, which involves the complete description and exact history of each particular thing.

Here is the principal goal which must be kept in mind: A prefabricated method can be used as a convenience for studying, a means of mutual understanding. But the sole true means of advancing natural science is to labor at the description and history of the various things which are its objects.

Things, in relation to us, are nothing in themselves; nor does giving them a name call them into existence. But they begin to exist for us when we become acquainted with their relations to each other and their properties. Yet even by means of their relations we are unable to give things a definition. Now, a definition such as we can construct verbally is still no more than a very imperfect representation of the thing, and we are never able adequately to define a thing without describing it exactly. This difficulty of forming an adequate definition is found constantly in all systems and in all the epitomes which have been attempted in order to relieve the burden on the memory. It must also be said that in natural things nothing is well-defined but that which is exactly described. Now, in order to describe exactly, it is necessary to have seen, reviewed, examined, and compared the thing which one wishes to describe. And it is necessary to do all this without prejudging things and without an eye to systematization. Otherwise the description would not have the character of truth, which is the only characteristic called

The "Initial Discourse"

for. Even the style of the description ought to be simple, clear, and measured. The nature of the enterprise does not allow of grandeur of style, of charm, even less of digressions, pleasantries, or equivocation. The sole adornment permitted is nobility of expression, of choice, and of propriety in the use of terms.

Of the great number of authors who have written on natural history, very few have described things well. To depict things simply and clearly, without changing or oversimplifying them, and without adding anything to them from one's imagination, is a talent all the more praiseworthy the less it is paraded about, a talent which is only to be perceived in a small number of persons who are capable of the particular attention necessary in order to pursue things in their finest details. Nothing is more common than works encumbered with numerous and dessicated nomenclature or with tedious and hardly natural methods which such authors think will bring them renown. Nothing is so rare as to discover exactitude in descriptions, novelty in details, and subtlety in observations.

Aldrovandi,[4] the most hard-working and knowledgeable of all naturalists, after sixty years of labor has left behind his immense volumes on natural history, which have been printed successively, most of them after his death. They could be reduced to one tenth their present size if all those things which are useless and foreign to the subject were removed. Except for this prolixity (which, I confess, is overwhelming), his books ought to be considered the best contribution that there is to the whole spectrum of natural history. The plan of his work is good, his distributions show discretion, the divisions are well demarcated, and his descriptions are quite exact--monotonous to be sure, but accurate. Historically he is less adequate, often mixing in the fabulous and giving evidence of quite a penchant for credulity.

In going over Aldrovandi's works I have been struck by a fault, or rather an excess, which is almost always found in books printed one or two hundred years ago, and which still characterizes the German scholar today. I refer to the vast amount of useless erudition with which they purposively stuff their works, such that the subject which they treat is drowned in an ocean of foreign matter over which they argue with such self-satisfaction and carry on with so little consideration for the readers that they appear to have forgotten what it was they had to say to you, telling you only what others have said on the matter. It appears to me that a man like Aldrovandi, having once conceived the plan of outlining the whole of natural history, sits in his library and reads one after the other the Ancients, the Moderns, the philosophers, the theologians, the jurists, the historians, the explorers, and the poets; and he reads them without any other end than that of seizing upon all the words and phrases which are directly or distantly related to his object. He himself copies and has others copy down all these remarks, arranges them alphabetically, and, after having filled several portfolios with notes of all kinds--often taken without scrutiny or discretion--he begins to work on a particular subject, wishing to let nothing that he has gathered go unused. Thus, when writing a natural history of cocks or oxen, he tells you everything that has ever been said about cocks or oxen, everything the Ancients have thought about them, everything that has been imagined about their qualities, their character, their courage; all the things they have been used for; all the old wives' tales about them; all the miracles attributed to them in various religions; all the superstitious stories they have occasioned; all the comparisons poets have drawn from them; all the attributes which certain people have ascribed to them; all the representations of them found in hieroglyphics or on coats-of-arms--in a word, all the stories and all the fables which have ever been noticed about cocks or oxen. How much natural history can one expect to find in this hodge-podge of writing? Indeed, if the author had not put natural history into some articles separated from others, the natural history would not be discernible, or at least not worth the pain of searching for.

This particular failing has been completely eliminated in this century. The order and precision which characterizes present writing has made the sciences more pleasant and easy to come by, and I am persuaded that this difference of style contributes perhaps as much to their advancement as the spirit of research which reigns today. For our predecessors searched as we do, but they gathered everything that they happened upon, whereas we reject that which appears to us to have little value. And we prefer a closely reasoned brief work to a huge volume of miscellaneous scholarship. One thing is to be feared in this connection, however, and this is that, coming to distrust erudition, we may also come to imagine that the mind is able to provide everything and that science is only an empty name.

Reasonable people will nevertheless always feel that the sole true science is the knowledge of facts. The mind itself is unable to provide this, and facts are in the sciences what experience is in ordinary life. The sciences might thus be divided into two principal classes which would contain all that is suitable for man to know. The first class encompasses the history of man in society, and the second, natural history. Both are founded upon facts which it is often important and always pleasant to know. The first is the inquiry of statesmen, and the second that of philosophers. And although the usefulness of the latter may not be as immediate as that of the former, it is certain that natural history is the source of the other physical sciences and the mother of all the arts. How many excellent medical remedies have been taken from certain productions of nature previously unknown! Furthermore, all the ideas of the arts have their models in the productions of nature. God created, and man imitates. All the inventions of men, whether they be necessities or conveniences, are only grossly executed imitations of that which nature makes with the utmost finesse.

But without insisting at further length on the utility that may be drawn from natural history, whether in connection with other sciences or in connection with the arts, let us return to our main object, the manner of studying and expounding natural history. The precise description and the accurate history of each thing is, as we have said, the sole end which ought to be proposed initially. So far as the description is concerned, one ought to show form, size, weight, colors, positions of rest and of movement, location of organs, their connections, their shape, their action, and all external functions. If there could be joined to all this an exposition of internal organs, the description would be all the more complete. But care must be taken against losing one's way in such a number of minor details or dwelling too long on some organs of minor importance, while considering too lightly the main and essential things. The history ought to follow the description, and it ought to treat only relations which the things of nature have among themselves and with us. The history of an animal ought to be not only the history of the individual, but that of the entire species. It ought to include their conception, the time of gestation, their birth, the number of young, the care shown by the parents, their sort of education, their instinct, the places where they live, their nourishment and their manner of procuring it, their customs, their instinctual cleverness, their hunting, and, finally, the services which they can render to us and all the uses which we can make of them. And when any of the internal organs of an animal are worthy of note, whether because of their striking configuration or because of the uses to which they might be put, they ought to be added either to the description or to the history of the species. But it would be foreign to the purposes of natural history to enter into a very detailed anatomical examination. At least, this is not its principal object. These details should be reserved for some memoirs on comparative anatomy.

This general plan ought to be followed and completed with all possible exactness. And in order to avoid falling into monotonous repetitions the form of descriptions ought to be varied and the thread of history changed as it appears necessary. And, likewise, for the sake of making descriptions less dry, it is wise to blend into them facts, comparisons, and reflections upon the uses of various organs—in a word, to write so that you can be read without boredom as well as without contention.

With regard to the general order and the method of distribution of the various subjects of natural history, this could be considered purely arbitrary. Consequently, one is certainly free to choose what seems either the most convenient or the most commonly accepted. But before giving reasons which would lead to the adoption of one system rather than another, it is still necessary to make a few reflections by means of which we shall try to make the reader aware of how far the divisions which we have made of natural productions correspond to reality.

In order to recognize this we must dismantle our prejudices for a moment and even abstract from our ideas. Let us imagine a man who indeed has forgotten everything, or who awakens to completely strange surroundings. Let us set this man in a field where animals, birds, fishes, plants, and stones appear successively to his eyes. This man, upon first perceiving them, would distinguish nothing and confound everything. But allow his ideas to become gradually more settled by means of repeated sensations from the same objects, and soon he will form a general idea of animated matter which he will easily distinguish from inaminate matter. And shortly thereafter he will distinguish quite accurately between animated matter and vegetative matter, and he will naturally arrive at that first great division, <u>Animal</u>, <u>Vegetable</u>, and <u>Mineral</u>. And since at the same time he will have come to a clear idea of those great and quite diverse objects, <u>Earth</u>, <u>Air</u>, and <u>Water</u>, he will come shortly to form a particular idea of the animals who inhabit the earth, of those who live in the water, and of those who take to the air. And, consequently, he will easily make that second division between <u>Four-footed</u> <u>Animals</u>, <u>Birds</u>, and <u>Fishes</u>. Likewise, in the vegetable kingdom he will distinguish trees and plants with facility, whether it be by their size, their substance, or their shape. This is what simple observation must necessarily show him, and what with the very least attention he could not fail to recognize. That is what we also must recognize as real, what we must respect as a division given by nature herself. Next, let us put ourselves in the place of that man, or let us suppose that he may have acquired as much knowledge and experience as we have on this matter. He will come to judge the objects of natural history by the connections which they have with his own life. Those which are the most necessary or useful to him will hold the first rank--for example, he will give preference in the order of animals to the horse, the dog, oxen, etc., and he will always know more about those which are most familiar to him. Next, he will occupy himself with those which, without being familiar, nevertheless inhabit the same places and climates as he does--such as deer, hares, and all the wild animals. And only after acquiring all these details will his curiousity lead him to inquire into what the animals of foreign regions may be like--those such as elephants, dromedaries, etc. The case will be the same with fishes, birds, insects, shellfish, plants, minerals, and all the other productions of nature. He will study them in proportion to their usefulness; he will consider them to the extent that they are familiar to him, and he will rank them in his mind relative to the order of his acquaintance with them, because that is indeed the order according to which

The "Initial Discourse"

he experienced them and according to which it is important to him to preserve them.

This order, the most natural of all, is what we believe ought to be followed. Our method of distribution is no more mysterious than that which we have just observed. We start from general divisions such as we have just indicated, divisions which are incontestable. Next we take those objects which interest us the most owing to the connections which they have with us. From this point we pass little by little to those which are most distant, and which are foreign to us. And we believe that this simple and natural manner of considering things is preferable to more recondite and complex methods because there is not one of them, whether of those which have been constructed, or of all those which might be constructed, which would not have more of an arbitrary element in them than our method. From every point of view, it is easier, more agreeable, and more useful to consider things in relation to us rather than from another point of view.

I foresee that two objections could me made to this. The first is that these great divisions which we regard as real are perhaps not exact, that, for example, we are not sure that it is possible to draw a line of separation between the animal kingdom and the vegetable kingdom, or indeed between the vegetable kingdom and the mineral, and that in nature it is possible to find things which partake equally in the properties of the one and the other, and which, consequently, one cannot register in either one or the other of these divisions.

To that I reply that if there exist things which are exactly half animal and half plant, or half plant and half mineral, etc., such things are presently unknown to us, so that in fact the division is complete and precise. And it is evident that the more general the divisions are, the less risk there will be of coming across mixed objects which might partake of the nature of the two things included in these divisions. In this manner, that same objection which we have used with advantage against particular distributions has no place when there is a question of divisions as general as this one is, particularly if these divisions are not made exclusive, and if there is no pretension of thereby including without exception not only all known beings but, further, all those which might be discovered in the future. Moreover, if the matter is considered carefully, it will surely appear that our general ideas are only composed of particular ideas. They are relative to a continuous chain of objects. It is only the central area of each link of this chain which we perceive clearly, while the two extremes continually evade our efforts to delineate them. In this manner, the more we pursue them the more they escape us, in such a way that we never seize anything except the general outlines of things. And, consequently, we ought not believe that our ideas, however general they may be, will ever encompass the detailed conception of all existing and possible things.

The second objection which will doubtless be made to us is that by following in our work the order that we have indicated we shall fall into the error of placing together objects which are quite different. For example, in the study of animals, if we begin with those which are the most useful and familiar to us, we shall be obliged to give the history of the dog after or before that of the horse, and this does not appear to be natural, since these animals are so different in other respects that they appear in no way made to be placed so close to each other in a treatise on natural history. And it will perhaps be added that there would be more value in following the former method of dividing animals into <u>Solipedes</u>, <u>Fissipeds</u>, and <u>those with cloven hoofs</u>, or else the new method of dividing animals on the basis of their teeth, or their mammary glands, etc.

This objection, which might appear attractive at first, disappears upon examination. Is it not better to arrange objects, not only in a treatise on natural history, but even in a scientific display, or anywhere else, in the order and position that they ordinarily occupy, rather than to force them together on the basis of a supposition? Isn't it better to have the horse, which is soliped, followed by the dog, which is fissiped, and which indeed customarily follows it, rather than by the zebra, which is little known to us, and which perhaps has no other connection with the horse than being soliped? Moreover, isn't there the same disadvantage in the latter arrangements as in ours? Doesn't a lion, because it is fissiped, resemble a rat, which is also fissiped, more than a horse resembles a dog? Doesn't a soliped elephant also resemble more closely a soliped donkey than does a deer, which is split-hoofed? If one wishes to make use of the new method in which the teeth and mammary glands are the specifying characters, and upon which the divisions and distributions are based, will it not be found that a lion more closely resembles a bat than a horse resembles a dog? Or, indeed, in order to make our comparison more exact, doesn't a horse resemble a pig more than a dog, or a dog resemble a mole more than a horse?[5] And since there are more disadvantages and just as great differences in these methods of arrangement as there are in ours, and since, besides, these methods do not have the same advantages as ours, and since they are much more distant from the ordinary and natural manner of considering things, we believe we have had sufficient reason for giving the preference to our method. Our divisions are based solely upon the relations which things seem to have with us.

We shall not examine in detail all the artificial methods which have been devised for the purpose of dividing animals. They are all more or less subject to the disadvantages of which we have spoken in connection with the systems of botany. It seems to us that an examination of only one of these systems suffices in order to discover the faults of the others. Consequently, we shall limit ourselves here to the examination of that system of M. Linnaeus, which is the newest of such

The "Initial Discourse"

methods, so that the reader might be in a position to judge whether or not we have been right to reject this system and to hold fast only to the natural order in which all men have customarily viewed and considered things.

M. Linnaeus divides all animals into six classes: namely, Quadrupeds, Birds, Amphibians, Fishes, Insects, and Worms. This initial division is, as anyone can see, quite arbitrary and very incomplete, for it gives us no idea of certain kinds of animals which are, however, quite numerous and very widespread--snakes, for example, and shellfish, and crustaceans. It appears at first glance that they have been overlooked. For one would not at first imagine that snakes might be amphibians, crustaceans insects, and shellfish worms. Instead of making only six classes, if this author had made twelve or more of them, and if he had spoken of quadrupeds, birds, reptiles, amphibians, cetacean fish, oviparous fish, dipnoan fish, crustaceans, shellfish, land insects, sea insects, freshwater insects, etc., he would have spoken more clearly, and his divisions would have been more accurate and less arbitrary. For, in general, the more one augments the number of divisions of the productions of nature, the more one approaches the truth, since in nature only individuals exist, while genera, orders, and classes only exist in our imagination.

If the general characters which M. Linnaeus employs and the manner in which he makes his particular divisions are examined, even more essential shortcomings will appear. For example, a general character such as the mammary glands, which is taken for purposes of identifying the quadrupeds, should at least belong to all quadrupeds. However, it has been known since Aristotle that the horse has no mammary glands whatsoever.[6]

M. Linnaeus divides the class of quadrupeds into five orders: the first, Anthropomorpha; the second, Ferae; the third, Glires; the fourth, Jumenta; and the fifth, Pecora. And, according to him, these five orders comprise all four-footed animals. It will become evident by the exposition, and even by the enumeration, of these five orders that his division not only is arbitrary, but, further, is quite poorly thought out. For the author places in the first order man, the ape, sloths, and scaly lizards [pangolins]. One must indeed have a mania for classification in order to put together beings as different as man and the sloths, or the ape and scaly lizards. Passing to the second order which he calls Ferae, the savage beasts, he indeed begins with the lion and the tiger, but he then continues with the cat, the weasel, the otter, the seal, the dog, the bear, the badger, and ends with the hedgehog, the mole, and the bat. Would one ever have believed that the appellation of Ferae (in Latin), savage or ferocious beasts (in French), was applicable to the bat, the mole, or the hedgehog? Or that domestic animals such as the dog and the cat might be savage beasts? Isn't that just as careless a use of ideas as it is of the words that represent them? But let us pass to the third order, <u>Glires</u>, the mice: these mice of M. Linnaeus are the porcupine, the hare, the squirrel, the beaver, and the rats. I confess that of all these I see only one species of rat which might indeed be a mouse. The fourth order is that of the <u>Jumenta</u>, or beasts of burden. These beasts of burden are the elephant, the hippopotamus, the shrew-mouse, the horse, and the pig--another assemblage, as it appears, which is so gratuitous and bizarre that it appears the author had designed it with just these ends in mind. Finally we have the fifth order, <u>Pecora</u>, or the cattle, which includes the camel, the deer, the goat, the sheep, and the bullock. But isn't there quite a difference between a camel and a sheep, or between a deer and a goat? And what reason can there be for claiming that these are animals of the same order, except that, if one wishes at all costs to create orders, and only a small number of them at that, it becomes quite necessary to cram beasts of all kinds into such categories? After examining the last division of animals into particular species, one finds that the lynx is only a species of cat, the fox and the wolf species of dog, the civet cat but a species of badger, the guinea pig but a species of hare, the water rat a species of beaver, the rhinoceros a species of elephant, the ass a species of horse, etc. And all that because there are some small resemblances between the number of mammary glands and teeth of these animals, or some slight resemblances in the form of their horns.

There is, however, a means of limiting this system of nature for the quadrupeds that suffers no omissions. Wouldn't it be more simple, more natural, and more true to call an ass an ass, a cat a cat, than to wish, without knowing why, that an ass might be a horse, and a cat a lynx?

The rest of the system can be judged by this example. Serpents, according to this author, are amphibians, crayfish are insects--and not only insects, but insects of the same order as lice and fleas; and all shellfish, crustaceans, and dipnoan fish are worms. Oysters, mussels, sea urchins, starfish, squid, and so forth, are, according to this author, only worms. Is it necessary to go any further to make it apparent that all these divisions are arbitrary and this method is not justifiable?

We reproach the Ancients for not having constructed systems, and the Moderns believe themselves to be quite a bit above them because they have constructed a great number of these methodological arrangements and these lists of classified objects of which we have just spoken. They have convinced themselves that this alone is sufficient to prove that the Ancients did not have nearly the knowledge of natural history that we have. However, the complete opposite of this is the case, and we shall have a thousand occasions in the continuation of this work to show that the Ancients were far more advanced and knowledgeable than we are, not in physics, but in

The "Initial Discourse"

the natural history of animals and minerals, and that the facts of that history were far more familiar to them than they are to us who should have profited from their discoveries and comments. Until such time as we see detailed examples of this, we shall be content with indicating here the general reasons which should be sufficient to make us think this to be the case, even should we not have particular proofs of it.

The Greek language is one of the most ancient of languages, and the one that has been in use for the longest time. Before and after Homer, Greek was written and spoken until the thirteenth or fourteenth centuries. And even today Greek, corrupted by foreign idioms, does not differ as much from ancient Greek as Italian does from Latin. This language, which may be regarded as the most perfect and the richest of all, as early as the time of Homer had been brought to a high degree of perfection, which of necessity supposes a considerable antiquity for it even before the century in which that great poet wrote. For one can judge the ancient or recent origin of a language by the greater or less quantity of words in use, and the more or less subtle variety of its constructions. Now we have in this language the names of a very great number of things which have no name in either Latin or French. The rarest of animals, certain species of birds or fish or minerals which one comes across only with great difficulty and quite rarely, have names, and in fact established names, in Greek. This obviously shows that these objects of natural history were known, and that the Greeks not only knew them, but even had a precise idea of what they were, which they would not have been able to acquire except by a study of these same objects--a study which necessarily supposes observations and formal comments. They even have names for varieties, and what we are only able to represent by a phrase is identified in that language by a single substantive. This abundance of words, this treasury of clear and precise expressions, does it not suppose the same abundance of ideas and knowledge? Isn't it obvious that peoples who have named far more things than we consequently knew far more about them? However, the Greeks did not create systems and arbitrary arrangements as we do. They thought that true science consisted in knowledge of facts. In order to acquire this knowledge, it was necessary to become familiar with the productions of nature; to give each thing its name in order to make it recognizable, in order to be able to talk about such things, in order to describe to oneself more often ideas of rare and singular things, and thus to multiply knowledge, which without this process would perhaps have disappeared, nothing being more subject to oblivion than that which has no name. Whatever is not in common usage can only be sustained by the aid of representations.

Moreover, the Ancients who have written on natural history were great men, men who did not restrict themselves to that field of study alone. They had learned minds, broad and thorough knowledge, and comprehensive views. And if it seems to us at first glance that they were somewhat imprecise in certain details, it is easy to recognize, in reading them with reflection, that they did not think that minute details merited as great attention as is given to them today. And whatever reproach the Moderns are able to lay at the doorstep of the Ancients, it seems to me that Aristotle, Theophrastus, and Pliny, who were the first naturalists, are also the greatest in certain respects. Aristotle's history of animals is perhaps still today the best that we have of this genre, and it is greatly to be desired that he had left us something as complete on vegetables and minerals. But the two books of plants which some authors attribute to him do not resemble his other works, and are indeed not his.[7] It is true that botany was not in great honor in his time. The Greeks and even the Romans did not regard it as a science which ought to exist in its own right, and which ought to be made a particular object of investigation. They considered it only in connection with agriculture, gardening, medicine, and the arts. And although Theophrastus, a disciple of Aristotle, knew more than five hundred kinds of plants, and although Pliny cites more than a thousand of them, they speak of them only so that we might know how to grow them, or in order to tell us that some are involved in the composition of drugs, that others may serve to decorate our gardens, etc. In a word, they only considered plants according to the use which might be made of them, and they did not bother to describe them with any precision.

The history of animals was better known to them than that of plants. Alexander gave orders and made quite considerable expenditures in order to gather specimens of animals, caused them to be brought from all lands, and placed Aristotle in a position to be able to observe them well. It appears from the latter's work that he perhaps knew them better and viewed them under more adequate categories than is the case today. Finally, although the Moderns have added their discoveries to those of the Ancients, I do not see that we have many works in natural history which can be placed above those of Aristotle and Pliny. But since the natural prepossession which one has for his own century might persuade one that what I have said is proposed rashly, I shall say a few words in exposition of the plan of their works.

Aristotle begins his history of animals by establishing the general differences and resemblances between various kinds of animals. Instead of dividing them on the basis of small special characteristics such as the Moderns do, he gathers historically all the facts and all the observations which bear on the general relations and the sensible characteristics. He draws these characteristics from the form, color, size, and all the exterior qualities of the whole animal, as well as from the number and position of its organs, from the size, movement, and form of its limbs, and from the likenesses or dissimilarities which are found in a comparison of these same parts. And he everywhere gives examples in order to make himself better

The "Initial Discourse"

understood. He also considers differences among animals in their style of life, their actions and their habits, their places of habitation, etc. He speaks of organs which are common to all animals and essential to them, and of those which they may lack and which are indeed missing in many kinds of animals. The sense of touch, he says, is the only thing which ought to be regarded as necessary, and which cannot be absent in any animal. And since this sense is common to all animals, it is impossible to give a name to that part of their bodies wherein the faculty of sensation resides. The most essential organs are those by which the animal secures its nourishment, those which receive and digest this food, and those by which it rids itself of what remains. He next examines the various methods of generation among animals and the variety of their different organs which are used for purposes of motion and for their natural functions. These general preliminary observations form a picture all the parts of which are interesting. And this great philosopher also tells us that he has presented such observations in this manner in order to give a foretaste of what is to follow and to stimulate the attention which the particular history of each animal, or rather of each thing, requires.

He begins with man, and describes him first, rather because he is the best-known animal than because he is the most perfect. And in order to make his description less dry and more piquant, he tries to draw ethical considerations while going over the physical connections of the human body. He indicates the characteristics of men by means of their facial traits. To be well versed in physiognomy would indeed be useful knowledge for whoever acquired it. But can this be gained from natural history? After this he describes man by means of all his parts, interior and exterior, and this description is the only one which can be considered to be complete. Rather than describing each animal in particular, he presents animals in terms of the relations which all the parts of their bodies have with those of man's body. When he describes the human head, for example, he compares the head of various species of animals with it, and does the same with all the other parts of the body. In the description of the lung of man, he brings in historically everything that is known about the lungs of animals, and he notes the history of those species which lack lungs. He does the same with the organs of generation. He draws upon all the varieties of animals according to their manner of copulating, generating, bearing and giving birth, etc. So far as blood is concerned, he gives the history of the bloodless animals. And so, following out this plan of comparison in which, as we see, man serves as the model, and giving only the differences which separate animals from man, he purposively eliminates all particular description. He avoids thus all repetition, he gathers facts, and he writes down not one useless word. Thus there is comprised in a small volume an almost limitless number of facts, and I do not believe that it would be possible to put in fewer words all that he has had to say about that material, which appears so little susceptible to such precision that it takes a genius like Aristotle to thus maintain in it both order and clarity at the same time. This work of Aristotle's appears to me like a table of materials which might have been extended with the greatest care for many thousands of volumes filled with descriptions and observations of all kinds. It is the most learned abridgment that has ever been made, if science is, indeed, the history of facts. And even if one were to suppose that Aristotle had drawn from all the books of his time that which he put into his own, the plan of the work, its distribution, the choice of examples, the exactness of the comparisons, a certain form in the ideas, which I shall gladly describe as philosophic in character, all this does not leave one in doubt for even an instant that he was himself far richer than those from whom he supposedly borrowed.

Pliny worked on an even greater project, one that was, perhaps, too vast. He wished to encompass everything, and he appears to have taken the measure of nature and to have found her still too small for the scope of his spirit. His <u>Natural History</u> takes in, exclusive of the history of animals, plants, and minerals, the history of heaven and earth, medicine, commerce, navigation, the history of the liberal and mechanical arts, the origin of customs, and finally all the natural sciences and all the skills of man. And what is astonishing is that in each part of the work Pliny is equally great. The elevation of thought, the nobility of style further reveal his profound erudition. Not only was he familiar with everything that it was possible to know in his time, but he had that facility for comprehensive thought which causes science to grow. He had the acuity of reflection upon which elegance and taste depend, and he transmits to his readers a certain freedom of mind, a boldness of thought, which is the germ of philosophy. His work, which matches nature in its variety, always shows nature in its best light. His work is, if you prefer, a compilation of everything that was written on the subject before him, a copy of everything that had been done before him which was excellent and useful to know. But this copy is drawn with such broad strokes, this compilation contains things arranged in so novel a manner, that it is preferable to most of the original works which consider the same material.

We have said that the faithful history and the exact description of each thing were the two sole objects which one ought to set oneself initially in the study of natural history. The Ancients have fulfilled the first well, and are perhaps as much superior to the Moderns in this respect as they are inferior to them in the second. For the Ancients have treated quite adequately the history of the life and habits of animals, the culture and uses of plants, the properties and uses of minerals, and at the same time they appear to have consciously neglected the description of each thing. It is not that they were not capable of doing so quite well. They obviously scorned to write about things which they regarded as useless. This fashion of thinking emphasized comprehensive views, and was not as unreasonable as one might think it to be. And, at the same time, authors such as Pliny were scarcely even capable of thinking otherwise. In the first place, they put a premium on brevity, and put in their works only those facts which were essential and useful, for they did not have the facility for multiplying and enlarging books that we do. In the second place, their concern with all the sciences always revolved around their usefulness, and they were much less devoted to vain curiosity than we are. Anything that was not of interest to society, or of use for health, or for the arts, they neglected. They related everything to the ethical side of man, and did not believe that things which had hardly any use were worth bothering with. A useless insect whose maneuvers our observers admire, a plant without healing qualities whose stamens our botanists consider, were for them merely an insect and a plant. One could cite, for example, the twenty-seventh book of Pliny's <u>Reliqua herbarum genera</u>, where he puts together all the plants on which he places no high value, and which he is content to mention alphabetically, indicating only some of their general characteristics and their uses for medicine. All this shows what little taste the ancients had for natural science. Or to speak more exactly, it shows that they had no idea of what we call particular and experimental natural science. They did not think that any advantage could be had from the scrupulous examination and exact description of all the parts of a plant or of a small animal, and they did not see the connections which that process might have with the explication of the phenomena of nature.

However, this object is the most important, and it is not necessary to imagine even today that, in the study of natural history, one ought to limit oneself solely to the making of exact descriptions and the ascertaining of particular facts. This is, in truth, and as has been pointed out, the essential end which ought to be proposed at the outset. But we must try to raise ourselves to something greater and still more worthy of our efforts, namely: the combination of observations, the generalization of facts, linking them together by the power of analogies, and the effort to arrive at a high degree of knowledge. From this level we can judge that particular effects depend upon more general ones; we can compare nature with herself in her vast operations; and, finally, we are able to open new routes for the further perfection of the various branches of natural philosophy. A vast memory, assiduity, and attention suffice to arrive at the first end. But more is needed here. General views, a steady eye, and a process of reasoning informed more by reflection than by study are what is called for. Finally, that quality of mind is needed which makes us capable of grasping distant relationships, bringing them together, and making out of them a body of reasoned ideas after having precisely determined their nearness to truth and weighed their probabilities.

Here there is need for a methodical approach to guide the mind, not for that [artificial method] of which we have spoken, for that only serves to arrange words arbitrarily, but for that method which sustains the very order of things, guides our reasoning, enlightens our views, extends them, and prevents us from being led astray.

The greatest philosophers have felt the need for such a method, and they have indeed attempted to give us principles and samples of it. But some of them have left us only the history of their thoughts, while others have left only the story of their imagination. And if some have risen to the elevated stations of metaphysics from which the principles, connections, and totality of the sciences can be viewed, none of them have communicated their ideas to us on these subjects or given us any advice concerning them; and the manner of properly conducting one's mind in the sciences is yet to be found. In the absence of precepts, examples have been substituted; in place of principles, definitions have been used; instead of authenticated facts, risky suppositions have been supplied by guesswork.

Even in our own century, when the sciences seem to be cultivated with care, I believe that it is easy to perceive that philosophy is neglected, perhaps more so than in any other century. The skills which one would like to call scientific have taken its place. The methods of calculus and geometry, those of botany and natural history--formulas, in a word, and dictionaries--occupy almost everyone. We think that we know more because we have increased the number of symbolic expressions and learned phrases. We pay hardly any attention to the fact that all these skills are only the scaffolding of science, and not science itself. We ought to use them only when we cannot do without them, and we ought always to be careful lest they happen to fail us when we wish to apply them to the edifice of science itself.

Truth, that metaphysical entity of which everyone believes himself to have a clear idea, seems to me to be confounded with such a great number of strange objects to which its name is applied that I am not at all surprised that it is hard to recognize. Prejudices and false applications are multiplied in proportion as our hypotheses have become more learned, more abstract, and more perfected. It is thus more difficult than ever to recognize what we can know, and to distinguish clearly what we ought to ignore. The following reflections will serve at least as advice on this important subject.

The word truth gives rise to only a vague idea; it never has had a precise definition. And the definition itself, taken in a general and absolute sense, is but an abstraction which exists only by virtue of some supposition. Instead of trying to form a definition of truth, let us rather try to make an enumeration of truths. Let us look closely at what are commonly called truths and try to form clear ideas of them.

There are many kinds of truths, and customarily placed in the first order are those of mathematics, which are, however, only truths of definition. These definitions are concerned with simple but abstract suppositions, and all the truths of this sort are nothing more than the worked-out and always abstract consequences of these definitions. We have made the suppositions, and we have combined them in all sorts of ways. The body of combinations that results is the science of mathematics. There is, then, no more in that science than what we have put into it, and the truths which are drawn from it can only be different expressions under which the suppositions which we have used are presented. Thus, mathematical truths are only the exact repetitions of definitions or suppositions. The last consequence is true only because it is identical with that which preceded it, and this latter in its turn with its antecedent. Thus one may proceed backward right to the first presupposition. And since definitions are the sole principles upon which everything is established, and since they are arbitrary and relative, all the consequences which can be deduced from them are equally arbitrary and relative. Hence, that which we call mathematical truth is thus reduced to the identity of ideas, and has nothing of the real about it. We make suppositions, we reason on the basis of our suppositions, we draw the consequences of them, we come to conclusions. The conclusion, or the last consequence, is a proposition which is true in proportion as our supposition was true. But the truth of this proposition cannot exceed that of the supposition itself. This is hardly the place to discourse on the methods of the science of mathematics, or on the abuse of such methods. It is sufficient for our purposes to have proved that mathematical truths are only truths of definition or, if you prefer, different expressions of the same thing, and that they are only truths in relation to the very definitions with which we started. It is for this reason that they have the advantage of always being precise and conclusive, but abstract, intellectual, and arbitrary.

Physical truths, on the other hand, are in no way arbitrary, and in no way depend on us. Instead of being founded on suppositions which we have made, they depend only on facts. A sequence of similar facts or, if you prefer, a frequent repetition and an uninterrupted succession of the same occurrences constitute the essence of this sort of truth. What is called physical truth is thus only a probability, but a probability so great that it is equivalent to certitude. In mathematics, one supposes. In the physical sciences, one sets down a claim and establishes it. There, one has definitions; here, there are facts. One goes from definition to definition in the abstract sciences, but one proceeds from observation to observation in the real sciences. In the first case one arrives at evidence, while in the latter the result is certitude. The word "truth" is used for both, and consequently corresponds to two different ideas. Its signification is vague and complicated, and it thus has not been possible to define the term in a general way. It has been necessary, as we have just seen, to distinguish the kinds of truth in order to form a clear idea of it.

I shall not speak of other orders of truths: those of the moral order, for example, which are in part real and in part arbitrary, would demand a lengthy discussion which would take us away from our goal, and that more especially because they have as object and end only decorum and probabilities.

Mathematical evidence and physical certitude are thus the only two aspects under which we ought to consider truth. As soon as it withdraws from one or another of these, it is no more than appearance and probability. Let us then examine what we can know through evident or certain science, after which we shall see what we can come to know only by conjecture; and finally we shall see what we ought to ignore.

We know, or we can know, in evident science, all the characteristics or, rather, all the relationships of numbers, lines, surfaces, and of all the other abstract quantities. We shall be able to know them in a more complete manner to the extent that we train ourselves to solve new problems, and more surely to the extent that we search out the causes of the difficulties which arise. Since we are the creators of this sort of knowledge, and since it takes under consideration absolutely nothing except what we ourselves have already imagined, it is impossible to have therein either obscurities or paradoxes which may be actual or impossible of resolution. A solution will always be found for these apparent difficulties through a careful examination of the premises, and by following all the steps which have been taken to arrive at the solution. Since the combination of these principles and the ways in which they can be used are innumerable, there is in mathematics a vast field of acquired knowledge to be gained of which we shall always be the masters, cultivating it when we wish, and from which we shall always garner the same abundance of truths.

But these truths would have been perpetually matters of pure speculation, of simple curiosity, and entirely useless if the means had not been found of conjoining them to physical truths. Before considering the advantages of that union, let us see what we could hope to know in this area.

The phenomena which offer themselves daily to our eyes, which follow one another and repeat themselves without interruption and uniformly, are the foundation of our physical knowledge. It is enough that a thing always happens in the same way for it to become a certainty or a truth for us. All the facts of nature which we have observed, or which we could observe, are just so many truths. And thus we are able to increase the number as much as we please by multiplying our observations. Our science is limited in this case only by the dimensions of the universe.

But when, after having determined the facts through repeated observations; when, after having established new truths through precise experiments, we wish to search out the reasons for these same occurrences, the causes of these effects, we find ourselves suddenly baffled, reduced to trying to deduce effects from more general effects, and obliged to admit that causes are and always will be unknown to us, because our senses, themselves being the effects of causes of which we have no knowledge, can give us ideas only of effects and never of causes. Thus we must be content to call cause a general effect, and must forego hope of knowing anything beyond that.

These general effects are for us the true laws of nature. All the phenomena that we recognize as holding to these laws and depending on them will be so many accountable facts, so many truths understood. Those phenomena which we are unable to associate with these general effects will be simple "occurrences" which we must keep in reserve until such time as a greater number of observations and a more extended experience make us aware of other facts and bring to light their physical cause, that is to say, the general effect from which these particular effects derive. It is here that the union of the two sciences of mathematics and physics might result in great advantages. The one gives the "how many," the other the "how" of things. And since it is a question here of combining and estimating probabilities in order to judge whether an effect depends more on one cause than on another, when you have imagined by physics the how, that is to say, when you have seen that such and such an effect might well depend upon such and such a cause, you then apply mathematics in order to assure yourself as to how often this effect happens in conjunction with its cause. And if you find that the result accords with the observations, the probability that you have guessed correctly is so increased that it becomes a certainty. But in the absence of such corroboration, the relation would have remained a simple probability.

It is true that this union of mathematics and physics can be accomplished only for a very small number of subjects. In order for this to take place it is necessary that the phenomena that we are concerned with explaining be susceptible to being considered in an abstract manner and that their nature be stripped of almost all physical qualities. For mathematics is inapplicable to the extent that such subjects are not simple abstractions. The most beautiful and felicitous use to which this method has ever been applied is to the system of the world. We must admit that if Newton had only given us the physical conformations of his system without having supported them by precise mathematical evaluations they would not have had nearly the same force. But at the same time one ought to be aware that there are very few subjects as simple as this, that is to say, as stripped of physical qualities as the Newtonian universe. For the distance of the planets is so great that it is possible to consider them in reference to each other as being no more than points. And it is possible simultaneously, without being mistaken, to abstract from all the physical qualities of the planets and take into consideration only their force of attraction. Their movements are, moreover, the most regular that we know, and suffer no retardation from resistance. All of this combines to render the explanation of the system of the world a problem in mathematics, for the realization of which fortunately there was needed only one well-conceived physical idea, that idea being to have thought that the force which makes bodies fall to the surface of the earth might well be the same as that which holds the moon in its orbit.

But, I repeat: there are very few subjects in physics in which the abstract sciences can be applied so advantageously. And I scarcely see anything but astronomy and optics to which they might be of any great service: astronomy, for the reasons which we have just explained; and optics because light being a body almost infinitely small, whose effects operate in straight lines with almost infinite speed, its properties approximate those of mathematics, which allows one to apply to optics with some success arithmetic and geometric measurement. I shall not speak of mechanics, for <u>rational</u> mechanics is itself a mathematical and abstract science from which practical mechanics, or the art of making and designing machines, borrows only one single principle, by which one is able to judge all the effects by abstracting from friction and other physical qualities. Also, it has always appeared to me that there was a sort of abuse in the manner in which experimental physics is taught, the object of that science being by no means the one ordinarily attributed to it. Mechanical effects--such as the power of levers and, pulleys, the equilibrium of solids and fluids, the effect of inclined planes, of centrifugal forces, etc. -- belonging entirely as they do to mathematics, and capable of being grasped by the mind with the utmost clarity [<u>la derniere évidence</u>], it seems to me superfluous to represent demonstrations of these effects to the senses. The true goal of experimental physics is, on the contrary, to experiment with all things which we are not able to measure by mathematics, all the effects of which we do not yet know the causes, and all properties whose circumstances we do not know. That alone can lead us to new discoveries, whereas the demonstration of mathematical effects will never show us anything except what we already know.

But this abuse is as nothing in comparison with the inconveniences into which one stumbles when one wishes to apply geometry and arithmetic to quite complicated subjects of natural philosophy, to objects whose properties we know too little about to allow us to measure them. One is obliged in all such cases to make suppositions which are always contrary to nature, to strip the subject of most of its qualities, and to make of it an abstract entity which has no resemblance to the real being. After long reasoning and calculation on the connections and the properties of this abstract entity, and after having arrived at a conclusion equally abstract, when it appears that something real has been found, and the ideal result is transferred back into the real subject, this process produces an infinity of false consequences and errors.

Here then is the most delicate and the most important point in the study of the sciences: to know how to distinguish what is real in a subject from what we arbitrarily put there in considering it, to recognize clearly the properties which belong to it and those which we give to it. This appears to me to be the foundation of the true method of leading one's mind in the way of the sciences. And if this principle were always kept in mind, a false step would never be taken. One might thus avoid falling into learned errors which at times are taken as truths. Paradoxes and insoluble problems in the abstract sciences would begin to disappear. The prejudices and the doubts which we ourselves bring to the sciences of the real would become apparent, and agreement would be reached on the metaphysics of sciences. Disputes would cease, and all would unite to advance along the same path following experience. Finally, we would arrive at the knowledge of all the truths which are within the competence of the human mind.[8]

When the subjects are too complicated to allow the advantageous application of calculation and measurement, as is almost always the case with natural history and the physics of the particular, it seems to me that the true method of guiding one's mind in such research is to have recourse to observations, to gather these together, and from them to make new observations in sufficient number to assure the truth of the principal facts, and to use mathematics only for the purpose of estimating the probabilities of the consequences which may be drawn from these facts. Above all, it is necessary to try to generalize these facts and to distinguish well those which are essential from those which are only accessories to the subject under consideration. It is then necessary to tie such facts together by analogies, confirm or destroy certain equivocal points by means of experiment, form one's plan of explication on the basis of the combination of all these connections, and present them in the most natural order. This order can be established in two ways: the first is to ascend from particular effects to more general ones, and the other is to descend from the general to the particular. Both ways are good, and the choice of one or the other depends more on the bent of the author than on the nature of things, which always allows of being treated equally well by either method. We are going to give trial to this proposition in the discourses which follow, of the "Theory of the Earth," of the "Formation of Planets," and of the "Generation of Animals."

NOTES

*The text used for translation purposes is the "Premier Discours" from the Histoire naturelle, générale et particulière avec la description du cabinet du Roi. (Deux-Ponts: Sanson, 1785-1790), vol. I, pp. 7-69. This pirated edition was the last edition of Buffon's work to appear during his lifetime. It has been checked against the first edition (Paris: Imprimerie Royale, 1749) vol. I, as found in Jean Piveteau, ed., Oeuvres philosophiques de Buffon (Paris: Presses Universitaires de France, 1954). I would like to thank the University of Oklahoma Library for the use of their copy of the Deux-Ponts edition.

The present translation and introduction is a slightly modified version of my article, "The 'Initial Discourse' to Buffon's Histoire Naturelle: The First Complete English Translation," which appeared in The Journal of the History of Biology, vol. 9, no. 1 (Spring, 1976), pp. 133-181. Copyright ©1976 by D. Reidel Publishing Company, Dordrecht, Holland.

[1] [Konrad von Gesner (1516-1565), Swiss polymath and naturalist.]

[2] [Joseph Pitton de Tournefort (1656-1708), French botanist.]

[3] "This system, indeed, namely Linnaeus's, is not only far more vile and inferior to the already-known systems, but is further exceedingly forced, slippery, and fallacious; indeed, I would consider it childish, for it not only brings after itself enormous confusion with regard to the division and denomination of plants, but it is also to be feared that from this would come an almost complete clouding and disruption of the more solid botanical systems." Vaniloq. Botan. specimen refutatum a Siegesbek, Petropoli, 1741 (sic).

[4] [Ulysse Aldrovandi (1522-1605), Italian Naturalist.]

[5] See Linnaeus, Systema naturae, p. 65 ff.

[6] [On this puzzling comment, see above pp. 94n8.]

[7] See the commentary of Scaliger.

[8] [For a lengthy criticism of the matter in the paragraphs from the bottom of p. 122 to this point, see Malesherbes' essay, in Part III below.]

Figure 4 - A God-like being, probably meant to suggest the Spirit of Nature, hovers over the still-forming earth. A large fissure develops as land shrinks and forms ocean basin. In the background figures of the zodiac along the ecliptic.

7. The "Second Discourse" and "Proofs of the Theory of the Earth" from Buffon's *Histoire naturelle* (1749) (selected)

Phillip R. Sloan

No feature of Buffon's Natural History created such an immediate reaction, nor had a more lasting impact than his novel geological and cosmological speculations put forth in the long treatises of the first volume of the Histoire naturelle. Historical approaches to geology and cosmology, as we have noted in the General Introduction, were not unique to Buffon, and reached back through Leibniz, William Whiston, John Woodward, Nicholas Steno and Thomas Burnet to Descartes. But in spite of this long traditon of predecessors, Buffon had placed the issues on a new conceptual and epistemological footing that made his presentation of the issues a beginning point for a more critical eighteenth-century discussion.

Of immediate importance, as the reviews in Part III will illustrate, was the absence in Buffon's account of appeal to creationism or miraculous intervention in the natural order. Furthermore, no attempt is made to reconcile this account with the chronology of the revealed religious tradition. When read in the context of his larger epistemological principles, as expressed in the "Premier discours" which opened this volume, it is evident that Buffon is not proposing his accounts as simple Cartesian fictions, but is instead accepting them as at least close approximations to a "real and physical" account of the history of nature.

In common with Leibniz, but in opposition to Newton, Buffon's world is autonomous since its first beginnings in the chance collision of a comet with the sun, excluding all subsequent divine intervention to alter or repair the order of nature.[1] At the same time, Buffon appears to have utilized a Newtonian methodological principle with greater consistency than Newton himself had done. In the third book of his Principia, Newton had argued that science is to be governed by a strong principle of analogy. Properties, such as attraction, which are empirically discovered to hold between bodies accessible to observation can, by analogy, be asserted of all bodies whatever. This is fashioned by Buffon into a principle of geological and cosmological reasoning. The apparent natural causes of such events revealed in the present can also be extended, by analogy, into the remote past and future.[2]

This extension seems to depend in two important ways on his concept of "physical" truth. First, it requires explanations in terms of immanent and specifically-acting causal agents, rather than "abstract" and purely general causes and forces. Secondly, it would imply that through the observation of the <u>constant</u> and <u>recurrent</u> action of such causes, an epistemological certitude is generated that can be expressed by a probability calculus in some cases. Just as he will argue subsequently that the "reality" of organic species can be known with increasing certitude by the constant observation of the reproduction of like by like in the succession of time, the larger history of nature can be known with increasing certainty through the extension of the action of recurrent causal agencies operating at present, to periods inaccessible to immediate observation. Observing the effects produced by the periodic tidal motions, for example, gives Buffon a basis upon which he can assert that the same specific and recurrent agency shaped the continents, formed the mountain ranges, and sculptured the sea bottom.[3]

In asserting that such recurrent causes can, however, act under a historically changing set of initial conditions, Buffon's geology is not a steady state or "actualist" system, such as would be articulated in the next century by Charles Lyell in his of <u>Principles of Geology</u> (1830-1833).[4] But Buffon's system is, at the same time, a strongly "uniformitarian" account, with unique events, such as the passage of a comet, appealed to only to account for the beginning of the solar system.

Although in his assertion of the complete autonomy of a natural order referred to in almost organismic terms as "nature,"[5] paralleling Leibniz in important ways, at one point we detect a divergence from Leibniz with important ramifications for the subsequent tradition. Difficult to locate in Buffon's accounts is any concept of teleological directedness, not simply in the external sense encountered in Newton and the British natural theological tradition generally, but also in the inherent and immanent sense of inner directedness to a larger harmony expounded by Leibniz and Wolff. As Buffon would eventually develop this point in his <u>Vues de la Nature</u> (1764, 1765), and the <u>Époques de la Nature</u> (1778), the world system, with both kinds of teleology denied, can only be a degenerating and ultimately dying system, gradually losing its necessary heat, momentum, and life-sustaining power. To his contemporaries and successors, particularly those in the German tradition, no feature of Buffon's thought was more in need of philosophical repair than the historical pessimism which seemed to be a consequence of this scientific cosmology. Immanuel Kant, who would only shortly afterwards propose his own account of the formation of the cosmos by the bi-polar action of Newtonian forces, nevertheless departs most markedly from Buffon in asserting an inherent teleology behind the self-organization of the universe.[6] This problem will also be raised in a biological context in the review by Albrecht von Haller in Part III.[7]

The following selections are taken from the first volume of the <u>Histoire naturelle</u>. The opening discussion, signed with the date of October 3, 1744, is taken from the long "Second

The "Second Discourse"

discours" of this volume. The second is excerpted from the same volume, and is dated September 20, 1745. Buffon's notes, and those of editor to both selections follow the "Proofs of Theory of the Earth," p. 160.

NOTES

[1] Newton's argument in this regard was expounded most explicitly in Query 31 to his Opticks. See Isaac Newton, Opticks 4th ed. (reprint New York: Dover, 1952), p. 397. Leibniz' counter forms an important issue in the Leibniz-Clarke debates that opened in print in 1717.

[2] See text below, pp. 148-49.

[3] Ibid.

[4] The distinction between "actualism" and "uniformitarianism" is maintained by recent historians of geology, although it is not always clear what is intended. In drawing this distinction, Martin Rudwick, for example, (The Meaning of Fossils [New York and London: MacDonald and Elezevier, 1972] p. 110), considers "uniformitarianism" to apply to the content of a geological theory, and "actualism" to denote the methodological thesis of inferring to the past from observations on present states and processes. As intended in our work, "actualism" is restricted to a methodological thesis like Lyell's, which asserts the extrapolation of similar causes and conditions observable in the present to the past. "Uniformitarianism" is a much less restrictive thesis, which affirms the action only of observed causes, but allows for radical changes in conditions and rates of action. The latter would be compatible with progressivist and degenerative models of the world.

[5] See especially his De la nature: Première vue opening volume 12 of the Histoire naturelle (1764). There he describes "nature" as an "immense living power animating and embracing all things, and subordinate to the First Being. . . ."

[6] Kant's indebtedness to Buffon's work, both in its general features and also for specific details and calculations, in his own Allgemeine Naturgeschichte des Himmels of 1755 is apparent at many points, with an important difference being Kant's even greater reliance on the polarity of forces, utilizing attraction and repulsion, in place of Buffon's attraction and inertial impulsion. A larger philosophical divergence is Kant's adherence to the Leibnizian concept of the inner telos of nature, which constitutes his chief reply to an Epicurean cosmology.

[7] See below, especially pp. 323-24.

Figure 5 - Original woodcut prefacing *Second discours*

SECOND DISCOURSE: THE HISTORY AND

THEORY OF THE EARTH*

Translated by J. S. Barr

> I have myself seen what once was solid land changed into sea; and again I have seen land made from the sea. Sea-shells have been seen lying far from the ocean, and an ancient anchor hs been found on a mountain-top. What once was a level plain, down-flowing waters have made into a valley; and hills by the force of the floods have been washed into the sea
>
> Ovid, Metamorphoses,[1]

Neither the figure of the earth,[2] its motion, nor its external connections with the rest of the universe pertain to our present investigation. It is the internal structure of the globe, its composition, form, and manner of existence which we propose to examine. The general history of the earth should doubtless precede that of its productions, as a necessary study for those who wish to be acquainted with nature in her variety of shapes, and the detail of facts relative to the life and manners of animals, or to the culture and vegetation of plants belong not, perhaps, so much to natural history, as to the general deductions drawn from the observations that have been made upon the different materials which compose the terrestrial globe: as its heights, depth, and inequalities of its form; the motion of the sea, the direction of mountains, the situation of rocks and quarries, the rapidity and effects of currents in the ocean, etc. This is the history of nature in its most ample extent, and these are the operations by which every other effect is influenced and produced. The theory of these effects constitutes what may be termed a primary science, upon which the exact knowledge of particular appearances as well as terrestrial substances entirely depends. This description of science may fairly be considered as appertaining to physics; but does not all physical knowledge, in which no system is admitted, form part of the history of nature? In a subject of great magnitude, whose relative connections are difficult to trace, and where some facts are but partially known, and others uncertain and obscure, it is more easy to form a visionary system than to establish a rational theory; thus it is that the theory of the earth has only hitherto been treated in a vague and hypothetical manner; I shall therefore but slightly mention the singular notions of some authors who have written upon the subject.

The first hypothesis I shall allude to, deserves to be mentioned more for its ingenuity than its reasonable solidity; it is that of an English astronomer, (Whiston) versed in the system of Newton, and an enthusiastic admirer of his philosophy;[3] convinced that every event which happens on the terrestrial globe depends upon the motions of the stars, he endeavours to prove, by the assistance of mathematical calculations, that the tail of a comet has produced every alteration the earth has ever undergone.

The next is the formation of an heterodox theologian, (Burnet) whose brain was so heated with poetical visions, that he imagined he had seen the creation of the universe.[4] After explaining what the earth was in its primary state when it sprung from nothing; what changes were occasioned by the deluge; what it has been and what it is, he then assumes a prophetic style, and predicts what will be its state after the destruction of human race.

The third comes from a writer (Woodward) certainly a better and more extensive observer of nature than the two former, though little less irregular and confused in his ideas;[5] he explains the principal appearances of the globe, by an immense abyss in the bowels of the earth, which in his opinion is nothing more than a thin crust that serves as a covering to the fluid it encloses.

The whole of these hypotheses is raised on unstable foundations and has given no light upon the subject, the ideas being unconnected, the facts confused, and the whole confounded with a mixture of physics and fable; and consequently has been adopted only by those who implicitly believe opinions without investigation, and who, incapable of distinguishing probability, are more impressed with the wonders of the marvellous than the relation of truth.

What we shall say on this subject will doubtless be less extraordinary, and appear unimportant, if put in comparison with the grand systems just mentioned, but it should be remembered that it is an historian's business to describe, not invent; that no suppositions should be admitted upon subjects that depend upon facts and observation; that his imagination ought only to be exercised for the purpose of combining observations, rendering facts more general, and forming one connected whole, so as to present to the mind a distinct arrangement of clear ideas and probable conjectures; I say probable because we must not expect to give exact demonstration on this subject, that being confined to mathematical sciences, while our knowledge in physics and natural history depends solely upon experience, and is confined to reasoning upon inductions.

In the history of the earth, we shall therefore begin with those facts that have been obtained from the experience of time, together with what we have collected by our own observations.

This immense globe exhibits upon its surface heights, depths, plains, seas, lakes, marshes, rivers, caverns, gulfs, and volcanos; and upon the first view of these objects we cannot discover in their disposition either order or regularity. If we penetrate into its internal part, we shall there find metals, minerals, stones, bitumens, sands, earths, waters, and matters of every kind, placed as it were by chance, and without the smallest apparent design. Examining with a more strict attention we discover sunken mountains,[6] caverns filled, rocks split and broken, countries swallowed up, and new islands rising from the ocean; we shall also perceive heavy substances placed above light ones, hard bodies surrounded with soft; in short we shall there find matter in every form, wet and dry, hot and cold, solid and brittle, mixed in such a sort of confusion as to leave room to compare them only to a mass of rubbish and the ruins of a wrecked world.

We inhabit these ruins, however, with a perfect security. The various generations of men, animals, and plants, succeed each other without interruption; the earth [abundantly supplies] their sustenance: the sea has its limits; its motions and the currents of air are regulated by fixed laws:[7] the returns of the seasons are certain and regular; the severity of the winter being constantly succeeded by the beauties of the spring: every thing appears in order, and the earth, formerly a <u>chaos</u>, is now a tranquil and delightful abode, where all is animated and regulated by such an amazing display of power and intelligence as fills us with admiration, and elevates our minds with the most sublime ideas of an all-potent and wonderful Creator.

Let us not then draw any hasty conclusions upon the irregularities of the surface of the earth, nor the apparent disorders in the interior parts, for we shall soon discover the utility, and even the necessity, of them; and, by considering

The "Second Discourse"   137

them with a little attention, we shall perhaps find an order of
which we had no conception, and a general connection that we
could neither perceive nor comprehend by a slight examination:
but in fact, our knowledge on this subject must always be confined. There are many parts of the surface of the globe with
which we are entirely unacquainted,[8] and have but partial ideas
of the bottom of the sea, which in many places we have not been
able to fathom. We can only penetrate into the [crust] of the
earth; the greatest caverns[9] and the deepest mines[10] do not
descend [beyond] the eighth-thousandth part of its diameter; we
can therefore judge only of the external and mere superficial
part; we know, indeed, that bulk for bulk the earth weighs four
times heavier than the sun, and we also know the proportion its
weight bears with other planets; but this is merely a relative
estimation; we have no certain standard nor proportion; we are
so entirely ignorant of the real weight of the materials, that
the internal part of the globe may be a void space, or composed
of matter a thousand times heavier than gold, nor is there any
method to make further discoveries on this subject; and it is
with the greatest difficulty any rational conjectures can be
formed thereon.[11]

We must therefore confine ourselves to a correct examination and description of the surface of the earth, and to those
trifling depths to which we have been enabled to penetrate.
The first object which presents itself is that immense quantity
of water which covers the greatest part of the globe; this
water always occupies the lowest ground, its surface always
level, and constantly tending to equilibrium and rest; nevertheless it is kept in perpetual agitation by a powerful
agent,[12] which opposing its natural tranquility impresses it
with a regular periodical motion, alternately raising and
depressing its waves, producing a vibration in the total mass,
by disturbing the whole body to the greatest depths. This
motion we know has existed from the commencement of time, and
will continue as long as the sun and moon, which are the causes
of it.

By an examination of the bottom of the sea, we discover
that to be fully as irregular as the surface of the earth;[13] we
there find hills[14] and valleys, plains and cavities, rocks and
soils of every kind; we shall perceive that islands are only
the summits[15] of vast mountains whose foundations are at the
bottom of the ocean; and we shall find other mountains whose
tops are nearly on a level with the surface of the water, and
rapid currents[16] which run contrary to the general movement:
they sometimes run in the same direction,[17] at others their
motions are retrograde, but never exceeding their bounds, which
appear to be as fixed and invariable as those which confine the
rivers of the earth. In one part we meet with tempetuous
regions, where the winds blow with irresistible fury; where the
sea and the heavens, equally agitated, join in contact with
each other, are mixed and confounded in the general shock; in
others, violent intestine motions, tumultuous swellings,[18]

water-spouts,[19] and extraordinary agitations, caused by volcanos, whose mouths, though a considerable depth under water, yet vomit fire from the midst of the waves, and send up to the clouds a thick vapour, composed of water, sulphur, and bitumen. Further we perceive dreadful gulfs or whirlpools,[20] which seem to attract vessels merely to swallow them up. On the other hand, we discover immense regions, totally opposite in their natures, always calm and tranquil,[21] yet equally dangerous; where the winds never exert their power, where the art of the mariner becomes useless, and where the becalmed voyager must remain until death relieves him from the horrors of despair. In conclusion, if we turn our eyes towards the northern or southern extremities of the globe, we there perceive enormous [sheets] of ice separating themselves from the polar regions,[22] advancing like huge mountains into the more temperate climes, where they dissolve and are lost to the sight.[23]

Exclusive of these principal objects the vast empire of the sea abounds with animated beings, almost innumerable in numbers and variety. Some of them covered with light scales move with astonishing celerity; others loaded with thick shells drag heavily along, leaving their tract in the sand; on others nature has bestowed fins, resembling wings, with which they raise and support themselves in the air, and fly to considerable distances; while there are those to whom all motion has been denied, who live and die immovably fixed to the same rock: every species, however, finds abundance of food in this their native element. The bottom of the sea and the shelving sides of the various rocks, produce great abundance of plants and mosses of different kinds; its soil is composed of sand, gravel, rocks and shells; in some parts a fine clay, in others a solid earth, and in general it has a complete resemblance to the land which we inhabit.

Let us now take a view of the earth. What prodigious differences do we find in different climates? What a variety of soils? What inequalities in the surface? But upon a minute and attentive observation we shall find the greatest chains of mountains[24] are nearer the equator than the poles; that in the Old Continent their direction is more from the east to west than from the north to south, and that on the contrary in the new world they extend more from north to south than from east to west; but what is still more remarkable, the form and direction of those mountains, whose appearance is so very irregular,[25] correspond so directly[26] that the prominent angles of one mountain are always opposite to the concave angles of the neighboring mountain, and are of equal dimensions whether they are separated by a small valley or an extensive plain. I have also observed that opposite hills are nearly of the same height, and that in general mountains occupy the middle of continents, islands, and promontories, which they divide by their greatest lengths.[27]

The "Second Discourse"

In following the courses of the principal rivers I have likewise found that they are almost always perpendicular with those of the sea [coasts] into which they empty themselves; and that in the greatest part of their courses they proceed nearly in the direction of the mountains from which they derive their source.[28]

The sea shores are generally bounded with rocks, marble, and other hard stones, or by earth and sand which has accumulated by the waters from the sea, or been brought down by the rivers; and I observe that opposite coasts, separated only by an arm of the sea, are composed of similar materials, and the beds of the earth are exactly the same.[29] Volcanos[30] I find exist only in the highest mountains, that many of them are entirely extinct; that some are connected with others by subterraneous passages,[31] and that their explosions frequently happen at one and the same time. There are similar correspondences between certain lakes and neighboring seas; some rivers suddenly disappear,[32] and seem to [rush down] into the earth. We also find internal, or mediterranean seas, constantly receiving an enormous quantity of water from a number of rivers without ever extending their bounds, most probably discharging by subterraneous passages all their superfluous supplies. Lands which have been long inhabited are easily distinguished from those new countries, where the soil appears in a rude state, where the rivers are full of cataracts, where the earth is either overflowed with water, or parched up with drought, and where every spot upon which a tree will grow is covered with uncultivated woods.

Pursuing our examination in a more extensive view, we find that the upper strata,[33] that surrounds the globe, is universally the same, [and] that this substance which serves for the growth and nourishment of animals and vegetables, is nothing but a composition of decayed animal and vegetable bodies reduced into such small particles that their former organization is not distinguishable. Penetrating a little further we find the real earth, beds of sand, limestone, argol, shells, marble, gravel, chalk, etc. These beds[34] are always parallel to each other[35] and of the same thickness throughout their whole extent. In neighbouring hills beds of the same materials are invariably found upon the same levels, though the hills are separated by deep and extensive intervals. All beds of earth, even the most solid strata,[36] as rocks, quarries of marble etc. are uniformly divided by perpendicular fissures; it is the same in the largest as well as smallest depths, and appears a rule which nature invariably pursues.

In the very bowels of the earth, on the tops of mountains,[37] and even the most remote parts from the sea, shells, skeletons of fishes, marine plants, etc. are frequently found; and these shells, fishes, and plants, are exactly similar to those which exist in the ocean. There are a prodigious quantity of petrified shells to be met with in an infinity of places, not only enclosed in rocks, masses of marble, limestone, as well as in earths and clays, but are actually incorporated and filled with the very substance which surrounds them. In short, I find myself convinced, by repeated observations [...] that marbles, stones, chalks, marles, clay, sand, and almost all terrestrial substances, wherever they may be placed, are filled with shells[38] and other [debris from] the sea.

These facts being enumerated, let us now see what reasonable conclusions are to be drawn from them.

The changes and alterations which have happened to the earth in the space of the last two or three thousand years are very inconsiderable indeed when compared with those important revolutions which must have taken place in those ages which immediately followed the creation; for as all terrestrial substances could only acquire solidity by the continued action of gravity, it would be easy to demonstrate that the surface of the earth was much softer at first than it is a present, and consequently the same causes which now produce but slight and almost imperceptible changes during many ages, would then effect great revolutions in a very short space. It appears to be a certain fact, that the earth which we now inhabit, and even the tops of the highest mountains were formerly covered with the sea, for shells and other marine productions are frequently found in almost every part; it appears also that the water remained a considerable time on the surface of the earth, since in many places there have been discovered such prodigious banks of shells that it is impossible so great a multitude of animals could exist at the same time[39]: this fact seems likewise to prove, that although the materials which composed the surface of the earth were then in a state of softness that rendered them easy to be disunited, moved and transported by the waters, yet that these removals were not made at once; they must indeed have been successive, gradual, and by degrees, because these kind of sea-productions are frequently met with more than a thousand feet below the surface, and such a considerable thickness of earth and stone could not have accumulated but by the length of time. If we were to suppose that at the deluge all the shell-fish were raised from the bottom of the sea, and transported over all the earth; besides the difficulty of establishing this supposition,[40] it is evident, that as we find shells incorporated in marble and in the rocks of the highest mountains, we must likewise suppose that all these marbles and rocks were formed at the same time, and that too at the very instant of the deluge; and besides, that previous to this great revolution there were neither mountains, marble, nor rocks, nor clays, nor matters of any kind similar to those we are at present acquainted with, as they almost all contain shells and other productions of the sea. Besides, at the time of the deluge the earth must have acquired a considerable degree of solidity, from the action of gravity, for more than sixteen centuries, and consequently it does not appear

The "Second Discourse"

possible that the waters, during the short time the deluge lasted, should have overturned and dissolved its surface to the greatest depths we have since been enabled to penetrate.

But without dwelling longer on this point, which shall hereafter be more amply discussed, I shall confine myself to well-known observations and established facts. There is no doubt but that the waters of the sea at some period covered and remained for ages upon that part of the globe which is now known to be dry land; and consequently the whole continents of Asia, Europe, Africa, and America, were then the bottom of an ocean abounding with similar productions to those which the sea at present contains: it is equally certain that the different strata which compose the earth are parallel and horizontal,[41] and it is evident their being in this situation is the operation of the waters which have collected and accumulated by degrees the different materials, and given them the same position as the water itself always assumes. We observe that the position of strata is almost universally horizontal: in plains it is exactly so, and it is only in the mountains that they are inclined to the horizon, from their having been originally formed by a sediment deposited upon an inclined base. Now I insist that these strata must have been formed by degrees, and not all at once, by any revolution whatever, because strata composed of heavy materials are very frequently found placed above light ones, which could not be, if, as some authors assert, the whole had been mixed with the waters at the time of the deluge,[42] and afterwards precipitated; in that case everything must have had a very different appearance to that which now exists. The heaviest bodies would have descended first, and each particular stratum would have been arranged according to its weight and specific gravity, and we should not see solid rocks or metals placed above light sand any more than clay under coal.

We should also pay attention to another circumstance; it confirms what we have said on the formation of the strata; no other cause than the motions and sediments of water could possibly produce so regular a position of [strata], for the highest mountains are composed of parallel strata as well as the lowest plains, and therefore we cannot attribute the origin and formation of mountains to the shocks of earthquakes, or eruptions of volcanos. The small eminences which are sometimes raised by volcanos or convulsive motions of the earth[43] are not by any means composed of parallel strata, they are a mere disordered heap of matters thrown confusedly together; but the horizontal and parallel position of the strata must necessarily proceed from the operations of a constant cause and motion always regulated and directed in the same uniform manner.

From repeated observations, and these incontrovertible facts, we are convinced that the dry part of the globe, which is now habitable, has remained for a long time under the waters of the sea, and consequently this earth underwent the same fluctuations and changes which the bottom of the ocean is at present actually undergoing. To discover therefore what formerly passed on the earth, let us examine what now passes at the bottom of the sea, and from thence we shall soon be enabled to draw rational conclusions with regard to the external form and internal composition of that which we inhabit.

From the creation the sea has constantly been subject to a regular flux and reflux: this motion, which raises and [lowers] the waters twice in every twenty-four hours, is principally occasioned by the action of the moon, and is much greater under the equator than in any other climates. The earth performs a rapid motion on its axis, and consequently has a centrifugal force, which is also the greatest at the equator; this latter, independent of actual observation, proves that the earth is not perfectly spherical, but that it must be more elevated under the equator than at the poles.

From these combined causes, the ebbing and flowing of the tides, and the motion of the earth, we may fairly conclude, that although the earth was a perfect sphere in its original form, yet its diurnal motion, together with the constant flux and reflux of the sea, must, by degrees, in the course of time, have raised the equatorial parts, by carrying mud, earth, sand, shells, etc. from other climes, and there depositing of them. Agreeable to this idea the greatest irregularities must be found, and, in fact, are found near the equator. Besides, as this motion of the tides is made by diurnal alternatives, and [has] been repeated, without interruption from the commencement of time, is it not natural to imagine, that each time the tide flows the water carries a small quantity of matter from one place to another, which may fall to the bottom like a sediment, and form those parallel and horizontal strata which are everywhere to be met with? For the whole motion of the water, in the flux and reflux being horizontal,[44] the matters carried away with them will naturally be deposited in the same parallel direction.

But to this it may be said, that as the flux and reflux of the waters are equal and regularly succeed, the two motions would counterpoise each other, and the matters brought by the flux would be returned by the reflux, and of course this cause of the formation of the strata must be chimerical; that the bottom of the sea could not experience any material alteration by two uniform motions, wherein the effects of the one would be regularly destroyed by the other, much less could they change the original form by the production of heights and inequalities.

To which it may be answered that the alternate motions of the waters are not equal, the sea having a constant motion from the east to the west; besides, the agitation, caused by the winds, [is opposed to the flux and reflux of the tides]. It will also be admitted that by every motion of which the sea is susceptible, particles of earth and other matters will be carried from one place and deposited in another; and these collections will necessarily assume the form of horizontal and parallel strata, from the various combinations of the motions of the sea always tending to move the earth, and to level these materials wherever they fall in the form of a sediment. But this objection is easily obviated by the well-known fact, that upon all coasts, bordering the sea, where the ebbing and flowing of the tide is observed, the flux constantly brings in a number of things which the reflux does not carry back. There are many places upon which the sea insensibly gains and gradually covers over,[45] while there are others from which it recedes, narrowing as it were its limits, by depositing earth, sands, shells, etc. which naturally take an horizontal position; these matters accumulate by degrees in the course of time, and being raised to a certain point gradually exclude the water, and so become part of the dry land for ever after.

But not to leave any doubt upon this important point, let us strictly [inquire] into the possibility of a mountain's being formed at the bottom of the sea by the motions and sediments of the waters. It is certain that on a coast which the sea beats with violence during the agitation of its flow, that every wave must carry off some part of the earth; for wherever the sea is bounded by rocks, it is a plain fact, that the water by degrees wears away those rocks,[46] and consequently carries away small particles every time the waves retire; these particles of earth and stone will necessarily be transported to some distance, and [to a specific place] where the agitation of the water is abated, and left to their own weight, they precipitate to the bottom in form of a sediment; and there form a first stratum either horizontal or inclined according to the position of the surface upon which they fall; this will shortly be covered by a similar stratum produced by the same cause, and thus will a considerable quantity of matter be almost insensibly collected together, and the strata of which will be placed parallel to each other.

This mass will continue to increase by new sediments, and by gradually accumulating, in the course of time become a mountain at the bottom of the sea, exactly similar to those we see on dry land, both as to outward form and internal composition. If there happen to be shells in this part of the sea where we have supposed this deposit to be made, they will be filled and covered with the sediment, and incorporated in the deposited matter, making a part of the whole mass, and they will be found situated in the parts of the mountain according to the time they had been there deposited; those that lay at the bottom previous to the formation of the first stratum, will be found in the lowest, and so according to the time of their being deposited, the latest in the most elevated parts.

So likewise, when the bottom of the sea, at particular places, is [disturbed] by the agitation of the water, there will necessarily ensue, in the same manner, a removal of the earth, shells, and other matters, from the [disturbed parts to the others]; for we are assured by all divers,[47] that at the greatest depths they descend, i.e. 20 fathom[s], the bottom of the sea is so troubled by the agitation of the waters that the mud and shells are carried to considerable distances. Consequently transportations of this kind are made in every part of the sea, and this matter falling must form eminences, composed like our mountains, and in every respect similar; therefore the flux and reflux, by the winds, the currents, an all the motions of the water, must inevitably create inequalities at the bottom of the sea.

Nor must we imagine that these matters cannot be transported to great distances, because we daily see grain, and other productions of the East and West Indies, arriving on our own coasts.[48] It is true these bodies are specifically lighter than water, whereas the substances of which we have been speaking are specifically heavier, but, however, being reduced to an impalpable powder, they may be sustained a long time in the water so as to be conveyed to considerable distances.

It has been supposed that the sea is not troubled at the bottom, especially if it is very deep, by the agitations produced by the winds and tides; but it should be recollected that the whole mass, however deep, is put in motion by the tides, and that in a liquid globe this motion would be communicated to the very center; that the power which produces the flux and reflux is a penetrating force, which acts proportionably upon every particle of its mass, so that we can determine by calculation the quantity of its force at different depths; but in short, this point is so certain that it cannot be contested but by refusing the evidence of reason.

Therefore, we cannot possibly have the least doubt that the tides, the winds, and every other cause which agitates the sea, must produce eminences and inequalities at the bottom, and these heights must ever be composed of horizontal or equally inclined strata. These eminences will gradually increase until they become hills, which will rise in situations similar to the waves that produce them; and if there is a long extent of soil, they will continue to augment by degrees; so that in course of time they will form a vast chain of mountains. Being formed into mountains they become an obstacle to and interrupt the common motion of the sea, producing, at the same time, other motions which are generally called currents. Between two neighbouring heights at the bottom of the sea a current will necessarily be formed[49] which will follow their common direction, and, like a river, form a channel, whose angles will be alternatively opposite during the whole extent of its course. These heights will be continually increasing, being subject only to the motion of the flux, for the waters, during the flow, will leave the common sediment upon their ridges; and those waters which are impelled by the current will force along with them, to great distances, those matters which would be deposited between both, at the same time hollowing out a valley with corresponding angles at their foundation. By the effects of these motions and sediments the bottom of the sea, although originally smooth, must become unequal, and abounding with

The "Second Discourse"

hills and chains of mountains, as we find it at present. The soft materials of which the eminences are originally composed will harden by degrees with their own weight; some forming parts purely angular, produce hills of clay; others, consisting of sandy and crystalline particles, compose those enormous masses of rock and flint from whence crystal and other precious stones are extracted; those [beds] formed with stony particles, mixed with shells, form those of lime-stone and marble, wherein we [today can find shells incorporated]; and others, compounded of matter more shelly, united with pure earth, compose all our beds of marle and chalk. All these substances are placed in regular beds, and all contain heterogeneous matter; marine productions are found among them in abundance, and nearly according to the relation of their specific weights; the lightest shells in chalk, and the heaviest in clay and lime-stone: these shells are invariably filled with the matter in which they have been enclosed, whether stones or earth; an incontestable proof that they have been transported with the matter that fills and surrounds them, and that this matter was at that time in an impalpable powder. In short, all those substances whose horizontal situations have been established by the level of the waters of the sea, will constantly preserve their original position.

But here it may be observed, that most hills, whose summits consist of solid rock, stone, or marble, are formed upon small eminences of much lighter materials, such for instance as clay, or strata of sand, which we commonly find extended over the neighbouring plains; upon which it may be asked, how, if the foregoing theory be just, this seemingly contradictory arrangement happens? To me this phenomenon appears to be very easy and naturally explained. The waters at first act upon the upper stratum of coasts, or bottom of the sea, which commonly consists of clay or sand, and having transported this, and deposited the sediment, it of course composes small eminences, which form a base for the more heavy particles to rest upon. Having removed the lighter substances, it operates upon the more heavy, and by constant attrition reduces them to an impalpable powder, which it conveys to the same spot, and where, being deposited, these stony particles, in the course of time, form those solid rocks and quarries which we now find upon the tops of hills and mountains. It is not unlikely that as these particles are much heavier than sand or clay, that they were formerly a considerable depth under a strata of that kind, and now owe their high situations to having been last raised up and transported by the motion of the water.

To confirm what we here assert, let us more closely investigate the situation of those materials which compose the superficial outer part of the globe, indeed the only part with which we have any knowledge. The different beds of strata in stone quarries are almost all horizontal, or regularly-inclined; those whose foundations are on clays or other solid matters are clearly horizontal, especially in plains. The quarries wherein we find flint, or brownish grey free-stone, in detached portions, have a less regular position, but even in those the uniformity of nature plainly appears, for the horizontal or regularly-inclined strata is apparent in quarries where these stones are found in great masses. This position is universal, except in quarries where flint and brown free-stone are found in small detached portions, the formation of which we shall prove to have been posterior to those we have just been treating of; for granite, vitrifiable sand, [clay], marble, calcareous stone, chalk, and marbles, are always deposited in parallel strata, horizontally or equally-inclined; the original formation of these are easily discovered, for the strata are exactly horizontal and very thin, and are arranged above each other like the leaves of a book. Beds of sand, soft and hard clay, chalk, and shells, are also either horizontal or regularly-inclined. Strata of every kind preserve the same thickness throughout [their] whole extent, which often occupy the space of many miles, and might be traced still farther by close and exact observations. In a word, the materials of the globe, as far as mankind [has] been enabled to penetrate, are arranged in an uniform position, and are exactly similar.

The strata of sand and gravel which have been washed down from mountains must in some measure be excepted; in valleys they are sometimes of a considerable extent, and are generally placed under the first strata of the earth; in plains, they are as even as the most ancient and interior strata, but near the bottom and upon the ridges of hills they are inclined, and follow the inclination of the ground upon which they have flowed. These being formed by rivers and rivulets, which are constantly in valleys changing their beds, and dragging these sands and gravel with them, they are of course very numerous. A small rivulet flowing from the neighbouring heights, in the course of time, will be sufficient to cover a very spacious valley with a stratum of sand and gravel; and I have often observed in hilly countries, whose base, as well as the upper stratum was hard clay, that above the source of the rivulet the clay is found immediately under the vegetable soil, and below [the source] there is the thickness of a foot of sand upon the clay and which extends itself to a considerable distance. These strata formed by rivers are not very ancient, and are easily discovered by the inequality of their thickness which is constantly varying, while the ancient strata preserves the same dimensions throughout; they are also to be known by the matter itself, which bears evident marks of having been smoothed and rounded by the motion of the water. The same may be said of the turf and perished vegetables which are found below the first strata of earth in marshy grounds; they cannot be considered as ancient but entirely produced by successive heaps of decayed trees and other plants. Nor are the strata of slime and mud which are found in many countries to be considered as ancient productions, having been formed by stagnated waters or inundations of rivers, and are neither so horizontal, nor equally-inclined, as the strata anciently produced by the regular motions of the sea. In the strata formed by rivers we constantly meet with river but scarcely ever sea shells, and the few that are found are broken and irregularly placed; whereas in the ancient strata there are no river shells; the sea shells are in great quantities, well preserved, and all placed in the same manner, having been transported at the same time and by the same cause. How are we to account for this astonishing regularity? Instead of regular strata why do we not meet with the matters that compose the earth jumbled together, without all kind of order? Why are not rocks, marbles, clays, marles, etc. variously dispersed, or joined by irregular or vertical strata? Why are not the heaviest bodies uniformly found placed beneath the lightest? It is easy to perceive that this uniformity of nature, this organization of earth, this connection of different materials, by parallel strata, without respect to their weights, could only be produced by a cause as powerful and constant as the motion of the sea, whether occasioned by the regular winds, or by that of the flux and reflux, etc.

These causes act with greater force under the equator than in other climates, for there the winds are more regular and the tides run higher; the most extensive chains of mountains are also near the equator. The mountains of Africa and Peru are the highest known, they frequently extend themselves through whole provinces, and stretch to considerable distances under the ocean. The mountains of Europe and Asia, which extend from Spain to China, are not so high as those of South America and Africa. The mountains of the North, according to the relation of travellers, are only hills in comparison with those of the Southern countries. Besides, there are very few islands in the northern seas, whereas in the torrid zone they are almost innumerable, and as islands are only the summits of mountains, it is evident that the surface of the earth has many more inequalities towards the equator than in the northerly climes.

It is therefore evident that the prodigious chains of mountains which run from the West to the East in the old continent, and from the North to the South in the new, must have been produced by the general motion of the tides; but the origin of all the inferior mountains must be attributed to the particular motions of currents, occasioned by the winds and other irregular agitations of the sea: they may probably have been produced by a combination of all those motions, which must be capable of infinite variations, since the winds and different positions of islands and coasts change the regular course of the tides, and compel them to flow in every possible direction: it is therefore not in the least astonishing that we should see considerable eminences whose courses have no determined direction. But it is sufficient for our present purpose to have demonstrated that mountains are not the produce of earthquakes, or other accidental causes, but that they are the effects resulting from the general order of nature, both as to their organization, and the position of the materials of which they are composed.

But how has it happened that this earth which we and our ancestors have inhabited for ages, which, from time immemorial, has been an immense continent, dry and removed from the reach of the waters, should, if formerly the bottom of the ocean, be actually larger than all the waters, and raised to such a height as to be distinctly separated from them? Having remained so long on the earth why have the waters now abandoned it? What accident, what cause could produce so great a change? Is it possible to conceive one possessed of sufficient power to produce such an amazing effect?

These questions are difficult to be resolved, but as the facts are certain and incontrovertible, the exact manner in which they happened may remain unknown, without prejudicing the conclusions that may be drawn from them; nevertheless by a little reflection we shall find at least plausible reasons for these changes.[50] We daily observe the sea gaining ground on some coasts, and losing it on others; we know that the ocean has a continued regular motion from east to west; that it makes loud and violent efforts against the low lands and rocks which confine it; that there are whole provinces which human industry can hardly secure from the rage of the sea; but there are instances of islands rising above, and others being sunk under the waters. History speaks of much greater deluges and inundations. Ought not this to incline us to believe that the surface of the earth has undergone great revolutions, and that the sea may have quitted the greatest part of the earth which it formerly covered? Let us but suppose that the old and new worlds were formerly but one continent, and that the Atlantis of Plato, was sunk by a violent earthquake: the natural consequence would be, that the sea would necessarily have flowed in from all sides, and formed what is now called the Atlantic Ocean, leaving vast continents dry, and possibly those which we now inhabit. This revolution therefore might be made [suddenly] by the opening of some vast cavern in the interior part of the globe, which an universal deluge must inevitably succeed: or possibly this change was not effected at once but required a length of time, which I am rather inclined to think: however these conjectures may be, it is certain the revolution has occurred, and in my opinion very naturally, for to judge of the future, as well as the past, we must carefully attend to what daily happens before our eyes. It is a fact clearly established by repeated observations of travellers[51] that the ocean has a constant motion from the east to west; this motion, like the trade winds, is not only felt between the tropics but also throughout the temperate climates, and as near the poles as navigators have gone; of course the Pacific Ocean makes a continual effort against the coasts of Tartary, China, and India: the Indian Ocean acts against the east coast of Africa, and the Atlantic in like manner against all the eastern coasts of America; therefore the sea must have always [gained land on the east and lost it on the west], and still continues to do so; and this alone is sufficient to prove the possibility of the change of earth into sea, and sea into land. If in fact, such are the effects of the sea's motion from east to west, may we not very reasonably suppose that Asia and the eastern continent are the oldest countries in the world, and that Europe and part of Africa, especially the western coasts of these continents, as Great Britain, France, Spain, Mauritania, etc. are of a more modern date? Both history and physics agree in confirming this conjecture.

There are, however, many other causes which concur with the continual motion of the sea from east to west, in producing these effects.

In many places there are lands lower than the level of the sea, and which are only defended from it by an isthmus of rocks, or by banks and dikes of still weaker materials; these barriers must gradually be destroyed by the constant action of the sea, when the lands will be overflowed, and constantly make part of the ocean. Besides, are not mountains daily decreasing by the rains which loosen the earth, and carry it down into the valleys?[52] It is also well known that floods wash the earth from the plains and high grounds into the small brooks and rivers, who in their turn convey into the sea. By these means the bottom of the sea is filling up by degrees, the surface of the earth lowering to a level, and nothing but time is necessary for the sea's successively changing places with the earth.

I speak not here of those remote causes which stand above our comprehension; of those convulsions of nature, whose least effects would be fatal to the world; the near approach of a comet, the absence of the moon, the introduction of a new planet, etc. are suppositions on which it is easy to give scope to the imagination. Such causes would produce any effect we chose, and from a single hypothesis of this nature, a thousand physical romances might be drawn, any of which the authors might term the <u>Theory of the Earth</u>. As historians we reject these vain speculations; they are mere possibilities which suppose the destruction of the universe, in which our globe, like a particle of forsaken matter, escapes our observation and is no longer an object worthy [of] regard; but to preserve consistency, we must take the earth as it is, closely observing every part and by inductions judge of the future from what exists at present; in other respects we ought not to be affected by causes which seldom happen, and whose effects are always sudden and violent; they do not occur in the common course of nature; but effects which are daily repeated, motions which succeed each other without interruption, and operations that are constant, ought alone to be the ground of our reasoning.

. . . .

Figure 6 - God creates the solar system by a collision of a passing comet with a rotating double star system. Matter hurled out from sun along the comet's path condenses to form the planets.

Figure 7 - Inspired man, probably Newton, who through geometry has understood the motion of the pendulum, the law of the inclined plane, and the force of attraction between earth and moon by a single set of principles.

PROOFS OF THE THEORY OF THE EARTH

ARTICLE I.

ON THE FORMATION OF THE PLANETS.

Translated by J.S. Barr

Our subject being Natural History, we would willingly dispense with astronomical observations; but as the nature of the earth is so closely connected with the heavenly bodies, and such observations being calculated to illustrate more fully what has been said, it is necessary to give some general ideas of the formation, motion, and figure of the earth and other planets.
The earth is a globe of about three thousand leagues diameter; it is situated one thousand millions of leagues from the sun, around which it makes its revolution in three hundred and sixty-five days. This revolution is the result of two forces, the one may be considered as an impulse from right to left, or from left to right, and the other an attraction from above downwards, or beneath upwards, to a common center. The direction of these two forces, and their quantities, are so nicely combined and proportioned, that they produce an almost uniform motion in an ellipse, very near to a circle. Like the other planets the earth is opaque, it throws out a shadow; it receives and reflects the light of the sun, round which it revolves in a space of time proportioned to its relative distance and density. It also turns round its own axis once in twenty-four hours, and its axis is inclined 66 [1/2] degrees on

the plane of the orbit. Its figure is spheriodal, the two axes of which differ about [160 1/15th] part from each other, and the smallest axis is that round which the revolution is made.

These are the principal phenomena of the earth, the result of discoveries made by means of geometry, astronomy, and navigation. We shall not here enter into the detail of the proofs and observations by which those facts have been ascertained, but only make a few remarks to clear up what is still doubtful, and at the same time give our ideas respecting the formation of the planets, and the different changes through which it is possible they have passed before they arrived at the state [in which we see them today].

There have been so many systems and hypotheses framed upon the formation of the terrestrial globe, and the changes which it has undergone, that we may presume to add our conjectures to those who have written upon the subject; especially as we mean to support them with a greater degree of probability than has hitherto been done; and we are the more inclined to deliver our opinion upon this subject, from the hope that we shall enable the reader to pronounce on the difference between an hypotheses drawn from possibilities, and a theory founded on facts; between a system such as we are here about to present, on the formation and original state of the earth, and a physical history of its real condition, which has been given in the preceding discourse.

Galileo [discovered] the laws of falling bodies, and Kepler observed, that the area described by the principal planets in moving round the sun, and those of the satellites round the planets to which they belong, are proportional to the time of their revolutions, and that such periods were also in proportion to the square roots of the cubes of their distances from the sun, or principal planets. Newton found that the force which caused heavy bodies to fall on the surface of the earth, extended to the moon, and retained it in its orbit; that this force diminished in the same proportion as the square of the distance increases, and consequently that the moon is attracted by the earth; that the earth and planets are attracted by the sun; and that, in general, all bodies which revolve round a center, and describe areas proportioned to the times of their revolution, are attracted towards that point. This power, known by the name of **gravity**, is therefore diffused throughout all matter; planets, comets, the sun, the earth, and all nature, is subject to its laws, and it serves as a basis [for] the general harmony which reigns in the universe. Nothing is better proved in physics than the actual [and individualized] existence of this power in every material substance. Observation has confirmed the effects of this power, and geometrical calculations have determined the quantity and relations of it[....]

This general cause being known, the effects would easily be deduced from it, if the action of the powers which produce it were not too complicated. A single moment's reflection upon the solar system will fully demonstrate the difficulties that have attended this subject; the principal planets are attracted by the sun, and the sun by the planets; the satellites are also attracted by their principal planets, and each planet attracts all the rest, and is attracted by them. All these actions and reactions vary according to the quantities of matter and the distances; and produce great inequalities and irregularities: how is so great a number of connections to be combined and estimated? It appears almost impossible in such a crowd of objects to follow any particular one; nevertheless, those difficulties have been surmounted, and calculation has confirmed the suppositions of [reason], each observation [has] become a new demonstration, and the systematic order of the universe is laid open to the eyes of all those who can distinguish truth from error.

[One thing causes us pause, and is in fact independent of this theory. This is the force of impulsion.] We evidently see the force of attracton always draws the planets towards the sun, [and] they would fall in a perpendicular line on that planet, if they were not repelled by some other power that obliges them to move in a straight line, and which impulsive force would compel them to fly off the tangents of their respective orbits, if the force of attraction ceased one moment. The force of impulsion was certainly communicated to the planets by the hand of the Almighty, when he gave motion to the universe; but we ought, as much as possible, to abstain in [natural philosophy] from having recourse to supernatural causes; and it appears that a probable reason may be given for this impulsive force, perfectly accordant with the laws of mechanics, and not by any means more astonishing than the changes and revolutions which may and must happen in the universe.

The sphere of the sun's attraction does not confine itself to the orbs of the planets, but extends to a remote distance, always decreasing in the same ratio as the square of the distance increases; it is demonstrated that the comets which are lost to our sight, in the regions of the sky, obey this power, and by it their motions, like that of the planets, are regulated. All these stars, whose tracks are so different, move round the sun, and describe areas proportioned to the time; the planets in ellipses more or less approaching a circle, and the comets in narrow ellipses of a great extent. Comets and planets move, therefore, by virtue of the force of attraction and impulsion, which continually acting at one time obliges them to describe these courses; but it must be remarked that comets pass over the solar system in all directions, and that the inclinations of their orbits are very different, insomuch, that although subject, like the planets, to the force of attraction, they have nothing in common with respect to their progressive or impulsive motions, but appear, in this respect, independent of each other: the planets, on the contrary, move round the sun in the same direction, and almost in the same plane, never exceeding [7 1/2] degrees of inclination in their planes, the most distant from their orbits. This conformity of position and direction in the motion of the planets, necessarily implies that their impulsive force has been communicated to them by one and the same cause.

May it not be imagined, with some degree of probability, that a comet falling into the body of the sun, will displace and separate some parts from the surface, and communicate to them a motion of impulsion, insomuch, that the planets may formerly have belonged to the body of the sun, and been detached therefrom by an impulsive force, which they still preserve?

This supposition appears to be at least as well founded as the opinion of Leibnitz, who supposes that the earth and planets have formerly been suns; and his system, of which an account will be given in the fifth article, would have been more comprehensive and more agreeable to probability, if he had raised himself to this idea. We agree with him in thinking that this effect was produced at the time when Moses said that God divided light from darkness; for, according to Leitnitz, light was divided from darkness when the planets were [darkened]; but in our supposition there was a physical separation, since the [opaque matter composing the bodies of the planets was actually separated] from the luminous matter which composes the sun.

This idea of the cause of the impulsive force of the planets will be found much less objectionable, when an estimation is made of the analogies and degrees of probability, by which it may be supported. In the first place, the motion of the planets is in the same direction, from West to East, and therefore, according to calculation it is sixty-four to one that such would not have been the case if they had not been indebted to the same cause for their impulsive forces.

This probably will be considerably augmented by the second analogy, viz. that the inclination of the planes of the orbits do not exceed 7 1/2 degrees; for by comparing the spaces, we shall find [it to be] twenty-four to one, that two planets [would be] found in their most distant planes at the same time, and consequently [24^5], or 7,692,624 to one, that all six would by chance be thus placed; or what amounts to the same, there is a great degree of probability that the planets have been impressed with one common moving force, which has given them this position. But what can have bestowed this common impulsive motion, but the force and direction of the bodies by which it was originally communicated? It may therefore be concluded, with great likelihood, that the planets received their impulsive motion by one single stroke. This probability, which is almost equivalent to a certainty, being established, I seek to know what moving bodies could produce this effect, and I find nothing but comets capable of communicating a motion to such vast bodies. By examining the course of comets, we shall be easily persuaded that it is almost necessary for some of them occasionally to fall into the sun. That of 1680 approached so near, that at its perihelion it was not more distant from the sun than a sixteenth part of [the solar] diameter, and if it returns, as there is every appearance it will, in 2255, it may then possibly fall into the sun. That must depend on the re-encounters it will meet with in its road, and on the retardation it suffers in passing through the atmosphere of the sun.[53]

We may therefore presume, with the great Newton, that comets sometimes fall into the sun; but this fall may be made in different directions. If they fall perpendicular, or in a direction not very oblique, they will remain in the sun, and serve for food to the fire which [consumes that star], and the motion of impulsion which they will have communicated to the sun, will produce no other effect than that of [displacing] it more or less, according as the mass of the comet will be more or less considerable; but if the fall of the comet is in a very oblique direction, which will most frequently happen, then the comet will only graze the surface of the sun, or slightly furrow it; and in this case, it may drive out some parts of matter to which it will communicate a common motion of impulsion, and these parts so forced out of the body of the sun, and even the comet itself, may then become planets, and turn round this luminary in the same direction, and in almost the same plane. We might perhaps calculate, what quantity of matter, velocity, and direction a comet should have to impel from the sun an equal quantity of matter to that which the six planets and their satellites contain; but it will be sufficient to observe here, that all the planets with their satellites, do not make the 650th part of the mass of the sun,[54] because the density of the large planets, Saturn and Jupiter, is less than that of the sun, and although the earth be four times, and the moon near five times more dense than the sun, they are nevertheless but as atoms in comparison with [the mass of this star].

However inconsiderable the 650th part may be, yet it certainly at first appears to require a very powerful comet to separate even that much from the body of the sun; but if we reflect on the prodigious velocity of comets in their perihelion, a velocity so much the greater, as they approach nearer the sun; if besides, we pay attention to the density and solidity of the matter of which they must be composed to suffer, without being destroyed, the inconceivable heat they endure; and consider the bright and solid light which shines through their dark and immense atmospheres, which surround and must obscure it, it cannot be doubted that comets are composed of extremely solid and dense matters, and that they contain a great quantity of matter in a small compass; that consequently a comet of no extraordinary bulk may have sufficient weight and velocity to displace the sun, and give a projectile motion to a quantity of matter, equal to the 650th part of the mass of this luminary. This perfectly agrees with what is known concerning the density of planets, which always decreases as their distance from the sun is increased, they having less heat to support; so that Saturn is less dense than Jupiter, and Jupiter much less than the Earth; therefore, if the density of the planets be as Newton asserts, proportional to the quantity of heat which they have to support, Mercury will be seven times more dense than the earth, and twenty eight times denser than the sun; and the comet of 1680 would be 28,000 times denser than the earth, or 112,000 times denser than the sun, and by supposing it as large as the earth, it would contain nearly an equal quantity of matter to the ninth part of the sun, or by giving it only the 100th part of the size of the earth, its mass would still be equal to the 900th part of the sun. From whence it is easy to conclude, that such a body, though it would be but a small comet, might separate and drive off from the sun a 900th or a 650th part, particularly if we attend to the immense velocity with which comets move when they pass in the vicinity of the sun.

Besides this, the conformity between the density of the matter of the planets [and] that of the sun deserves some attention. It is well known that both on and near the surface of the earth there are some matters [14,000 or 15,000] times denser than others. The densities of gold and air are nearly in this relation. But the internal parts of the earth and planets are composed of a more uniform matter, whose comparative density varies much less; and the conformity in the density of the planets and that of the sun is such, that of 650 parts which compose the whole of the matter of the planets, there are more than 640 of the same density as the matter of the sun, and only ten parts out of these 650 which are of a greater density, for Saturn and Jupiter are nearly of the same density as the sun, and the quantity of matter which these planets contain, is at least 64 times greater than that of the four inferior planets, Mars, the Earth, Venus and Mercury. We must, therefore, admit, that the matter, of which the planets are generally composed, is nearly the same as that of the sun, and that consequently the one may have been separated from the other.

. . . .

The comet, therefore, by its oblique fall upon the surface of the sun, having driven therefrom a quantity of matter equal to the 650th part of its whole mass; this matter, which must be considered in a liquid state, will at first have formed a torrent, the grosser and less dense parts of which will have been driven the farthest, and the smaller and more dense having received only the like impulsion, will remain nearest its source; the force of the sun's attraction would inevitably act upon all the parts detached from [it], and constrain them to circulate around [its] body, and at the same time the mutual attraction of the particles of matter would form themselves into globes at different distances from the sun, the nearest of which necessarily moving with greater rapidity in their orbits than those at a distance.

But, another objection may be [raised], and it may be said, if the matter which composes the planets had been separated from the sun, they, like [it], would have been burning and luminous bodies, not cold and opaque, for nothing resembles a globe of fire less than a globe of earth and water, and by comparison, the matter of the earth and planets is perfectly different from that of the sun.

To this it may be answered, that in the separation, the matter changed its form, and the light or fire was extinguished by the stroke which caused this motion of impulsion. Besides, may it not be supposed that if the sun, or a burning star, moved with such velocity as the planets, that the fire would soon be extinguished: and that is the reason why all luminous stars are fixed, and that those stars which are called new, and which have probably changed places, are frequently extinguished, and lost? This remark is somewhat confirmed by what has been observed on comets; they must burn to the center when they pass to their perihelion. Nevertheless they do not become luminous themselves, they only exhale burning vapours, of which they leave a considerable part behind them in their course.

I own, that in a medium where there is very little or not resistance, fire may subsist and suffer a very great motion without being extinguished: I also own, that what I have just said extends only to the stars which totally disappear, and not to those which have periodical returns and appear and disappear alternatively without changing place in the heavens. The phenomena of these stars has been explained in a very satisfactory manner by M. de Maupertuis, in his discourse on the figure of the [heavenly bodies].[55] But the stars [étoiles] which appear and afterwards disappear entirely, must certainly have been extinguished, either by the velocity of their motion, or some other cause. We have not a single example of one luminous star [astre] revolving round another; and among the number of planets which compose our system, and which move round the sun with more or less rapidity, there is not one luminous of itself.

It may also be added, that fire cannot subsist so long in the small as in large masses, and that the planets must have burnt for some time after they were separated from the sun, but were at length extinguished for want of combustible matter, as probably will be the sun itself, and for the same reason; but in a length of time as far beyond that which extinguished the planets, as it exceeds in quantity of matter. Be this as it may, the matter of which the planets are formed being separated from the sun, by the stroke of a comet, that appears as a sufficient reason for the extinction of their fires.

The earth and planets, at the time of their quitting the sun, were in a state of total liquid fire: in this state they remained only as long as the violence of the heat which had produced it; and which heat necessarily underwent a gradual decay: it was in this state of fludity that they took their circular forms, and that their regular motions raised the parts of their equators, and lowered their poles. [This figure, which accords so well with the laws of hydrostatics, necessarily supposes that the earth and planets have been in a fluid state, and I am here in agreement with Mr. Leibnitz.[56] This state of fluidity would be caused by violent heating, and] the internal part of the earth must be a vitrifiable matter, of which sand, granite, etc. are the fragments and scoria.

It may, therefore, with some probability be thought, that the planets [originated from] the sun, that they were separated by a single stroke which gave to them a motion of impulsion, and that their position at different distances from the sun, proceeds only from their different densities. It now only remains, to complete this theory, to explain the diurnal motion of the planets, and the formation of the satellites; but this, far from adding difficulties to my hypothesis, seems on the contrary, to confirm it.

For the diurnal motion, or rotation, depends solely on the obliquity of the stroke, an oblique impulse therefore on the surface of a body will necessarily give it a rotative motion; this motion will be equal and always the same, if the body which receives it is homogeneous, and it will be unequal if the body is composed of heterogeneous parts, or of different densities; hence we may conclude, that in all the planets the matter is homogeneous, since their diurnal motions are equal, and regularly performed in the same period of time. [This is another proof of the separation of parts of greater and lesser density when the planets were formed].

But the obliquity of the stroke might be such as to separate from the body of the principal planet a small part of matter, which would of course continue to move in the same direction; these parts would be united, according to their densities, at different distances from the planet, by the force of their mutual attraction, and at the same time follow its course around the sun, by revolving about the body of the planet, nearly in the plane of its orbit. It is plain that those small parts so separated are the satellites: thus the formation, position, and direction of the motions of the satellites perfectly agree with our theory; for they have all the same motion in concentrical circles round their principal planet; their motion is in the same direction, and that nearly in the plane of their orbits. All these effects, which are common to them, and which depend on an impulsive force, can proceed only from one common cause, which is, impulsive motion, communicated to them by one and the same oblique stroke.

What we have just said on the cause of the motion and formation of the satellites, will acquire more probability if we consider all the circumstances of the phenomena. The planets which turn the swiftest on their axes, are those which have satellites. The earth turns quicker than Mars in the relation of about 24 to 15; the earth has a satellite, but Mars has none. Jupiter, whose rapidity round its axis is five to six hundred times greater than that of the earth, has four satellites, and there is a great [likelihood] that Saturn, which has five, and a ring, turns still more quickly than Jupiter.

It may even be conjectured with some foundation that the ring of Saturn is parallel to the equator of the planet, so that the plane of the equator of the ring, and that of Saturn, are nearly the same; for by supposing, according to the preceding theory, that the obliquity of the stroke by which Saturn has been set in motion [would have been very great, the velocity of axial rotation that would have resulted from this oblique blow would at first have been such that the centrifugal force exceeded that of gravity, and there would] be detached from its equator, and neighboring parts, a considerable quantity of matter, which will necessarily have taken the figure of a ring, whose plane must be nearly the same as that of the equator of the planet; and this quantity of matter having been detached from the vicinity of the equator of Saturn, must have lowered the equator of that planet, which [is the cause] that, notwithstanding its rapidity, the diameters of Saturn cannot be so unequal as those of Jupiter, which differ from each other more than an eleventh part.

However great the probability of what I have advanced on the formation of the planets and their satellites may appear to me, yet, every man has his particular measurement to estimate probabilities of this nature, and as this measurement depends on the strength of the understanding to combine more or less distant relations, I do not pretend to convince the incredulous. I have not only thought it my duty to offer these ideas, because they appear to me reasonable, and calculated to clear up a subject, on which, however important, nothing has hitherto been written, but because the impulsive motion of the planets enters at least one half in the composition of the universe, which gravity alone cannot unfold. I shall only add the following questions to those who are inclined to deny the possibility of my system.

1. Is it not natural to imagine, that a body in motion has received that motion by the stroke of another body?

2. Is it not very probable, that when many bodies move in the same direction, that they have received this direction by one single stroke, or by many strokes directed in the same manner?

3. Is it not more probable, that when many bodies have the same direction in their motion and are placed in the same plane, that they received this direction and this position by one and the same stroke, rather than by a number?

4. At the time a body is put in motion by the force of impulsion, is it not probable that it receives it obliquely, and consequently is obliged to turn on its axis so much the quicker, as the obliquity of the stroke will have been greater? If these questions should not appear unreasonable, the theory, of which we have presented the outlines, will cease to appear an absurdity. . . .

NOTES

*Buffon's Natural History, Containing a Theory of the Earth, a General History of Man, of the Brute Creation, and of Vegetables, Minerals, Etc., trans. J.S. Barr (London: J.S. Barr, 1792) vol. 1, pp. 1-40; 69-82; 91-99. Buffon's notes and the quote from Ovid as given in the original 1749 Imprimerie Royale edition of the Histoire naturelle, have been restored to this text. Spellings have been modernized, and occasional minor grammatical changes have been inserted for clarity. For a detailed criticism of Buffon's theory of the earth, see first review from the Bibliotheque raisonee, Part III, below.

[1][Ovid, Metamorphoses, bk. xv, line 262-267, trans. by Frank J. Miller (Loeb Classical Library; New York & London: Heinemann, 1933) II, 383. This is given in Latin in Buffon's original text.]

[2]See the "Proofs of the Theory of the Earth," art. 1, below.

[3]See the "Proofs of the Theory of the Earth," art. 2, below. [Not included. Reference is to William Whiston, A New Theory of the Earth, from its Original to the Consummation of all Things. . . (London, 1696)].

[4]See Proofs of the Theory of the Earth," art. 3 [not included. Burnet's work is the Sacred Theory of the Earth: Containing an Account of it Original Creation, and of all the General Changes which it hath Undergone (1st. English edition London, 1684)].

[5]See "Proofs. . ., art. 4 [not included. Reference is to John Woodward, An Essay Towards a Natural History of the Earth (London, 1695)].

[6]See Seneca, Quaestiones, lib. 6, cap. 21; Strabo, Geographica, lib. 1; Orosius, [Historiarum] lib. 2, cap. 18; Pliny [Naturalis historia] lib. 2, cap. 19; Histoire de l'académie des sciences, year 1708, p. 23.

[7]See "Proofs. . .," art. 14. [Not included].

Proofs of the Theory of the Earth   161

   [8] See "Proofs...," art. 6. [Not included].

   [9] See *Philosophical Transactions*, Abridged ed., vol. II, p. 323.

   [10] See Boyle's *Works*, vol. III, p. 232.

   [11] See "Proofs...," art. 1. [Given below].

   [12] See "Proofs...," art. 12. [Not included].

   [13] See "Proofs...," art. 13. [Not included].

   [14] See the map prepared by Mr. Buache in 1737 on the depths of the ocean between Africa and America.

   [15] See Varenius, *Geographica generalis*, p. 218.

   [16] See "Proofs...," art. 13. [Not included].

   [17] See Varenius, p. 140. See also the *Voyage* of [François] Pyrard, p. 137.

   [18] See the *Voyages* of Shaw, vol. 2, p. 56.

   [19] See "Proofs...," art. 16. [Not included].

   [20] The Maelstrom in the Norwegian Sea.

   [21] The calms and the gyres of the Ethopian Sea.

   [22] See "Proofs...," art. 6 and 10. [Not included].

   [23] See the Map of the expedition of Mr. Bouvet, prepared by Mr. Buache in 1739.

   [24] See "Proofs...," art. 9. [Not included].

   [25] See "Proofs...," art. 9 and 12. [Not included].

   [26] See the *Lettres philosophiques* of Bourguet, p. 181.

   [27] See Varenius, *Geographica*, p. 69.

   [28] See "Proofs...," art. 10. [Not included].

   [29] See "Proofs...," art. 7. [Not included].

   [30] See "Proofs...," art. 16. [Not included].

   [31] See Kircher, *Mundus subterraneus*, "Preface."

[32]See Varenius, Geographica, p. 43.

[33]See "Proofs...," art. 7. [Not included].

[34]See "Proofs...," art. 7. [Not included].

[35]See Woodward, p. 41 ff.

[36]See "Proofs...," art. 8. [Not included].

[37]See "Proofs...," art. 8. [Not included].

[38]See Steno, Woodward, Ray, Bourguet, Scheuchzer, the Philosophical Transactions, the Memoires de l'Académie, etc.

[39]See "Proofs..." art. 8. [Not included].

[40]See "Proofs...," art. 5. [Not included].

[41]See "Proofs...," art. 7. [Not included].

[42]See "Proofs...," art. 4. [Not included].

[43]See "Proofs...," art. 17. [Not included].

[44]See "Proofs...," art. 12. [Not included].

[45]See "Proofs...," art. 19. [Not included].

[46]See the Voyages of Shaw, vol. 2, p. 69.

[47]See Boyle's Works, vol. 3, p. 232.

[48]Particularly on the coasts of Scotland and Ireland.

[49]See "Proofs...," art. 13. [Not included].

[50]See "Proofs...," art. 19. [Not included].

[51]See Varenius, Geographica generalis, p. 119.

[52]See Ray's Discourses, p. 226; [Robert] Plot, Natural History [of Oxfordshire], etc.

[53]See Newton [Principia mathematica] 3rd ed., p. 525.

[54]See Newton [op. cit.], p. 405.

[55][Pierre de Maupertuis, Discours sur les differens figures des astres ... (Paris, 1732)].

[56] *Protogaea, Acta eruditorum Lipsiae*, [(January, 1693). This was just a short summary of the text published in full only at Göttingen in 1749 as *Protogaea sive de prima facie Telluris* . . . .].

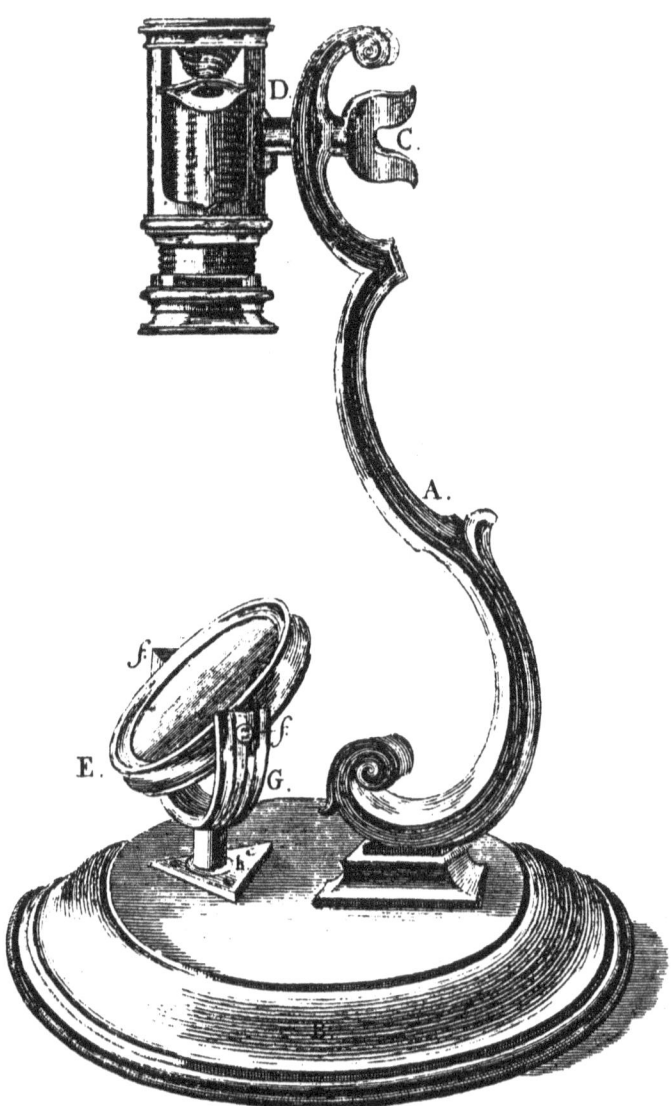

Figure 8 - The scroll-mounted Wilson screw-barrel simple microscope manufactured by John Cuff in the 1740's. A condensing lens and aperture diaphragms attach at the lower end. An ivory slide-holding mount is inserted in lenticular opening at D. Interchangeable lenses of varying focal lengths screw at top. This microscope, described by Henry Baker, exactly matches the specifications of the microscope used by Buffon and Needham.

8. Buffon on the Generation of Animals (1749), (selected)
Phillip R. Sloan

The best known, and most controversial of all of Buffon's biological speculations were undoubtably those concerned with the issue of plant and animal generation, which were expounded in the second volume of the Histoire naturelle (1749). In the first selection which follows, taken from the lengthy "Histoire générale des animaux," and dated February 6, 1746, Buffon proposes his theory of the organic molecules and the "internal" molds, and explains how these are to account for major biological phenomena, such as growth, nutrition and reproduction. Read in light of his "Reflections on the Law of Attraction," it can be observed that the "internal" molds represent not simply Newtonian microforces, such as those encountered in Newton's Queries to the Opticks, but are also a concretizing of the abstract agency of attraction, in such a way that these forces can govern, supposedly, the formation of the embryo, the process of nutrition and metabolism, and even the identity of organic species over time.

Uncertainty persists over whether Buffon was proposing in this treatise simply a homogenous organic matter, differentiated only by the action of the internal molds, or advocating different kinds of organic molecules themselves, whose own specificity manifests itself in resultant unique molds.

The second and third selections, dated from the Spring of 1748, present the most critical of his controversial experiments, and Buffon's summary reflections on them, which established to Buffon's satisfaction, on empirical grounds, the main predictive claims of the organic molecule theory. These experiments, carried out with the assistance of the French anatomist Marie-Louis Daubenton, the botanist Thomas-François Dalibard, and the English microscopist John Turberville Needham, seemed to prove both the existence of universally-distributed molecular bodies in all organic tissues, and also the surprising existence of identical moving bodies in the "semen" of male and female mammals. The results of these experiments are attested to independently by Needham in his own writings.[1]

In view of the controversial nature of these observations, it is of significant interest to resolve some of the ambiguities surrounding both the instruments used and the nature of the observations described here. Traditionally, it has been assumed that the instrument brought by Needham to France, and referred to in Buffon's discussion as the one utilized in these experiments, was a compound microscope. From the scope pictured in the included plate prefacing the discussion on generation, and the subsequent claims of Lazzaro Spallanzani in his attack on Buffon's and Needham's experiments,[2] it has been presumed that the scope employed was the popular compound microscope manufactured by John Cuff. This microscope, while providing ease of focus and a wide field of vision, also displays poor resolving power, and marked spherical aberration at higher powers.[3] As a consequence, it has been common to attribute the alleged observations of organic molecules to optical defects of the lenses, poor resolution that prevented the discrimination of spermatozoa and microbes, the inexperience of the observers, or other technical difficulties in the experiment.[4]

In view of the fact that Needham was one of the finest microscopists of the late eighteenth century, however, the assumption of simple incompetence in the experiments is somewhat strained, and it would seem that another account is possible.

Several lines of evidence strongly suggest that the microscope used in these experiments was not the compound Cuff scope, but a modified Wilson Screw-Barrel microscope, also manufactured by John Cuff, that is a more powerful and optically-superior simple microscope, mounted vertically on a scroll stand, with a concave reflecting mirror for illumination. This particular scope is described by Needham's friend Henry Baker in 1742,[5] and the optical specifications for this microscope, with one exception, exactly match those supplied by Buffon in 1748 for the microscope being used to make the controversial observations. In discussing the quality of this scroll-mounted microscope, Henry Baker was comparing it directly to a variety of Leeuwenhoek's simple microscopes available to him, and considered it a superior instrument.[6]

The specifications supplied for the Wilson-Cuff microscope give a maximal magnification of 400X at a focal length of 1/50th of an inch. Assuming quality lenses, which on tested antique simple microscopes have given resolutions often approaching theoretical values,[7] the observations are not so much complicated by the crudity of the instrument as by the greater probability that they could be reaching the level of the observation known to affect suspended particles in any solution termed Brownian motion. Significantly, the original observations by Robert Brown in 1827 of this phenomenon were made with a simple microscope of 1/32 inch focal length, which would give a theoretical magnification of only 320X. It is of interest to quote Brown's conclusions on his own observation:

> Reflecting on all the facts with which I had now become acquainted, I was disposed to believe that the minute spherical particles or molecules of apparently uniform size. . . ,were in reality the supposed constituent or elementary molecules of organic bodies, first so considered by Buffon and Needham, then by Wrisberg with greater precision, soon after and still more particularly by Muller, and, very recently, by Milne Edwards. . . .[8]

Brown describes uniformly-shaped molecular bodies in constant oscillatory motion, which are to be found not only in infusions of all organic materials, but also in most inorganic. Consequently, it is possibly the superiority of the Buffon-Needham microscope to those used by others that prevented the replication of these observations until the nineteenth century.[9] This would explain some of the tenacity with which Buffon adhered to the organic molecule theory in all his writings, in spite of the criticisms of contemporaries.

NOTES

[1] J.T. Needham, "A Summary of Some Late Observations Upon the Generation, Composition, and Decompositon of Animal and Vegetable Substances," Phil. Trans. Roy. Soc. Lond. 45, No. 490 (1748), pp. 642-44.

[2] Lazzaro Spallanzani, "Observations and Experiments upon the Seminal Vermiculi of Man and Other Animals, with an Examination of the Celebrated Theory of Organic Molecules," in: Tracts on the Nature of Animals and Vegetables, Tranlated by J. G. Dalyell (Edinburgh: A. Constable, 1799; first published Moderna, 1776), pp. 132-3.

[3] See S. Bradbury, "The Quality of the Image Produced by the Compound Microscope: 1700-1840," in: Historical Aspects of Microscopy, ed. S. Bradbury and G. L'E. Turner (Cambridge: Heffer, 1967), pp. 151-73.

[4] See, for example, Luigi Belloni, "Micrografia illusoria e 'animalcula'," Physis rivista di storia dela scienza 4 (1962), pp. 65-73. Belloni assumes the microscope employed is the compound Cuff pictured in the woodcut. See plates below, pp. 170, 191. I wish to express my indebtedness to Sister Elaine Abels for this reference and for other technical assistance in

this matter. I have also benefited greatly from access to antique and replicated microscopes through the assistance of Dr. James B. McCormick of Replica Rara Ltd. Mr. Gerard L'E. Turner of the Museum of the History of Science at Oxford University also kindly allowed me access to an antique specimen of the Scroll-mounted Wilson microscope.

[5]H. Baker, The Microscope Made Easy (London, 1742), pp. 14-15. See plate of this microscope, p. 164 above.

[6]Compare the table given by Baker in his "An Account of Mr. Leeuwenhoek's Microscopes," Phil. Trans. Roy. Soc. Lond. 51, No. 458, (1744), p. 513 and that in G.L. Buffon, "Découverte de la liqueur seminale dans les femelles vivipares et du reservoir qui la contient," Mem. de l'acad. roy. sci. 1748 (Paris, 1752), p. 228. Buffon has obviously made a misprint here in giving the value of the focal length of his highest power lens as 1/10 inch, rather than 1/50 as in Baker. It should also be observed that while Buffon speaks of a "double" microscope, he does not call it a "compound" microscope. Although the Cuff compound scope is sometimes referred to as a "Double-Constructed" microscope, it is more likely that Buffon is speaking of the double lens system of the Wilson Screw-Barrel scope, which has a condensing lens mounted below the slide stage, thus given it a markedly different construction than a Leeuwenhoek or Lieberkuhn simple microscope, while not altering the fact that the observations are made only with a single lens system. My personal observations on various infusions of organic materials with a precisely replicated Wilson-Edinburgh microscope, manufactured by Replica Rara Ltd., with an optical system essentially identical to the scroll-mounted Wilson scope, revealed it to be a remarkably high quality instrument, with excellent resolution and defects mainly in the greatly limited field of vision and clumsiness of focusing at higher powers.

[7]See P.H. Van Cittert, Descriptive Catalogue of the Collection of Microscopes in Charge of the Utrecht University Museum (Gröningen: Noordhoff, 1934), pp. 6-12.

[8]Robert Brown, "A Brief Account of Microscopical Observations . . . and on the General Existence of Active Molecules in Organic and Inorganic Bodies," Edin. New Phil. Jour. 5 (1828), p. 363. See also Brown's "Additional Remarks on Active Molecules," Edin. New Phil. Jour. 8 (1829-30), pp. 41-46. The theoretical magnification of a simple lens can be readily computed from the formula:

$$M = \frac{D}{f.l.}$$

where f.l. is the focal length, and D is a standardized distance, that of least distance of distinct vision. In the eighteenth century this was generally taken as 8 English inches, and is the value used by Buffon and Needham, for example. On nineteenth century scopes, this is more typically 10 inches. At present 250 mm. is the accepted international standard.

[9]Contemporaries often had great difficulty replicating Robert Brown's observations until using a simple microscope of similar construction to those employed by Brown.

Figure 9 - Commonly reproduced but highly stylized depiction of the Buffon experiments on seminal bodies. From left to right the figures have been sometimes identified as a Surgeon, Daubenton, probably Needham, and Buffon. Firm evidence on the identities is, however, lacking. The microscope depicted is a stylized version of the John Cuff compound scope. *See figure 11.*

THE GENERATION OF ANIMALS*

CHAPTER II

OF REPRODUCTION IN GENERAL

Translated by J.S. Barr

We shall now make a more minute inspection into this common property of animal and vegetable nature; this power of producing its resemblance; this chain of successive individuals, which constitutes the real existence of the species; and without attaching ourselves to the generation of man, or to that of any particular kind of animal, let us inspect the phenomena of reproduction in general; let us collect facts, and enumerate the different methods nature makes use of to renew organized beings. The first, and as we think the most simple method, is to collect in one body an infinite number of [similar] organic bodies, and so to compose its substance, that there is not a part of it which does not contain a germ of the same species, and which [cannot] consequently of itself become a whole, resembling that of which it constitutes a part. This [process] seems to suppose a prodigious waste, and to carry with it profusion; yet it is a very common magnificence of nature, and [one] which manifests itself even in the most common and inferior kinds, such as worms, polyps, elms, willows, gooseberry-trees, and many other plants and insects, each part of which contains a whole, [which] by the single effect of expansion alone may become a plant or an insect. By considering organized beings, in this point of view, an individual is a whole, uniformly organized in all its parts; a compound of an infinity of resembling figures and similar parts, an assemblage of germs, of small individuals of the same kind, which can expand in the same mode according to circumstances, and form new bodies, composed like those from whence they proceed.

By examining this idea thoroughly, we shall discover a connection between animals, vegetables, and minerals, which we could not expect. Salts, and some other minerals, are composed of parts resembling each other, and to all that composes them; a grain of salt is a cube, composed of an infinity of smaller cubes, which we may easily perceive by a microscope:[1] these are also composed of other cubes still smaller, as may be perceived with a better microscope; and we cannot doubt, but that the primitive and constituting particles of this salt are likewise cubes so exceedingly minute as to escape our sight, and our imagination. Animals and plants which can multiply by all their parts, are organized bodies, of which the primitive and constituting parts are also organic and similar, [and] of which we discern the aggregate quantity, but cannot perceive the primitive parts [except] by reason and analogy.

This leads us to believe that there is an infinity of organic particles actually existing and living in nature, the substance of which is the same with that of organized bodies, [just as there is, as we have recognized, an infinity of similar inert particles of inanimate bodies. And as it would perhaps be necessary for millions of small cubes of salt to accumulate in order to make the visible individual cube of sea-salt], so likewise millions of organic particles, like the whole, are required to form [a single] one out of that multiplicity of germs contained in an elm or a polyp; and as we must separate, bruise, and dissolve a cube of sea-salt to perceive, by means of crystallization, the small cubes of which it is composed; we must likewise separate the parts of an elm or polypus to discover, by means of vegetation and expansion, the small elms or polyps contained in those parts.

The difficulty of giving way to this idea arises from a prejudice strongly established, that there is no method of judging of the complex, except by the simple, and that, to conceive the organic constitution of a body we must reduce it to its simple and unorganized parts, and that it is more easy to conceive how a cube is composed of other cubes than how one polypus is composed of others; but if we attentively examine what is meant by simple and complex we shall then find that in this, as in every thing else, the plan of Nature is quite different from the very rough draft of it formed by our ideas.

Our senses, as is well known, do not furnish us with exact [and complete notions of the things we need to know], inasmuch that if we are desirous of estimating, judging, comparing, measuring, etc. we are obliged to have recourse to [outside] assistance, to rules, principles, instruments, etc. All these [aids are the works of the human mind, and more or less pertain to the reduction or] the abstraction of our ideas; this abstraction, therefore, is what is called the simple, and the difficulty of reducing them to this abstraction, the complex. Extension, for example, being a general and abstracted property from Nature, is not very complex, nevertheless, to form a judgment of it we have supposed extensions without depth, without breadth, and even points without any extension at all. All these abstractions have been invented for the support of our judgment, and the few definitions made use of in geometry have occasioned a variety of prejudices and false conclusions. All that can be reduced by these definitions are termed <u>simple</u>, and all that cannot be readily reduced are called <u>complex</u>; from hence a triangle, a square, a circle, a cube, etc. are simple subjects, as well as all curves, whose geometrical laws we are acquainted with; but all that we cannot reduce by these abstracted figures and laws are complex. We do not consider that these geometrical figures exist only in our imagination; that they are not to be found in nature, or, at least, if they are discoverable there, it is because she exhibits every possible form, and that it is more difficult and rare to find simple figures of an equilateral pyramid, or an exact cube in nature, than compounded forms of a plant or an animal. In every thing, therefore, we take the abstract for the simple, and the real for the complex. In nature, on the contrary, the abstract has no existence, every thing is compounded; we shall never, of course, penetrate into the intimate structure of bodies; we cannot, therefore, pronounce on what is complex in a greater of lesser degree, excepting by the greater or lesser [relation] each subject has to ourselves and to the rest of the universe; from which reason it is we judge that the animal is more compounded than the vegetable, and the vegetable more than the mineral. This notion is just with relation to us, but we know not, in reality, whether the animal, vegetable, or mineral, is the most simple or complex; and we are ignorant whether a globule, or a cube, is more indebted [to] an exertion of nature than a seed or an organic particle. If we would form conjectures on this subject we might suppose that the most common and numerous things are the most simple; but then animals would be the most simple, since the number of their kind far exceeds that of plants or minerals.

But without taking up more time on this discussion it is sufficient to have shown that the opinions we commonly have of the simple and complex are ideas of abstraction, that they cannot be applied to the compound productions of Nature, and that when we attempt to reduce every being to elements of a regular figure, or to prismatic, cubical, or globular particules, we substitute our own imaginations in the place of realities; that the forms of the constituting particles of different bodies are absolutely unknown to us, and that consequently, we can suppose, that an organized body is composed of organic particles as well as that a cube is composed of other cubes. We have no other rule to judge by than experience. We perceive that a cube of sea-salt is composed of other cubes, and that an elm consists of other smaller elms, because by taking an end of a branch, or root, or a piece of the wood separated from the trunk, or a seed, they will alike produce a new tree. It is the same with respect to polyps, and some other kinds of animals, which we can multiply by cutting off and separating any of the different parts; and since our rule for judging in both is the same why should we judge differently of them?

It therefore appears very probable, by the above reasons, that there really exists in nature [an infinity] of small organized beings, alike, in every respect, to the large organized bodies seen in the world; that these small organized beings are composed of living organic particles, which are common to animals and vegetables, and are their primitive and [incorruptible] particles; that the assemblage of these particles forms an animal or plant, and consequently that reproduction, or generation, is only a change of form made by the addition of these resembling parts alone, and that death or dissolution is nothing more than a separation of the same particles. Of the truth of this we apprehend there will not remain a doubt after reading the proofs we shall give in the following chapters. Besides, if we reflect on the manner in which trees grow, and consider how so considerable a volume can arise from so small an origin, we shall be convinced that it proceeds from the simple addition of small [similar] organized particles. A grain produces a young tree, which it contained in miniature. At the summit of this small tree a bud is formed, which contains the young tree for the succeeding year, and this bud is an organic part, resembling the young tree of the first year's growth. A similar bud appears the second year containing a tree for the third; and thus, successively, as long as the tree continues growing, at the extremity of each branch, new buds will form, which contain, in miniature, young trees like that of the first year. It is, therefore, evident, that trees are composed of small organized bodies, similar to themselves, and that the whole individual is formed by the assemblage of small resembling individuals.

But, it may be asked, were not all these organized bodies contained in the seed, and may not the order of their expansion be traced from that source? For the bud which first appeared was evidently surmounted by another similar bud, which was not expanded till the second year, and so on to the third: and consequently, the seed may be said really to contain all the buds, or young trees that would be produced for a hundred years, or till the dissolution of the tree itself. This seed, it is also plain, not only contained all the small organized bodies which one day must constitute the individual tree, but also every seed, every individual, and every succession of seeds and individuals, to the total destruction of the species.

This is the principal difficulty, and we shall examine it with the strictest attention. It is certain, that the seed produces, by the single expansion of the bud, or germ it contains, a young tree the first year, and that this tree existed in miniature in that bud, but it is not equally certain, that the bud of the second year, and those of succeeding, were all contained in the first seed, no more than that every organized body and seed, which must succeed to the end of the world, or to the destruction of the species [would be]. This opinion supposes a progress to infinity, and [makes each existing] individual, a source of eternal generations. The first seed, in that case, must have contained every plant of its kind which has existed or ever will exist; and the first man must actually and individually have contained in his loins every man which has or will appear on the face of the earth. Each seed, and each animal, agreeable to this opinion must have possessed within [it] an infinite posterity. But the more we suffer ourselves to wander into these kind of reasonings, the more we lose the sight of truth in the labyrinth of infinity; and, instead of clearing up and solving the question, we confuse and involve it in more obscurity; it is placing the object out of sight, and afterward saying it is impossible to see it.

Let us investigate a little these ideas of infinite progression and expansion. From whence do they arise? What do they represent? The ideas of infinity can only spring from an idea of that which is limited; for it is in that manner we have an idea of an infinity of succession, a geometrical infinity: each individual is a unit; many individuals compose a finite number, and the whole species is the infinite [number]. Thus in the same manner as a geometrical infinity may be demonstrated not to exist, so we may be assured, that an infinite progression or [development] does not exist; that it is only an abstract idea, a [suppression] of the idea of finity, of which we take away the limits that necessarily terminate all size;[2] and that, of course, we must reject from philosophy every opinion which leads to an idea of the actual existence of geometrical or arithmetical infinity.

The partisans, therefore, of this opinion must acknowledge that their infinity of succession and multiplication, is, in fact, only an indeterminate or indefinite number; a number greater than any we can have an idea of, but which is not infinite. This being granted, they will tell us, that the first seed of an elm, for example, which does not weigh a grain, really contains all the organic particles necessary for the formation of this and every other tree of same kind which ever shall appear. But what do they explain to us by this answer? Is it not cutting the knot instead of untying it, and eluding the question when it should be resolved?

When we ask, how [creatures are reproduced]? and it is answered that this multiplication was completely made in the first body, is it not acknowledging that they are ignorant how it is made, and renouncing the will of conceiving it? The question is asked, how one body produces its like? and it is answered, that the whole was created at once. Can we receive this as a solution? For whether one or a million of generations have passed the like difficulty remains, and so far from explaining the supposition of an indefinite number of germs, increases the obscurity, and renders it incomprehensible.

I own, that in this circumstance, it is easier to [raise] objections than to establish probabilities, and that the question of reproduction is of such subtle nature, as possibly never to be fully resolved; but then we should search whether it is totally inscrutable, and by that examination, we shall discover all that is possible to be known [on] the subject, or at least, why we must remain ignorant of it.

There are two kinds of questions, some [pertaining] to the first causes, the others only [to] particular effects; for example, if it is asked why matter is impenetrable? it must either remain unanswered, or be replied to by saying, matter is impenetrable, because it is impenetrable. It will be the same with respect to all the general qualities of matter, whether relative to gravity, extension, motion, or rest; no other reply can be given, and we shall not be surprised that such is the case, if we attentively consider, that in order to give a reason for a thing, we must have a different subject from which we may deduce a comparison, and therefore if the reason of a general cause is asked, that is, of a quality which belongs to all in general, and of which we have no subject to which it does not belong, we are consequently unable to reason upon it; from thence it is demonstrable [that] it would be useless to make such enquiries, since we should go against the supposition that quality is general and universal.

If, on the contrary, the reason of a particular effect depends immediately on one of the general causes above mentioned, and whether it partakes of the general effect immediately, or by a chain of other effects, the question will be equally solved, provided we distinctly perceive the dependence these effects have on each other, and the connections there are between them.

But if the particular effect, of which we [seek] the reason, does not appear to depend on these general effects, nor to have any analogy with other known effects; then, this effect being the only one of its kind, and having nothing in common with other effects at least known to us, the question is insolvable; because, not having, in this point, any known subject which has any connection with that we would explain, there is nothing from whence we can draw the reason sought after. When the reason of a general cause is demanded, it is unanswerable because it exists in every object and, on the other hand, the reason of a singular or isolated effect is not found, because not any thing known has the same qualities. We cannot explain the reason of a general effect, without discovering one more general; whereas the reason of an isolated effect may be explained by the discovery of some other relative effect, which although we are ignorant of at present, chance or experience may bring to light.

Besides these, there is another kind of question, which may be called the question of fact. For example, why do trees, dogs, etc. exist? All these fact questions are totally insoluble, for those who answer them by final causes, do not consider that they take the effect for the cause; the connection particular objects have with us having no influence on their origin. Moral affinity can never become a physical reason.

We must carefully distinguish those questions where the why is used, from those where the how is employed, and more so from those where the how many is mentioned. Why is always relative to the cause of the effect, or to the effect itself. How is relative to the mode from which the effect springs, and the how many has relation only to the proportionate quantity of the effect.

All these distinctions being explained, let us proceed to examine the question concerning the reproduction of bodies. If it is asked, why animals and vegetables reproduce? we shall clearly discover, that this being a question of fact, it is insolvable, and useless to endeavour at the solution of it. But if it is asked, how animals and vegetables reproduce? we reply by relating the history of the generation of every species of animals, and of the reproduction of each distinct vegetable; but, after having run over all the methods of an animal engendering its [likeness], accompanied even with the most exact observations, we shall find it has only taught us facts without indicating causes; and that the apparent methods which Nature makes use of for reproduction, do not appear to have any connection with the effects resulting therefrom; we shall be still obliged to ask what is the secret mode by which she enables different bodies to propagate their own species.

The question is very different from the first and second; it gives liberty of enquiry and admits the employment of imagination, and therefore is not insolvable, for it does not immediately belong to a general cause; nor is it entirely a question of fact, for provided we can conceive a mode of reproduction dependent upon, or not repugnant to original causes, we shall have gained a satisfactory answer; and the more it shall have a connection with other effects of nature, the better foundation will it be raised upon.

By the question itself it is therefore permitted to form hypotheses, and to select that which shall appear to have the greatest analogy with the other phenomena of nature. But we must exclude from the number, all those which suppose the thing already done; for example, such as suppose that all the germs of the same species were contained in the first seed, or that every reproduction is a new creation, an immediate effect of the Almighty's will; because these hypotheses are questions of fact, and [about] which it is impossible to reason. We must also reject every hypothesis which might have final causes for its object; such as, we might say, that reproduction is made in order for the living to supply the place of the dead, that the earth may be always covered with vegetables, and peopled with animals; that man may find plenty for his subsistence, etc. because these hypotheses, instead of explaining the effects by physical causes, are founded only on arbitrary connections and moral [conventions]. At the same time we must not rely on these absolute axioms and physical problems, which so many people have improperly made use of, as principles; for example, [that] there is no fecundation made apart from the body, <u>nulla foecondatio extra corpus</u>; [that] every living thing is produced from an egg; [that] all generation supposes sexes, etc. We must not take these maxims in an absolute sense, but consider them only as signifying things generally performed in one particular mode rather than in any other.

Let us, therefore, search after an hypothesis which has not any of these defects, and by which we cannot fall into any of these inconveniences; if, then, we do not succeed in the explanation of the [mechanism] nature makes use of to effect the reproduction of beings, we shall, at least, arrive at something more probable than what has hitherto been advanced.

As we can make moulds, by which we can give to the external parts of bodies whatever figure we please, let us suppose nature can form the same, by which she not only bestows on bodies the external figure but also the internal. Would not this be one mode by which reproduction may be performed?

Let us, then, consider on what foundation this supposition is raised; let us examine if it contains any thing contradictory, and afterwards we shall discover what consequences may be derived from it. Though our senses are only judges of the external parts of bodies, we perfectly comprehend external affection and different figures. We can also imitate nature, by representing external figures by different modes, as by painting, sculpture, and moulds; but although our senses are only judges of external qualities we know there are internal qualities, some of which are general, as gravity. This quality, or power, does not act relatively to surfaces but proportionably to the masses or quantities of matter; there [are], therefore, very active qualities in nature, which even penetrate bodies to the most internal parts; but we shall never gain a perfect idea of these qualities, because, not being external, they cannot fall within the compass of our senses; but we can compare their effects, and deduce analogies therefrom, to [give an account of] the effects of similar qualities.

If our eyes, instead of representing to us the surface of objects only, were so formed as to show us the internal parts alone, we should then have clear ideas of the latter, without the smallest knowledge of the former. In this supposition the internal moulds, which I have supposed to be made use of by nature, might be as easily seen and conceived as the moulds for external figures. In that case we should have modes of imitating the internal parts of bodies as we now have for the external. These internal moulds, although we [can never obtain them, can be possessed by nature, just as she has the property of] gravity, which penetrates to the internal particles of matter. The supposition of these moulds being formed on good analogies it only remains for us to examine if it includes any contradiction.

It may be argued tht the expression of an internal mould includes two contradictory ideas; that the idea of a mould can only be related to the surface, and that the internal, according to this, must have a connection with the whole mass, and, therefore, it might as well be called a massive surface, as an internal mould.

I admit, that when we are about to represent ideas which have not hitherto been expressed, we are obliged to make use of terms which seem contradictory; for this reason philosophers have often employed foreign terms on such occasions, instead of applying those in common use, and which have a received signification; but this artifice is useless, since we can show the opposition is only in the words, and that there is nothing contradictory in the idea. Now I affirm that a simple idea cannot contain a contradiction, that is, when we can form an idea of a thing, if this idea is simple it cannot be compounded, it cannot include any other idea, and, consequently, it will contain nothing opposite nor contrary.

Simple ideas are not only the primary apprehensions which strike us by the senses but also the primary comparisons which form from those apprehensions, for the first apprehension itself is always a comparison. The idea of the size of an object, or of its remoteness, necessarily includes a comparison with bulk or distance in general; therefore, when an idea only includes comparison it must be regarded as simple, and from that circumstance, as containing nothing contradictory. Such is the idea of the internal mould. There is a quality in nature, called gravity, which penetrates the internal parts of bodies. I take the idea of internal mould relatively to this quality, and, therefore, including only comparison, it [contains nothing opposed or contrary].

Let us now see the consequences that may be deduced from this supposition; let us also search after facts corresponding therewith, as it will become so much the more probable, as the number of analogies shall be greater. Let us begin by unfolding this idea of internal moulds, and by explaining in what manner we understand it, we shall be brought to conceive the modes of reproduction.

Nature, in general, seems to have a greater tendency to life than death, and to organize bodies as much as possible; the multiplication of germs, which may be infinitely increased, is a proof of it; and we may assert, with safety, that if all matter is not organized it is because organized beings destroy each other, for we can augment as much as we please the quantity of living and vegetating beings, but we cannot augment the quantity of stones or other inanimate matters. This seems to indicate that the most common work of nature is the production of the organic part, and in which her power knows no bounds.

To render this intelligible let us make a calculation of what a single germ might produce. The seed of an elm, which does not weigh the hundredth part of an ounce, at the end of 100 years will produce a tree whose volume will be [10 toises cubed]. At the tenth year this tree will have produced 1000 seeds, which being all sown [would yield 1000 trees, which], at the end of 100 years would each have also a volume equal to [10 toises cubed]. Thus in 110 years there is produced more than [10,000 cubic toises] of organized matter; 10 years more there will be 10,000,000 [cubic toises], without including the 10,000 increased every year, which would make 100,000 more; and 10 years after there will be [10 trillion cubic toises]; thus in 130 years a single shoot will produce a volume of organized matter which would fill up a space of 1000 cubic leagues; 10 years after it would comprehend a [million cubic leagues], and in 10 years more 1,000,000 times 1,000,000 cubic leagues; so that in 150 years the whole terrestrial globe might be entirely converted into one single kind of organized matter. In this production of organized bodies nature would know no bounds if it were not for the resistance of [kinds of matter] which are not susceptible of organization, and this proves that she does not incline to form inanimate, but organized beings, and that in this she never stops but when irresistible inconveniences are opposed thereto. What we have already said on the seed of an elm may be said of any other; and it would be easy to demonstrate, that if we were to hatch every egg produced by hens for the space of 30 years there would be a sufficient number of fowls to cover the whole surface of the earth.

[This] kind of calculation demonstrates that organic formation is the most common work of nature, and apparently that which costs her the least labour. But I will go farther; the general division which we ought to make of matter seems to me to be into living and dead matter instead or organized and brute; the brute is only that matter produced by the death of animals or vegetables; I could prove it by that enormous quantity of shells and other cast-off matters of living animals which compose the principal part of stones, marble, chalk, marle, earth, turf, and other substances, which we call brute matter, and which are only the ruins of dead animals or vegetables; but a reflection, which seems to me well founded, will, perhaps, make it better understood.

Having meditated on the activity of nature to produce organized bodies, and [having] seen that her power, in this respect, is not limited; having proved that infinity of organic living particles, which constitute life, must exist; having shown that the living body costs the least trouble of nature, I now search after the principal causes of death and destruction,

and I find that bodies in general, which have the power of
converting matter into their proper substance, and to
assimilate the parts of other bodies, are the greatest
destroyers. Fire, for example, turns into its own substance
almost every species of matter, and is the greatest means of
destruction known to us. Animals seem to participate in the
qualities of flame; their internal heat is a kind of fire;
therefore, after fire, animals are the greatest destroyers, and
they assimilate and convert into their own substance every
matter which may serve them for food: but although these two
causes of destruction are very considerable, and their effects
perpetually incline to the annihilation of organized beings,
the cause of reproduction is infinitely more powerful and
active; she seems to borrow, even from destruction itself,
means to multiply, since assimilation, which is one cause of
death, is, at the same time, a necessary means of producing
life.

To destroy an organized being is, as we have observed,
only to separate the organic particles of which it is composed;
these particles remain separated till they are reunited by some
active power. But what is this active power?--It is the power
which animals and vegetables have to assimilate the matter that
serves them for food; and is not this the same, or at least has
it not great connection with, that which is the cause of reproduction?

CHAPTER III

OF NUTRITION AND GROWTH

The body of an animal is a kind of internal mould, in which the nutritive matter assimilates itself with the whole in such a manner that, without changing the order and proportion of the parts, each receives an augmentation, and it is this augmentation of bulk which some have called expansion, because they imagined every difficulty would be removed by the supposition that the animal was completely formed in the embryo, and that it would be easy to conceive that its parts would expand, or unfold, in proportion as it would increase by the addition of accessory matter.

But if we would have a clear idea of this augmentation and expansion, how can it be done otherwise than by considering the animal body, and each of its parts, as so many internal moulds which receive the accessory matter in the order that results from the position of all their parts? This expansion cannot be made by the addition to the surfaces alone, but, on the contrary, by an intimate susception which penetrates the mass, and thus increases the size of the parts, without changing the form, from whence it is necessary that the matter which serves for this expansion should penetrate the internal part in all its dimensions; it is also as necessary that this penetration be made in a certain order and proportion, so that no one point can receive more than another. [Otherwise] some parts would expand quicker than others, and the form [would] be entirely changed. Now what can prescribe this rule to accessory matter, and constrain it to arrive perpetually and proportionally to every point of the external parts, [unless] we conceive an internal mould?

It therefore appears certain that the body of an animal or vegetable is an internal mould of a constant form, but [one in which the mass and the volume may be increased] proportionally, by the extension of this mould in all its external and internal dimensions, [and] that this extension also is made by the intus-susception of any accessory or foreign matter which penetrates the internal part, and becomes similar to the form, and identical [in] substance with, the matter of the mould [itself].

But of what nature is this matter which the animal or vegetable assimilates with its own substance? What can be the nature of that power which gives it the activity and necessary motion to penetrate the internal mould? And if such a power does exist must it not be similar to that by which the internal mould itself would be produced?

These three questions include all that can be desired on this subject, and seem to depend on each other so much that I am persuaded the reproduction of an animal or vegetable cannot be explained in a satisfactory manner if a clear idea of the mode of the operation of nutrition is not obtained; we must, therefore, examine these three questions separately, in order to compare the consequences resulting therefrom.

The first, which relates to the nutritive nature of this matter, is in part resolved by the reasons we have already given, and will be fully demonstrated in the succeeding chapter. We shall show that there exists an infinity of living organic particles in nature; that their production [is] of little expense to nature, since their existence is constant and invariable; and that the causes of death only separate without destroying them. Therefore the matter which the animal or vegetable assimilates is an organic matter of the same nature as the animal or vegetable itself, and which consequently can augment the size without changing the form or quality of the matter of the mould, since it is, in fact, of the same form and quality as that which it is constituted with. Thus, in the quantity of aliments which the animal takes to support life, and to keep its organs in play, and in the sap, which the vegetable takes up by its roots and leaves, there is a great part thrown off by transpiration, secretion, and other excretory modes, and only a small portion retained for the nourishment of the parts and their expansion. It is very probable, that in the body of an animal or vegetable [a separation is made of the inert and organic parts of the food]; that the first are carried off by the causes just mentioned. [Therefore] only organic particles remain, and the distribution of them is made by means of some active power which conducts them to every part in an exact proportion, insomuch that neither receive more nor less than is needful for its equal nutrition, growth, or expansion.

The second question, [is] what can be the active power which causes this organic matter to penetrate and incorporate itself with this internal mould? By the preceding chapter it appears, that there exist in nature powers relative to the internal part of matter, and which have no relation with its external qualities. These powers, as already observed, will never come under our cognizance, because their action is made on the internal part of the body, whereas our senses cannot reach beyond what is external; it is therefore evident, that we shall never have a clear idea of the penetrating powers, nor of the manner by which they act; but it is not less certain that they exist, than that by their means most effects of nature are produced; we must attribute to them, the effects of nutrition and expansion, which cannot be effected by any other means than the penetration of the most intimate recesses of the original mould; in the same mode as gravity penetrates all parts of matter, so the power which impels or attracts the organic particles of food, penetrates into the internal parts of organized bodies, and as those bodies have a certain form, which we call the internal mould, the organic particles, impelled by the action of the penetrating force, cannot enter therein but in a certain order relative to this form, which consequently it cannot change, but only augment its dimensions, and thus produce the growth of organized bodies; and if in the organized body, expanded by this means, there are some particles, whose external and internal forms are like that of the whole body, from those reproduction will proceed.

The third question: is it not by a similar power [that] the internal mould itself is reproduced? It appears, that it is not only a similar but the same power which causes [development] and reproduction, for in an organized body which [developes], if there is some particle like the whole, it is sufficient for that particle to become one day an organized body itself, perfectly similar to that of which it made a part. This particle will not at first present a figure striking enough for us to compare with the whole body; but when separated from that body, and receiving proper nourishment, it will begin to expand, and in a short time present a similar being, both externally and internally, as the body from which it had been separated: thus a willow or polyp, which contain more organic particles similar to the whole than most other substances, if cut into ever such a number of pieces, from each piece will spring a body similar to that from whence it was divided.

Now in a body [in which every particle is like every other], the organization is the most simple, as we have observed in the first chapter; for it is only the repetition of the same form, and a composition of similar figures, all organized alike. It is for this reason, that the most simple bodies, or the most imperfect kinds, are reproduced with the greatest ease, and in the greatest plenty; whereas, if an organized body contains only some few particles like itself, then, as such alone can [attain] the second [development], consequently, the reproduction will be more difficult, and not so abundant in number; the organization of these bodies will also be more compounded, because the more the organized parts differ from the whole, the more the organization of this body will be perfect, and the more difficult the reproduction will be.

Nourishment, expansion, and propagation, then are the effects of one and the same and cause. The organized body is nourished by the particles of aliments analogous to it; it expands by the intimate susception of organical parts which agree with it, and it propagates because it contains some original particles which resemble itself. It only remains to examine, whether these similar organic particles come into the organized body by nutriment, or whether they were there before, and have an independent existence. If we suppose the latter, we shall fall in with the doctrine of the infinity of parts, or similar germs contained one in the other; the insufficiency and absurdity of which hypothesis we have already shown; we must therefore conclude that similar parts are extracted from the food; and after what has been said, we hope to explain the manner in which the organic molecules are formed, and how the minute particles unite.

There is, as we have said, a separation of the parts in the nutriment; the organic, from those analogous to the animal or vegetable, by transpiration and other excretory modes; the organical remain and serve for the expansion and nutriment of the body. But these organic parts must be of various kinds, and as each part of the body receives only those similar to itself, and that in a due proportion, it is very natural to imagine, that the superfluity of this organic matter will be sent back from every part of the body into one or more places, where all these organical molecules uniting, form small organized bodies like the first, and to which nothing is wanting but the mode of expansion for them to become individuals of the same species; for every part of the body sending back organized parts, like those of which they themselves are composed, it is necessary, that from the union of all these parts, there should result organized bodies like the first. This being admitted, may we not conclude this is the reason why, during the time of expansion and growth, organized bodies cannot produce, because the parts which expand, absorb the whole of the organic molecules which belong to them and not having any superfluous parts, consequently are incapable of reproduction.

This explanation of nutrition and reproduction will not probably be received by those who admit [as the basis of their philosophy only] a certain number of mechanical principles, and reject [everything] which does not depend on them; [this is, they say, the great difference between the old philosophy and that of today: it is no longer permissible to postulate causes. It is necessary to give an account of everything by the laws of mechanics, and only those explanations are satisfactory which can be deduced from them. And as that account which you give of nutrition and reproduction do not depend on these, we ought not to admit them]. But I am quite of a different opinion from those philosophers; for it appears to me that, by admitting only a certain number of mechanical principles, they do not see how greatly they contract the bounds of philosophy, and that for one phenomenon that can be explained by a system so confined, a thousand would be found exceeding its limits.

The idea of explaining every phenomenon in nature by mechanical principles, was certainly a great and beautiful exertion, and [one] which Descartes first attempted. But this idea is only a project, and if properly founded have we the means of performing it? These mechanical principles are the extension of matter, its impenetrability, its motion, its external figure, its divisibility, and the communication of movement by impulsion, by elasticity, etc. The particular ideas of each of these qualifies we have acquired by our senses, and regard them as principles, because they are general and belong to all matter. But are we certain these qualities are the only ones which matter possesses, or rather, must we not think these qualities, which we take for principles, are only modes of perception; and that if our senses were differently formed, we should discover in matter, qualities different from those which we have enumerated? To admit only those qualities [in] matter which are known to us, seems to be a vain and unfounded pretension. Matter may have many general qualities which we shall ever be ignorant of; she may also have others that human assiduity may discover, in the same manner as has recently been done with respect to gravity, which [exists universally in all tangible] matter. The cause of impulsion, and such other mechanical principles, will always be as impossible to find out as that of attraction, or such other general qualities. From hence is it not very reasonable to say, that mechanical principles are nothing but general effects, which experience has pointed out to us in matter, and that every time a new general effect is discovered, either by reflection, comparison, measure, or experience, a new mechanical principle will be gained, which may be used with as much certainty and advantage as any we are now acquainted with?

The defect of Aristotle's philosophy was making use of particular effects as common causes; and that of Descartes in making use of only a few general effects as causes, and excluding all the rest. The philosophy which appears to me would be the least deficient, is that where general effects [only] are made use of for causes, [but] seeking to augment the number of them, by endeavouring to generalize particular effects.

In my explanation of [development] and reproduction, I admit the received mechanical principles, the penetrating force of [gravity], and by analogy I have strived to point out that there are other penetrating powers existing in organized bodies, which experience has confirmed. I have proved by facts, that matter inclines to organization, and that there exists an infinite number of organic particles. I have therefore only generalized some observations, without having advanced any thing contrary to mechanical principles, when that term is used as it ought to be understood, as denoting the general effects of nature.

. . . .

CHAPTER VI

EXPERIMENTS ON THE METHOD OF GENERATION

I often reflected on the above systems, and was every day more and more convinced that my theory was infinitely the most probable. I then began to suppose that, by a microscope, I might be able to attain a discovery of the living organic particles, from which I thought every animal and vegetable drew their origin. My first supposition was, that the spermatic animalcules, seen in the seed of every male, might, possibly, be these organic particles; on which I reasoned as follows:

If every animal and vegetable contain a quantity of living organic particles, these particles would be found in their seed, and in a greater quantity than in any other substance because the seed is an extract of what is most analogous to the individual, and the most organic; and the animalcule we see in the seed of males are, perhaps, only these same living organic molecules, or, at least, the first union, or assemblage of them. But if this is so, the seed of the female must also contain similar living organic molecules, and, consequently, we ought to find moving bodies there as well as in the male: and since the living organic particles are common both to animals and vegetables we should also find them in the seeds of plants, in the nectarium, and in the stamina, which are the most essential parts of vegetables, and which contain the organic molecules necessary for reproduction. I then seriously thought of examining the seminal liquors of both sexes, and the germs of plants, with a microscope. I thought, likewise, that reservoirs of the female seed might possibly be the cavities of the glandular bodies,[3] in which Valisnieri, and others, had uselessly sought for the egg; and at length determined to undertake a course of observations and experiments. I first communicated my ideas to Mr. Needham, a gentleman well known for his microscopical observations, [which he had published in 1745. This skillful man, deserving of commendation, had been recommended by Mr. Folkes, President of the Royal Society of London. Having formed a friendship with him, I believed that I could do no better than to communicate my ideas to him, and as he had an excellent microscope, superior to my own and more convenient, I asked him to let me make use of it for my experiments. I read him the entire part of my work that had been completed, and at the same time I told him that I believed I had found the true reservoir of the female semen, and that I did not doubt that the fluid contained in the cavity of the glandular body was the true seminal fluid of the female. I said I was persuaded that one would find under microscopic observation some spermatic animals in this fluid, as in the semen of the male, and that I was very much inclined to believe that there would also be discovered moving bodies in the most substantial parts of plants, such as in all the nectarium etc., and that there was a great likelihood that the spermatic animals discovered in the seminal fluid of males were only the first assemblage of the organic particles which must be in much greater number in this fluid than in all other substances constituting the animal body. Mr. Needham seemed to me to respect these ideas, and had the generosity to lend me his microscope, and even wished to be present at some of my experiments. At the same time I communicated to Messrs. Daubenton, Gueneau de Montbeillard and Dalibard my theory and project of experiments, and although I was very practiced in making optical observations and experiments, and I well knew how to distinguish reality from appearance in microscopic observation, I believed that I must not trust this to my eyes alone, and enlisted Mr. Daubenton to assist and observe with me. I cannot overemphasize how much I owe to his friendship in having been willing to abandon his ordinary tasks in order to pursue with me for several months these experiments which I recount. He made me notice a great number of things which would perhaps have escaped me. In such delicate matters, where it is easy to be deceived, one is very fortunate to find some one who would indeed wish not only to criticize, but also to assist. Messrs. Needham, Dalibard and Gueneau have seen a portion of the things I am going to recount, and Mr. Daubenton has seen them all as well as I have.

Persons who are not in the practice of using the microscope will find it valuable that I add some remarks here which will be of use when they wish to repeat these experiments or make new ones on this matter. Preference should be given to double microscope in which the object is viewed from above over the simple and double microscopes in which the object is viewed horizontally and by incident light. These double microscopes have a plane or concave mirror which illuminates the object from below. The concave mirror is to be preferred when observation is made with the strongest lens. Leeuwenhoek, who without doubt has been the greatest and most indefatigible of all microscopic observers, only made use of the simple microscope, it seems, with which he viewed objects by daylight or by that of a candle. If this is true, as the plate at the head of his book would seem to indicate, he has needed a diligence and inconceivable patience to err so little as he has on the almost infinite number of things he has observed by such clumsy means. He bequeathed all his microscopes to the Royal Society of London, and Mr. Needham has assured me that the best did not reveal so much detail as the strongest lens of the one I have used, and with which I made all my observations. If this is so, it is necessary to note that the majority of plates that Leeuwenhoek has given of microscopic objects, especially those of spermatic animals, shows them much larger and longer than they are in fact seen, which might lead one into error.

Furthermore, these alleged animals from man, dogs, rabbits, chickens etc. that are engraved in the Philosophical Transactions, No. 141 and in Leeuwenhoek, Vol. 1, p. 161, copied subsequently by Valisneri, Baker etc. appear much smaller under the microscope than in the plates representing them.]4 This renders the microscopes we speak of preferable to the horizontal, as they are more stable.

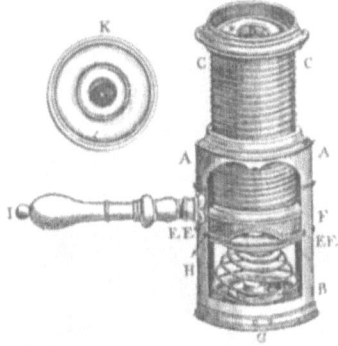

Figure 10 - The popular James Wilson-style screw-barrel single pocket microscope in its common hand-held version. Condensing lens and aperture diaphragms are screwed at *C*. Single lenses of varying focal lengths (*K*) are screwed at *G*, and ivory slide is inserted at *EE*. Ordinary viewing is with scope held in horizontal position with slide placed vertically.

The motion of the hand, with which the microscope is held, produces a little trembling, which causes the object to appear wavering, and never presents the same part for any time. Besides there is always a motion in the liquors caused by the agitation of the external air, at least, if we do not put the liquor between two plates of glass, or very fine talc, which diminishes somewhat of its transparency, and greatly lengthens the experiment; but the horizontal microscope, whose tables are vertical, has the still greater inconvenience, that the most ponderous parts of the drop of liquor fall to the bottom; consequently, there are three motions, that of the trembling of the hand, the agitation of the fluid by the action of the air, and also that of the parts of the liquor falling to the bottom, from the combination of which, certain small globules which we see in these liquors, may appear to move by their own motion and powers, while they only obey the compounded power of those three causes.

When we put a drop of liquor on the [slide] of the double microscope, although horizontally placed, and in the most advantageous situation, we still see one common motion in the liquor which forces all that it contains to one side. We must

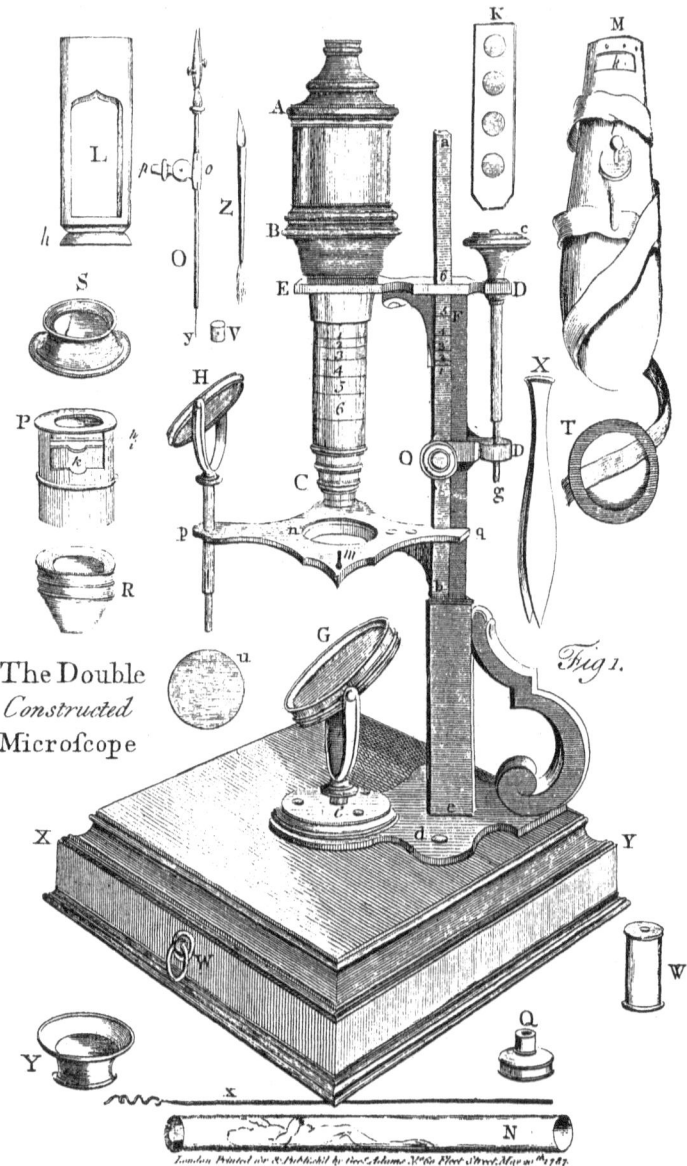

Figure 11 - The popular compound microscope manufactured by John Cuff, entering widespread scientific use in 1744. For illumination, breadth of field, and ease of use, it was superior to all previous compound microscopes, and was an excellent instrument at magnifications below 200X. Depicted with this are standard accessories: *L*, a sleeve-type Lieberkuhn mirror to improve illumination of opaque objects; *R*, a substage aperture diaphragm; *K*, a typical ivory slide; *M*, the famous "fish plate," for observing circulation of the blood in the fish tail; *Q*, interchangeable objective lenses, screwed at *C*.

wait till the fluid is in an equilibrium and at rest, before we make our observations; for it often occurs, that this motion of the fluid [drags along] many globules, and forms a kind of whirling motion which returns one of these globules in a very different direction to the others. The eye is then fixed on the globules, and seeing one take a different course from the rest, supposes it an animal, or at least a body, which moves of itself, whereas its motion is only owing to that of the fluid; and as the liquor is apt to dry and thicken in the circumference of the drop, endeavours must be made to fix the lens on the center of it. The drop should also be as large as possible, and contain as much liquor as will permit sufficient transparency to see perfectly what it contains.

Before we begin to make observations, we should have a perfect knowledge of our microscope. There is no glass whatsoever but in which there are some spots, bubbles, threads, and other defects, which should be nicely inspected, in order that such appearances should not be represented as real and unknown objects: we must also endeavour to learn what effect the imperceptible dust has which adheres to the glasses of the microscope; a perfect knowledge of which may be acquired by observing the microscope several times.

To make proper observations, the sight, or focus, of the microscope must not precisely fall on the surface of the liquor, but a little [beneath] it; as not so much reliance should be placed on what passes upon the surface, as what is seen in the body of the liquor. There are often bubbles on the surfaces which have irregular motions produced by the contact of the air.

We can see much better with the light of two short candles, than in the brightest day, provide that this light is not agitated, which is avoided by putting a small shade on the table, enclosing the three sides of the lights and the microscope.

It will often appear as though dark and opaque bodies become transparent, and even take different colors, or form concentrical and colored rings, or a kind of a rainbow on their surface; and other matters, which are seen at first sight transparent and clouded, become black and obscure; these changes are not real, but only depend on the obliquity [with which light falls on the body], and the height of the plane in which they are found.

When there are bodies in a liquor which seem to move with great swiftness, especially when they are on the surface, they form a furrowed motion in the liquor, which appears to follow the moving body, and which we might be inclined to mistake for a tail. This appearance deceived me at first, but I clearly perceived my error, when these little bodies met others which stopped them; for there was no longer any appearance of tails. These are the remarks which occurred during my experiments, and which I submit to those who would make use of the microscope for the observation of liquors.

EXPERIMENTS

I. I took from the seminal vessels of a man, who died a violent death, and whose body was still warm, all the liquor therein contained, and put it into a small bottle; of this I put a drop on the slide of a [very good double] microscope, without the addition of water or any other liquor. The first thing which [appeared], was a vapour which steamed from the liquor towards the lens, and obscured it. These vapours being dissipated, I perceived large filaments, [Fig. 12, 1] which in some places seemed to extend into different branches, and in others to intermingle together. These filaments clearly appeared to be internally agitated by an undulating motion, and looked like hollow tubes which contained some moving substance. I distinctly saw two of these filaments [Fig. 12, 2] were joined together, and had a vibration, nearly like that of two extended strings, which were tied at the two extremities, and pulled asunder in the middle. These filaments were composed of globules which touched each other, and resembled beads. I afterwards saw [Fig. 12, 3] filaments which swelled in certain parts, and I observed, that on the side so swelled small globules came out, which had a distinct motion like that of a pendulum [Fig. 12, 4]; these small bodies were fastened to the filaments by a small thread, which lenghtened gradually as the little body moved; and at last I saw their little bodies entirely separated from the large filament, carrying after them the small thread which connected them. As this liquor was very thick, and the filaments too near each other, I diluted another drop with rain water, in which I was assured there were no animals. I then saw [Fig. 12, 5] the filaments much separated, and very distinctly perceived the motion of these little bodies, which was now more free, and they swam much quicker; and if I had not seen them separate from the filaments, and carry along with them their thread, I should have taken the moving body in this second observation for an animal, and the thread for its tail. I then attentively observed one of the these filaments, that was much thicker than these small bodies, and I had the satisfaction of seeing two of those bodies which separated with difficulty, drag along with them a long a small thread, which obstructed their motion.

This seminal liquor was at first very thick; but by degrees it became more fluid; in less than an hour, it was almost transparent; and in proportion as this fluidity increased the phenomena changed, as I shall relate.

II. When the seminal liquor attained more fluidity, the filaments were no longer to be seen, but the little bodies appeared in great numbers [Fig. 12, 6]; they have for the most part a motion like that of a pendulum, and they draw after them a long thread, which it may clearly be perceived they want to get rid of; their motion forwards is very slow, vibrating to the right and left. The motion of a boat fastened in the midst of a rapid stream to one fixed point, pretty well represents the motion of these bodies, excepting that the boat remains in the same place, whereas they advance by degrees. [However], they do not always keep the same parts in the same direction; but at each vibration they take a considerable rolling motion; so that, besides their horizontal motion, they have one of a vertical balance, which proves that these bodies are of a globular figure, or, at least, that their lowest part is not sufficiently extended to maintain them in the same position.

III. At the end of two or three hours, when the liquor was more fluid, we saw [Fig. 13, 7] a greater quantity of these moving bodies. They seemed to be more free; the threads were shorter; their progressive motion was more direct, and their horizontal motion was greatly diminished; for the longer the threads are the greater is the angle of their vibration; and, in proportion as the these threads diminish in length, the vibratory motion lessens, and the progressive motion increases. The vertical balance still subsisted, and was always plainly perceptible.[5]

IV. In five or six hours the liquor attained its utmost fluidity. Most of these moving substances were entirely disengaged from their threads; they were of an oval figure, [Fig. 13, 8] and moved progressively with great swiftness, and by their various motions had a stronger resemblance than ever to real animals. Those who had their threads still adhering, were not so brisk as the others; and among those that had no threads, some seemed to change their shape and size, some were round, some oval, and others thicker at their extremities than in the middle; the balancing and rolling motion was still observable.

V. At the end of twelve hours a kind of gelatinous matter was settled to the bottom of the bottle: it was of an ash-colour and of [some firmness]; the liquor that swam above was almost as clear as water, with a kind of bluish tint, resembling water, in which a little soap had been dissolved; nevertheless it still preserved its viscosity. The moving bodies had then a great activity, were loosened from their threads, and moved in all directions. I saw some of them change their form, and from oval become round; and others separate and from one oval form two. As they became smaller, their activity increased.

VI. In twenty-four hours the liquor had deposited a great quantity of gelatinous matter. I diluted it with water, but it did not readily mix, and required a considerable time to dissolve. It then appeared composed of an infinite number of opaque tubes that formed a kind of network, in which no regular disposition nor the least motion could be seen: in the clear liquor some few small bodies were still moving. The next morning there were also a very few; but after that time I saw no more in this liquor than the globules, without any appearance of motion.

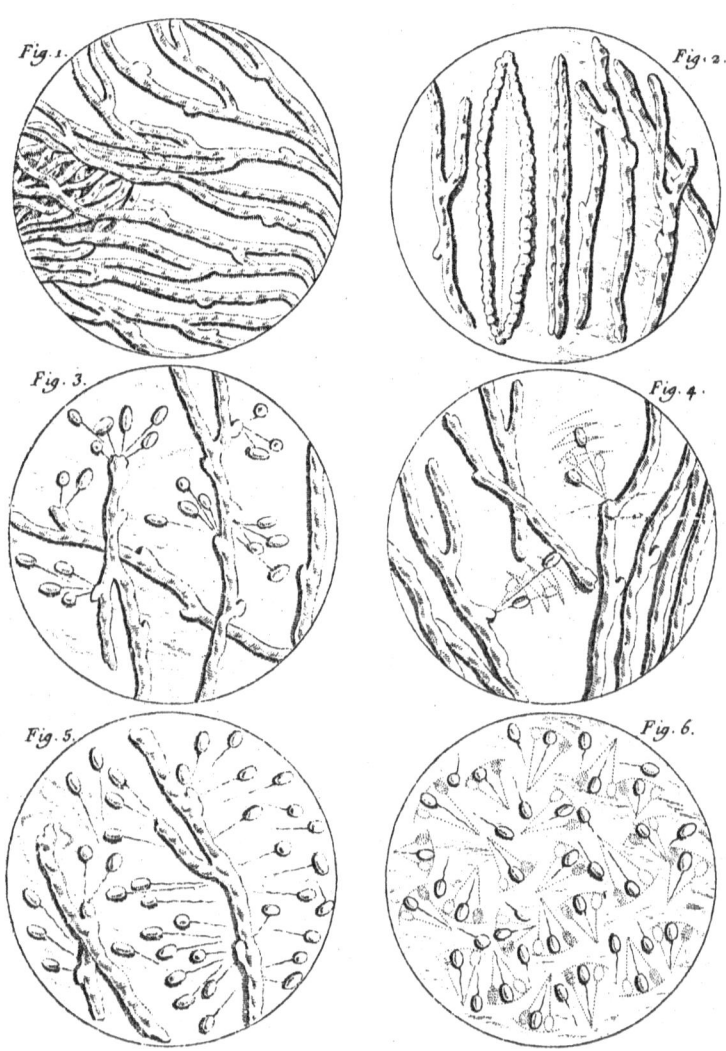

Figure 12 - Woodcut depiction of the spermatic bodies seen in the semen of a male cadaver shortly after death. Figs. 1 and 2 depict filamentous strands within which spherical bodies are seen to form. Figs. 3 and 5 show the tailed bodies detaching. Figs. 4 and 6 depict the free bodies beginning their pendular oscillations.

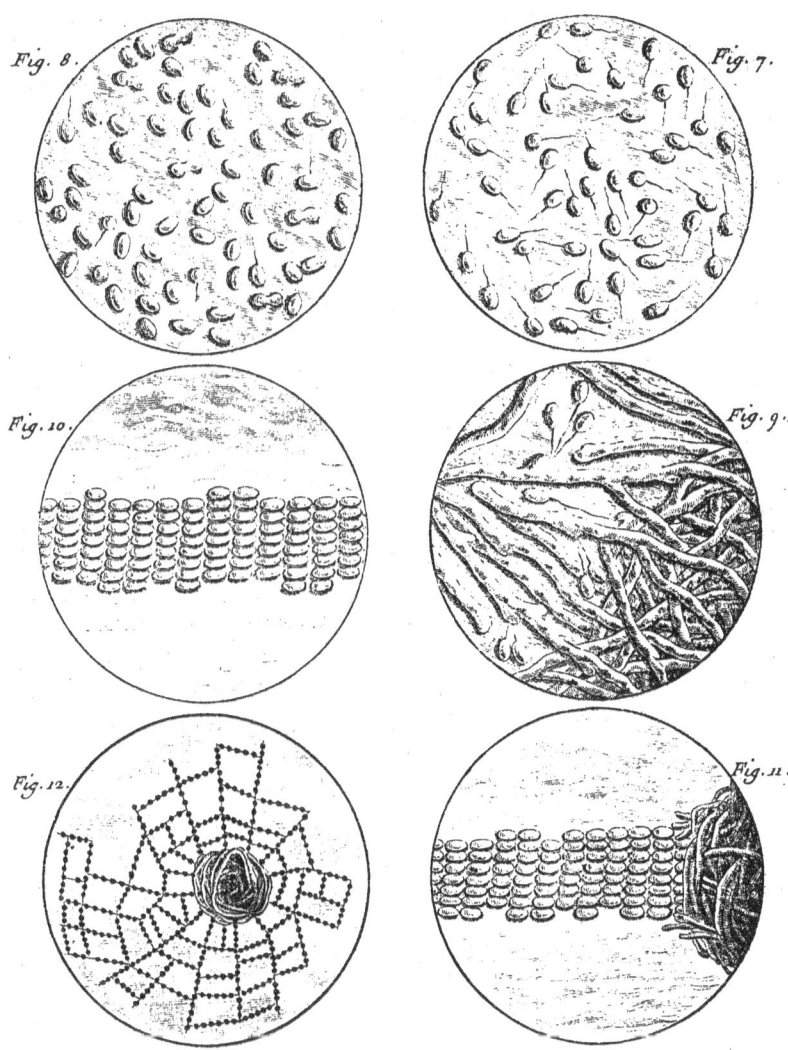

Figure 13 - Figs. 7 and 8 depict the observations on the seminal fluid of the cadaver two to six hours after death, indicating return of bodies to free-swimming condition. Figs. 9, 10 and 11 (not discussed in the selection) represent the formation of the bodies on filaments in the seminal fluid of a second cadaver and their coalescence into linear rows 10 to 12 hours after death. Fig. 12 shows the formation of rectangularly arrayed filaments composed of the tail-less spermatic bodies after several hours.

These experiments were repeated several times with the most possible exactness; and I am persuaded, that those threads above mentioned, are not tails, nor do they make any part of the individual body; for these threads have no proportion with the rest of the body; they are of different sizes, although the moving bodies are always nearly of the same [size] at the same time. The globule appears [obstructed] in its motion, as its tail is longer or shorter; sometimes it cannot advance, but move only from right to left, or from left to right, when the tail is very long; and it is clearly seen that they use great efforts to get rid of them.

. . . .

XXV. As I was persuaded, not only by my own theory, but also by the observations of all those who had made experiments before me, that the female, as well as the male, has a seminal and prolific liquor; and, as I had no doubt, but the reservoir of this liquor was the [cavity of the] glandular body of the testicle, where prejudiced anatomists attempted to find the egg, I purchased several dogs and bitches, and some male and female rabbits, which I kept separate from each other; and in order to have a comparative object with the liquor of the female, I again observed the seminal liquor of a dog, and discovered there [Fig. 15, 19] the same moving bodies [as described in the XIth experiment I have noted beforehand. These bodies draw after them some threads which resemble tails which they labor to free themselves of. Those with the shortest tails move themselves with more agility than the others. They all have a greater or lesser rocking or rolling motion, and generally their progressive motion although very evident and well marked, was not of great rapidity.][6]

XXVI. While I was thus occupied, a bitch was dissected which had been four or five days in heat, and had not received the [male] dog. The testicles [i.e. ovaries] were readily found, and on one of them I discovered a [prominent red glandular] body, about the size of a pea, which perfectly resembled a little nipple; on the outside was a visible orifice formed by two lips; one of which jutted out more than the other. Having introduced a small instrument into this orifice, a liquor dropped from it, which we carefully caught to examine with the microscope. The surgeon replaced the testicles in the body of the animal, which was yet alive, in order to keep them warm. I then examined this liquor with a microscope, and, at the first glance, had the satisfaction to see moving bodies with tails, exactly like those I just before saw in the seminal liquor of the [male] dog [Fig. 15, 20]. Messrs. Needham and Daubenton, who observed them with me, were so surprised at this resemblance, that they could scarcely believe but that these spermatic animals were the same, and thought I had forgotten to

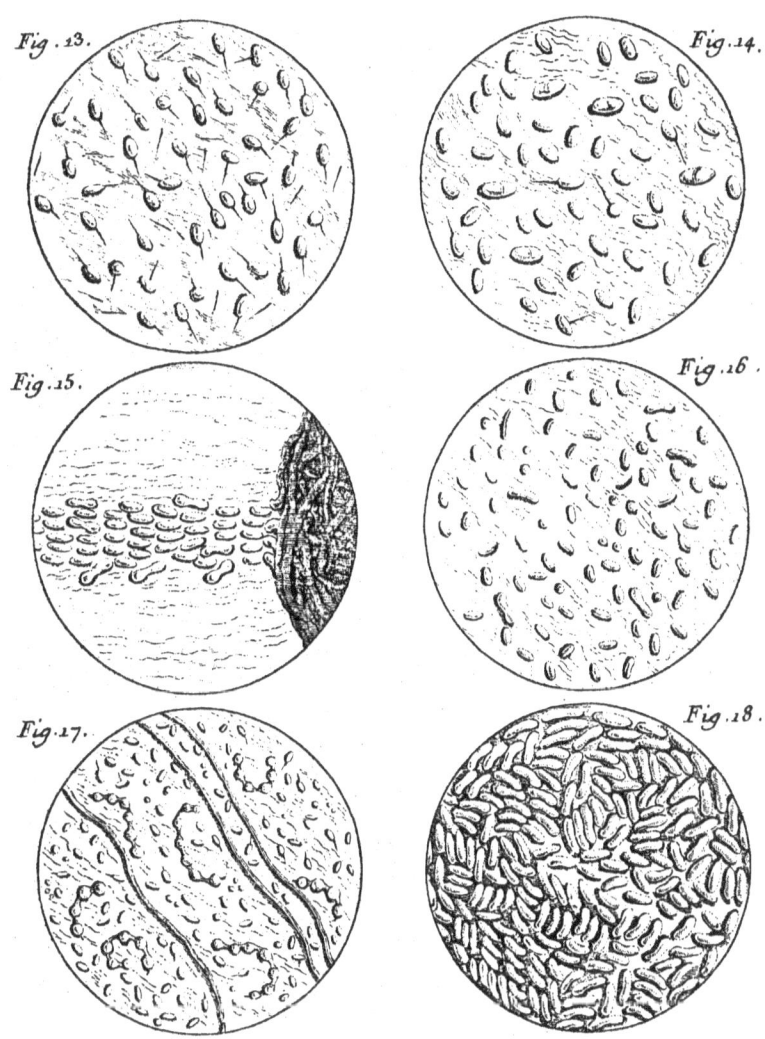

Figure 14 - (Not treated in the selection). Figs. 13 and 14 show tailed spermatic bodies, analogous to those in man, from a male dog. Long filament precursors are absent. Fig. 15 illustrates mucilagenous matter in the semen producing moving globules. Fig. 16 represents bodies in a water infusion of dog testicles after three days. Fig. 17 shows seminal bodies from a male rabbit, and fig. 18 those from a ram.

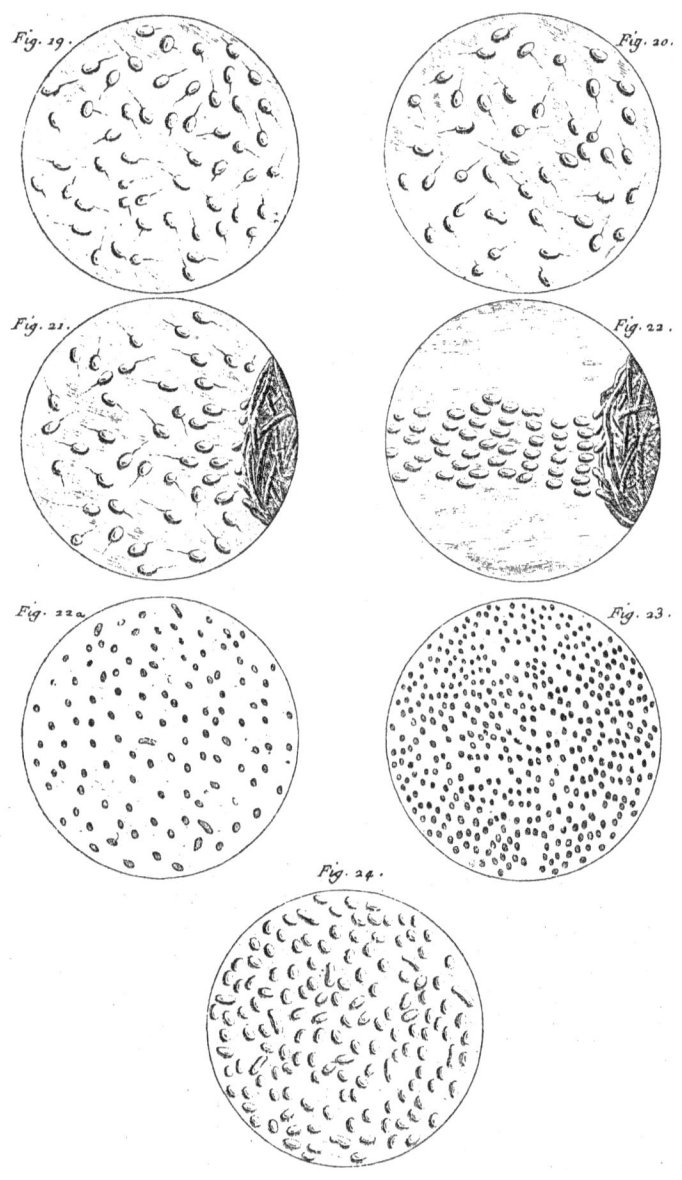

Figure 15 - Original woodcut depicting the controversial observation of seminal bodies found in fluid taken from the Graffian follicle of a female mammal. Fig. 19 depicts fresh observations on a male dog for comparison. Figs. 20 and 21 show identical bodies from a female dog. Fig. 22 illustrates the tendency of female spermatic bodies to coalesce in the same way as in the male. Fig. 22a (not discussed in the selection) shows the spermatic bodies seen in a similar sample from a cow. Fig. 23 depicts bodies forming in water infusion of cow "testicle." Fig. 24 shows spermatic bodies from another cow.

change the [slide] of the microscope, or that the instrument with which we had gathered the liquor of the female, might before have been used for the [male] dog. Mr. Needham then took different instruments, and having obtained some fresh liquor, he examined it first, and saw there the same kind of animals, and was convinced, not only of the existence of spermatic animals in the seminal liquor of the female, but likewise of their resemblance to those of the semen of the male. [We saw the same phenomenon ten times in a row, and on different drops, since there was a rather large quantity of fluid in this glandular body, into which the opening penetrated to a cavity almost three lines deep].

. . . .

CHAPTER XIII

REFLECTIONS ON THE PRECEEDING EXPERIMENTS

By the experiments we have just described, I was assured that females, as well as males, have a seminal liquor which contains moving [bodies]; that these [bodies] were not real animals, but only living organic particles; and that those particles exist, not only in the seminal liquors of the two sexes, but even in the flesh of animals, and in the germs of vegetables. To discover whether all the parts of animals, and all the germs of vegetables, contained living organic particles, I caused infusions of the flesh of different animals to be made, and of more than twenty kinds of seeds of different plants; and after they had infused four or five days, in vials closely stopped up, I had the satisfaction to see moving organic parts in them all; some appeared sooner, and others later; some preserved their motion for months together, while others were soon deprived of it; some directly produced large moving globules, that had the appearance of real animals, which changed their figures, separated, and became successively smaller: others produced only small globules, whose motions were very brisk; others produced filaments which lengthened and seemed to vegetate, swelled, and afterwards thousands of moving globules issued therefrom; but it is useless to detail my observations on the infusion of plants, since Mr. Needham [has pursued this with much greater care than I have been able to do myself, and this skillful naturalist will soon present to the public the collection of discoveries he has made on this subject.][7] I read the preceding treatise to that able naturalist, and often reasoned with him on the subject particularly on the probability that the germs of vegetables contained similar moving bodies to those in the seed of male and female animals. He thought those views sufficiently founded to deserve to be pursued; and therefore began to make experiments on all parts of vegetables; and I must own that the ideas I gave him on this subject have reaped greater profit under his hands than they would have done from me. I could quote many examples, but shall confine myself to one, because I indicated the circumstance I am going to relate.

To determine whether the moving [bodies] seen in the infusions of flesh were true animals, or only, as I supposed, moving organic particles, Mr. Needham [reasoned] that he had only to examine [the juice from] some roasted meat, because if they were animals the fire must destroy them; and if not animals, they might still be found there as well as when the meat was raw; having therefore taken the jelly of veal, and other roasted meat, he infused them for several days in water, closely corked up in vials, and, upon examination, he found in every one of them a great quantity of moving [bodies]. He showed me some of these infusions, and among the rest, that of the jelly of veal, in which there were [some kinds of moving bodies], perfectly like those in the seminal liquor of a man, a dog, and a bitch, when they have no threads, nor tails; and although we perceived them to change their figures, their motions so perfectly resembled those of an animal which swims, that whoever saw them, without being acquainted with what has been already mentioned, might certainly have taken them for real animals. I shall only add, that Mr. Needham assured himself, by a multiplicity of experiments, that all parts of vegetables contain moving organic particles, which confirms what I have said, and extends my theory on the composition of organized beings, and their reproduction.

[All animals, both male and female, all those lacking sexes, and all plants, of whatever species they might be, in short all living or vegetating bodies, are composed of living organic particles, demonstrable to all the world.] These organic parts, are in the greatest abundance in the seminal liquor of animals, and in seeds of vegetables. It is from the union of these organic parts returned from all parts of the animal or vegetable body, that reproduction [of a form always similar to the animal or plant in which it occurs, is accomplished.] The union of these organic parts cannot be made but by the means of an internal mould, in which the form of the animal or vegetable is produced. It is in this also, [that] the essence of the unity and continuity of the species consists, and will so continue while the great Creator permits their existence.

But before I draw general conclusions from the system I am establishing, I must endeavour to remove some objections which might be made, and mention some other circumstances which will serve to place this matter in a better light.

It will be asked, why I deny those moving [bodies] in the seminal liquors to be animals, since they have constantly been regarded as such by Leeuwenhoek, and every other naturalist, who has examined them? I may be also told, that living organic particles are not perfectly intelligible, [unless] they are to be looked upon as animalcules; and to suppose an animal is composed of a number of small animals, is nearly the same as saying that an organized being is composed of living organic particles. I shall therefore endeavour to answer these objections in a satisfactory manner.

It is certain that almost all naturalists agree in looking on the moving [bodies] in seminal liquors as real animals, [and there are only a few who, like Verheyen,[8] have not observed them with good microscopes, and have believed that the motion seen in these fluids could be originating from seminal spirits, which are presumed to be in great agitation] but it is no less certain, from my own observations, and those of Mr. Needham, on the seed of the calmar, that these moving substances are more simple and less organized beings than animals.

The word _animal_, in the acceptation we commonly receive it, represents a general idea formed of particular ideas drawn from particular animals. All general ideas include many different ones, which approach, or are, more or less, distant from each other, and, consequently, no general idea can either be exact or precise. The general idea which we form of an animal may be taken principally from the particular idea of a dog, a horse, and other beasts, which appear to us to act and move according to the impulse of their will, and which are besides composed of flesh and blood, seek after their food, have sexes, and the faculty of reproduction. The general idea, therefore, expressed by the word _animal_, must comprehend a number of particular ideas, not one of which constitutes the essence of the general idea, for there are animals, which appear to have no reason, will, progressive motion, flesh nor blood, and which only appear to be a congealed substance: there are some which cannot seek their food, but only receive it from the element they live in: there are some which have no sensation, not even that of feeling, at least in any sensible degree: there are some [that] have no sexes, or are both in one; there only belongs, therefore, to the animal a general idea what is common also to the vegetable, that is, the faculty of reproduction.

The general idea then is formed from the whole taken together, which whole being composed of different parts, there [are], consequently, between these parts degrees and [gradations]. An insect, in this sense, is something less of an animal than a dog; an oyster still less than an insect; a sea-nettle, or a fresh-water polyp, still less than an oyster; and as nature acts by insensible [gradations] we may find beings, which are still less animated than a sea-nettle, or a polyp. Our general ideas are only artificial methods to collect a quantity of objects in the same point of view; and they have, like the artificial methods we [have spoken of (Vol. I, Discourse 1)], the defect of never being able to comprehend the whole. They are, likewise opposite to the [progression] of nature, which is uniform, insensible, and always particular, insomuch, that by our endeavouring to comprehend too great a number of particular ideas in one single word we have no longer a clear idea of what that word conveys; because, the word being received, we imagine that it is a line drawn between the productions of nature; that all above this line is _animal_, and all below it _vegetable_, another word, as general as the first, and which is used as a line of separation between organized bodies and inanimate matter. But as we have already said, these lines of separation do not exist in nature; there are beings which are neither animals, vegetables, nor minerals, and which we in vain might attempt to arrange with either. For example, when Mr. Trembley, [the celebrated author who discovered some animals which multiply themselves by each of their detached, incised or separated parts], first observed the polyp he employed a considerable time before he could determine whether it was an animal or a plant; and possibly from this reason--that it is perhaps neither one nor the other, and all that can be said is, that is approaches nearest to an animal; and as we suppose every living thing must be either an animal or a plant, we do not credit the existence of an organized being, that cannot be referred to one of those general names; whereas there must, and in fact are, a great number of organized beings, which are neither the one nor the other. The moving [bodies] perceived in seminal liquors, in infusions of the flesh of animals, in seed, and other parts of plants, are all of this kind. We cannot call these animals, nor can we say they are vegetables, and certainly we can still less assert they are minerals.

We can therefore affirm, without fear of advancing too much, that the grand division of nature's productions into <u>Animals</u>, <u>Vegetables</u>, and <u>Minerals</u>, do not contain every material being: since there are some that exist which cannot be classed in this division. We have already observed, that nature passes by insensible [gradations] from the animal to the vegetable, but from the vegetable to the mineral, the passage is quick, and the distance considerable; from whence the law of nature's passing by imperceptible degrees appears [contradicted]. This has made me suppose, that by examining nature closely, we shall discover [first] intermediate organized beings, which, without having the power of reproduction, like animals and vegetables, would, nevertheless, have a kind of life and motion; [then] other beings which, without being either vegetables or animals, might possibly enter into the composition of both, and [then] likewise other beings which would be only the assemblage of the organic molecules I have spoken of in the preceding chapters.

. . . .

RECAPITULATION[9]

All animals procure nutriment from vegetables, or other animals which feed upon vegetables; there is, therefore, one common matter to both, which serves for the nutrition and [development] of every thing which lives or vegetates. This matter cannot perform [this] but by assimilating itself to each part of the animal, or vegetable, and by intimately penetrating the texture and form of these parts, which I have called the <u>internal mould</u>. When this nutritive matter is more abundant than is necessary to nourish and expand the animal, or vegetable, it is sent back from every part of the body, and deposited in one, or more reservoirs, in the form of a liquor; this liquor contains all the molecules analogous to all parts of the body, and, consequently, all that is necessary for the reproduction of a young being, perfectly resembling the first. Commonly this nutritive matter does not become superabundant, in most kinds of animals, till they have acquired the greatest part of their growth; and it is for this reason, that animals are not in a state of engendering before that time.

When this nutritive and productive matter, which is universally [distributed], has passed through the internal mould of an animal or vegetable, and has found a proper matrix, it produces an animal, or vegetable, of the same kind: but when it does not meet with a proper matrix, it produces organized beings different from animals and vegetables, as the moving and vegetating bodies seen in the seminal liquor of animals, in the infusion of the germ of plants, etc.

This productive matter is composed of organic particles, always active, the motion and action of which are fixed by the inanimate parts of matter in general, and particularly by oily and saline bodies, but as soon as they are disengaged from this foreign matter, they [resume] their action and produce different kinds of vegetations and other animated beings [which move progressively].

By the microscope, the effects of this productive matter may be perceived in the seminal liquors of animals of both sexes. The seed of the female viviparous animals is filtered through the glandular bodies which grow upon their testicles, and these glandular bodies contain a large quantity of seminal fluid in their internal cavities. Oviparous females have, as well as the viviparous, a seminal liquor, which is still more active than the viviparous. The seed of the female is, in general, like that of the male; when they are both in a natural state, they decompose after the same manner, contain similar organic bodies, and they alike offer the same phenomena.

All animal or vegetable substances include a great quantity of this organic and productive matter. To perceive it, we need only separate the inanimate parts in which the active particles of this matter are engaged. And this is done by infusing animal or vegetable substances in water. The salts will dissolve, the oils separate, and the organic particles will be seen by their putting themselves in motion. They are in greater abundance in the seminal liquors than in any other parts, or, rather, they are less entangled by the inanimate parts. In the beginning of this infusion, when the flesh is but slightly dissolved, the organic matter is seen under the form of moving bodies, which are almost as large as those of the seminal liquors: but, in proportion as the decomposition augments, these organic particles diminish in size and increase in motion; and when the flesh is entirely decomposed, or corrupted, these same particles are exceedingly minute, and their motion exceedingly rapid. It is then that their matter may become a poison, like that of the tooth of a viper, wherein Mr. Mead perceived an infinite number of small pointed bodies, which he took for salts, although they are only these same organic particles in a state of great activity. The pus which issues from wounds, abounds with [these particles], and it may take such a degree of corruption as to become one of the most subtle poisons; for every time this active matter is exalted to a certain point, which may be known by the rapidity and minuteness of the moving bodies it contains, it will become a species of poison. It is the same with the poison of vegetables. The same matter which serves to feed us when in its natural state, will destroy us when corrupted. [Blighted rye], for instance, throws the limbs of men and animals into a gangrene who feed on it. It is also [seen in] comparing the matter which adheres to our teeth, which is the residue of our food, with that from the teeth of a viper, or mad dog, which is only the same matter too much [excited] and corrupted to the last degree.

When this organic and productive matter is found collected in a great quantity in some part of an animal, where it is obliged to remain, it forms living beings which have been ever regarded as animals; the taenia, ascarides, all the worms found in the veins, liver, in wounds, in corrupted flesh, and pus, have no other origin; the eels in paste, vinegar, and all the [supposed] microscopical animals are only different forms which this active matter takes of itself, according to circumstances, and which invariably tends to organization.

In all animal and vegetable substances, decomposed by infusion, this productive matter manifests itself immediately under the form of vegetation. Filaments are seen to form, which grow and extend like plants. Afterwards these extremities and knots swell and burst, to give passage to a multitude of bodies in motion, which appear to be animals; so that it seems as if all nature began by a motion of vegetation. It is seen by microscopical objects, and likewise by the expansion or unfolding of the animal embryo; for the foetus at first has only a species of vegetable motion.

[Fresh] food does not furnish any of these moving molecules for a considerable time. Several days infusion in water is required for fresh meat, grain, kernels, etc., before they offer to our sight any moving bodies; but the more matters are corrupted, decomposed, or exalted, the more suddenly these moving bodies manifest themselves; they are all free from other matters in seminal liquors; but a few hours infusion is required to see them in pus, [blighted rye], honey, drugs, etc.

There exists, therefore, an organic matter, universally diffused in all animal and vegetable substances, which alike serves for their nutrition, their growth, and their reproduction. Nutrition is performed by the intimate penetration of this matter in all parts of the animal or vegetable body. Expansion or growth is only a kind of more extended nutrition, which is made and performed as long as the parts have sufficient ductility to swell and extend; and reproduction is made by the same matter when it superabounds in the body of the animal or vegetable; each part of the body sends back, to the appropriate reservoirs, the organic particles which exceed what are sufficient for their nourishment. These particles are absolutely analogous to each part from which they are sent back, because they were destined to nourish those parts from hence, [and] when all the particles sent back from [the entire body], collect together, they must form a body similar to the first, since each particle is like that part from which it was detached; thus it is, that reproduction is effected in all kinds of trees, plants, polyps, [aphids], etc., where one individual can produce its like; and it is also the first mode which nature uses for the reproduction of animals which have need of the communication of different sexes; for the seminal liquors of both sexes contain all the necessary molecules for reproduction; but something more is required for its effectual completion, which is the mixture of these two liquors in some place suitable to the [development] which must result therefrom, which place is the matrix of the female.

There are, therefore, no pre-existing germs, no germs contained one in the other, <u>ad infinitum</u>; but there is an organic matter perpetually active, and always ready to [mold itself], assimilate, and produce beings similar to those which receive it. [The species of] animals and vegetables, therefore, can never be extinct; so long as there subsists individuals the species will ever be new; they are the same at present as they were three thousand years ago, and will perpetually exist, by the powers they are endowed with, unless annihilated by the will of the Almighty Creator.

NOTES

*<u>Buffon's Natural History, Containing a Theory of the Earth, a General History of Man, of the Brute Creation, and of Vegetables, Minerals, Etc.</u>, translated by J.S. Barr (London: J.S. Barr, 1792), Vol. II, pp. 272-310; Vol. III, pp. 83-93;

109-111; 159-69; 309-316. We have restored in brackets substantial passages deleted from the Barr text. Detailed criticisms on Buffon's arguments are to be found below in review section, Part II.

[1][We have deleted a latin quotation from Antoine van Leeuwenhoeck's Arcana naturae, vol 1, p. 3 that appears as a note in Buffon's original, but is deleted in Barr.]

[2]My demonstration of this can be seen in the preface to the translation of Newton's Fluxions, page [45-46 above] and ff.

[3][Corps glanduleux in French. In Buffon's discussions, and in the subsequent reviews by Albrecht von Haller and the anonymous reviewer for the Bibliotheque raisoneé, both translated below, there is a persistent uncertainty concerning the anatomical structures under discussion, and a repeated confusion is evident between the ovary proper, the Graffian follicles, and what is termed the corpus luteum. (lit. "yellow body"). Consequently, while Buffon here refers to the corps glanduleux in this selection, his reviewers will commonly speak of the corps jaune or gelben Drusen, seeming to mean in fact the corpus luteum. The relation of these anatomical structures is as follows: the mammalian "ovary", accepted by many seventeenth and eighteenth century anatomists as the direct analog of the ovarium of the bird, and by Buffon as the female "testicle", forms on its surface prior to ovulation large fluid-filled vesicles--the so-called Graffian follicles. Originally these were considered by their discoverers to be the actual eggs of the mammal, and only in 1828 was it demonstrated by Von Baer that the actual mammalian egg was only a microscopic structure within the follicle. On bursting at the maturation of the egg, the follicle emits both its contained fluid and the accompanying egg cell. The ruptured Graffian follicle then undergoes a marked change in structure, with its internal cavity becoming gradually filled with yellow glandular tissue, and if pregnancy occurs, continues to enlarge. The function of this body has been determined to be the secretion of endocrine hormones (progesterone and estrogen) before, during and after pregnancy. See Marshall's Physiology of Reproduction, 3rd ed., edited A.S. Parkes (London: Longmans and Green, 1956), pp. 467-96. A useful review article with historical remarks is R.J. Harrison, "The Development and Fate of the Corpus Luteum in the Vertebrate Series," Biol. Rev. 23 (1948), pp. 296-331. From the description of the experiments, it would seem certain that Buffon is taking samples of the fluid of the enlarged and ripened Graffian follicles prior to rupture, and is not experimenting with the subsequent corpus luteum.]

4[Passage restored from the French original. The descriptions in this section are fully consistent with the working of the simple microscope discussed in the introduction to the section and pictured on page 164 above. This would explain Buffon's reference to the horizontal viewing of objects, since in the ordinary version of the Wilson Screw-Barrel microscope, it was mounted with a hand-held mount, and the slide would ordinarily be viewed holding the microscope horizontally directed toward a light source, with the slide itself arranged vertically. The stable vertical mounting made possible by the scroll mount and the use of a mirror would in this case place the slide in a horizontal position.]

5[The repeated references in these experiments to the rapid vibration and oscillation of the molecular bodies strongly suggest an observation of Brownian motion. See introductory remarks on this selection above, pp. 166-67. My own observations on various solutions of organic materials, prepared as described by Needham, ("A Summary of Some Late Observations Upon the Generation, Composition, and Decomposition of Animal and Vegetable Substances," Phil. Trans. Roy. Soc. Lond. 45, No. 490 (1748), pp. 615-66), with an achromatic microscope at a magnification of 550X revealed, after several days incubation, what were probably coccoid bacteria in several of the solutions, moving by Brownian motion.]

6[Passage restored from French original.]

7[Needham, op. cit. endorsed several of Buffon's claims about the observations, but in subsequent works his agreement with Buffon became much more uncertain. See on this J. Roger, Les sciences de la vie dans la pensée française du xviiie siècle 2nd ed. (Paris: Colin, 1971), pp. 494-520. Professor Shirley Roe of Harvard University is presently embarked on a detailed study of Needham and the controversy over the experiments.]

8[Phillippe Verheyen (1648-1710), Louvain anatomist.]

9[This is section from the final summary discussion closing the sections on generation. It ends in the original with the date 27 May, 1748.]

PART III:

The First Receptions

9. The *Journal de Trévoux* Reviews (1749–50)
John Lyon

The Memoires pour l'histoire des sciences et des beaux-arts, also known as the Journal de Trévoux, was founded under the auspices of Louis Auguste de Bourbon, the Duke of Maine. In 1695 the Duke had founded a printing establishment at Trevoux, in Dombes, and in 1701 asked the Jesuits of the College de Louis-le-Grand in Paris to assume the editorship of memoirs "useful to the history of the arts and sciences," and to the defense of religion, to be published at Trevoux.[1] The first volumes appeared that year, and ran with few interruptions until about 1782, when the periodical disappeared.[2]

The printing of the Journal moved from Trevoux to Lyons (1732), and then to Paris (1734), but the editorship remained under the control of the Jesuits until the suppression of that Society in France in 1762. Thereafter the Journal fell upon rather indifferent editorship until its demise. In general the policy of the editors was conciliatory and moderate, with a few notable exceptions. The Journal attempted to avoid unnecessary polemics and altercations, and devoted its pages to articles and reviews concerning the arts, letters, sciences, morals, religion, and politics. Though its position might be considered intransigent in matters of good morals and religion, it was not notably dogmatic on matters of politics.

The most significant of its editors was Guillaume-François Berthier (1704-1782), who controlled the Journal from 1745 to 1762, and the policies of moderation and conciliation noted above particularly characterized his editorship.³ Berthier had himself been educated by the Jesuits at Bourges, and entered the Jesuit Novitiate at the age of 18.

It was he who was the apparent author of the review of Buffon's Histoire naturelle in the issues of the Journal between September, 1749, and June, 1750, translated here.⁴ Berthier's moderate and tactful review was attacked by the Jansenist Nouvelles écclesiastiques, as will subsequently be obvious. Berthier's attitude toward Buffon's religious orthodoxy may appear naive and credulous in light of Herault de Sechelles' observations on this matter (see the Voyage à Montbard, later in this volume). Likewise, his willingness to write off Buffon's extravagant hypotheses as "effects of an over-active imagination" rather than deliberate impiety,⁵ and his praise for the "retraction" Buffon published in Volume IV of the Histoire naturelle (see the "Letters of the Deputies and Syndic of the Faculty of Theology of Paris," and Buffon's reply, later in this volume) published in the December, 1753 issue of the Journal de Trévoux⁶ may seem a bit misplaced.⁷

NOTES

¹Gustave Dumas, Histoire du journal de Trévoux (Paris: 1936); p. 20. Cited in John Nicholas Pappas, Berthier's Journal de Trévoux and The Philosophes. Ann Arbor, Mich.: University Microfilms (Doctoral Dissertation series, 1955), 1978, p. 2. See also Charles Ledre, Histoire de la Presse (Paris: Librairie Artheme Fayard; 1958), p. 85.

²Pappas, op. cit., p. 27

³Ibid., pp. 13, 14, 18, 42.

⁴The June, 1750 review is exclusively concerned with Volume III of the Histoire naturelle and with Daubenton's technical work. It has not been translated here.

⁵Ibid., p. 232

⁶Cited in Ibid., p. 230.

⁷Pappas seems to misread the case when he suggests that Berthier and Buffon might be bracketed together as "pioneers in the development of Modern Christian Apologetics." (p. 245)

REVIEW OF BUFFON'S HISTOIRE NATURELLE

September, 1749

THE JOURNAL DE TRÉVOUX*

Article CV

Translated by John Lyon

    This great work which we shall be considering in some detail is the result of the labors of MM. de Buffon and d'Aubenton [sic]. The former is the Intendant of the Jardin Royale des Plantes; the latter, Garde and Demonstrateur du Cabinet Royal d'Histoire Naturelle. M. de Buffon, the author of the first volume, begins with a discourse on the manner of studying and expounding natural history. This is a piece which deserves the attention of readers by reason of the principles which it contains, and by the ingenuous capacious, and pleasant fashion in which the details are set forth. We wish to explicate this discourse at some length.
    No one doubts that natural history, as well as many other sciences which are matters of everyday conversation, cannot but be presented in a didactic fashion, and accompanied by a display of precepts, method, and cumbersome observations. Now, this manner is the bane of the intellect the ruin of education, and of fair knowledge. Just as in society the multitude of laws indisposes hearts, so too in the empire of the sciences and the arts the excessive weight of regulations overwhelms genius and stifles talents. M. de Buffon avoids this danger with a charming sagacity. He knows that natural history presents infinite difficulty. But, without directing us to any details of its career; without discouraging us by setting forth a tiresome theory, he simply advises us to observe often and frequently to repeat our observations of the objects which adorn this universe: animals, plants, vegetables -- in a word, all the productions which relate to natural history.
    But what? Simply look, and look again? Is this the way to knowledge? Should the eyes be the sole judges in this matter, while judgement, reflection, and comparison shall count for nothing? Our author replies by this principle: When one has often observed then one generalises his ideas, forms a method of arrangement and systems of explication. Now who doesn't recognize [scait] how much more pleasant and advantageous for self-instruction this is, rather than to be bound slavishly to the instruction of masters? The discoveries which the mind makes effortlessly, or by spontaneous effort, little by little develop the taste for science. And this taste which

is so necessary for everything, but at the same time so rare, is not acquired through precepts. This is no doubt an excellent theory, and one which should be exported from natural history to all other disciplines of the arts and sciences!

We would be better educated, perhaps, if there were fewer pedagogues and textbooks or, rather (for there is no need to exaggerate here), if our pedagogues and those who explain textbooks to us could more readily engage our natural curiosity.

It is obvious that in this regard the time of youth is of paramount importance. M. de Buffon desires that the young be exposed to everything, that natural history be presented to them in its turn, that what is most singular in that science be pointed out to them -- but without first giving precise explications. Rather, curiosity should be aroused by an air of mystery, while allowing these young minds to inquire of themselves, or seek for the extremely utilitarian responses which they have wished and waited for.

Nor does the author condemn all methods. He is content to dismiss those in which he finds no advantage. But, in order to save the young from danger, he notes the defects of most modern methods, which have been set up, it is claimed, in order to facilitate the knowledge of natural hsitory, while in reality they serve better to store up words in the memory than ideas in the mind. An example will make this apparent.

Since the establishment of letters and arts botany has been broadly cultivated -- botany, that fine and extensive part of natural history. Among the great number of botanists who have offered their works to the public there are few who have not claimed to have made general systems, that is to say, to have related all species of plants to a few fixed and certain points in such fashion that their distribution into Classes and Genera generally comprises all that is meant by the vegetable kingdom. But success has not been as widespread as effort in this respect, and there is no occasion to be surprised by the disparity. For certain anomalous and intermediate species have always been found which cannot be classified according to the system nor related to [s'allier] other plants. And what has caused these systems to fail is the gratuity with which their authors have chosen a single part of plants (sometimes the leaf, sometimes the flower, sometimes the fruit, sometimes the stamens) to give their specific character to them, and to serve as the essential juncture which they have marked out as discriminating the formation of their genera and classes.

Now there is nothing less analogous to the extremely rich and varied plan upon which Nature has operated. Nor is anything less appropriate to convey clear ideas of things. What should we say of a man who wished to know the difference between animals based on the differences of their hair? He who dares to judge of the difference between plants solely by their leaves, or flowers, or fruits, or stamens, is almost in the same position. Both the one and the other deviate thus from the true end of science. Both at best acquire an arbitary language, and a means of mutual understanding; but this does not lead to any real understanding. And in this fashion are the various characteristics of botany reduced to a dry, abstract, complicated nomenclature, such that, says our author, a man should rather engrave in his memory the figures of all plants, and retain clear ideas of them, which is true botany, than retain all the names which the various methods give to these plants.

All this is but the rough sketch and outline of M. de Buffon's judicious critique. He shows us how the love of system creeps in, becomes acknowledged, and spreads in botany. He acknowledges the sagacity of M. de Tournefort who had felt the defects of all abstract methods and proposed his distributions and exceptions with infinite skill. But since his time a contemporary has dared to confound the most different things. This fellow has put in the same classes the mulberry and the nettle, the tulip and the barberry, the elm and the carrot, the rose and the strawberry, the oak and the bloodwort.

This attack is directed straight at the celebrated Linnaeus, an author, it is known, of a new system which destroys all previous ideas. According to him, plants should be classified by their stamens, and our author declares with firmness that this is to play with nature and with those who study it. He shows the inconvenience, the defects, even the ridiculousness of such a method. Those who read this section of the discourse will see that energy and freedom still remain among the writers of natural history. We add that despite our esteem for the learned Swedish botanist,[1] we are not very sorry to see his system disturbed. It has long seemed difficult for us to abandon the traces of our Tournefort and to return to new elements of botany, precisely because it pleased M. Linnaeus to overturn all common notions. By virtue of the discourse of M. de Buffon we are supported, so to speak, in our manner of thinking, and we acquire strong arms to defend us against the partisans of novelty.

Our author was not content to show the road which leads to the study of natural history. He felt obliged above all to mark out the route which leads to science. This route, according to him, consists in working at the description and at the history of the various things which science has as its objects. Thus, considering the matter in hand, when one has observed, re-observed, and examined animals, plants, vegetables, in a word, all the riches of natural history, it is then necessary to put oneself at the task of describing them, without prejudice, without systematic ideas, without deviation, without ornaments, and without equivocation. And this description ought to include the size, the weight, the colors, the condidtions of repose and movement, the position of the parts and their relations, their shape, their actions, and all their exterior functions, without neglecting the interior of which it is possible to gain some knowledge. "History ought to come after description, and ought to be uniquely concerned with the relations which the things of nature have between themselves and with us. The history of an animal ought not be the history of an individual but of the entire species of these animals. It ought to include their generation, the duration of pregnancy, the time of their labor, the number of young, the care fathers and mothers take of the young, their kind of education, their instinct, the places where they live, food and the manner in which it is procured, their manners, their means of deception, the hunting of them, and then the uses to which man may put them, etc."

We only try to set forth here the ideas of the author. We wish we were able to transcribe what he says of Aldrovandi, <u>the most diligent and erudite of all naturalists</u>. This section is crucial, and is quite well written. The gist of the passage is that Aldrovandi, like all the savants [Érudits] of his time often builtup science at the expense of taste and judgment. But in looking over the precise and orderly spirit which reigns today in the composition of books, M. de Buffon remarks quite sensibly <u>that it is to be feared that, coming to mistrust erudition, we shall also come to imagine that mind can supply everything, and that science is but an empty name</u>.

After the description and history of natural objects, it is necessary to portion it out into various subjects. Regarding this, it appears to our author that there is no order more appropriate than that in which all the beings appear which surround us. Thus, when we open our eyes we see animals, plants, and minerals. And since the earth, the air, and the waters are objects which strike the senses incessantly, we form for ourselves without difficulty an idea of those animals which live on the earth or in the waters, or those which take to the skies. And among these animals we particularly mark out those which have the closest relations with us, those which are most familiar to us and are found in the climate where we live. The same case prevails in the vegetable kingdom, and likewise in the mineral. We are led to study them in proportion to the usefulness which we can extract from them, and to the readiness with which they are present to us.

This order, the most natural of all, is that which has been employed in this great work which we announce. It is not hard to imagine the advantages it possesses. Nevertheless, M. de Buffon, with great appropriateness, refutes objections which could be proposed to this arrangement, and his replies show that he has a great knowledge of various sysems. This is particularly the case with the most modern of them, that is, the system of M. Linnaeus, who is the author of a system of classifying animals which is as peculiar as his botanical system. Our author attacks this method also. He points out how often it is arbitrary, inadequate, and ill-founded. He uses this occasion to declare his quite high esteem for the practice of the ancients, those very ancients whom we believe lived in the infancy of the Arts, the contemporaries of Homer, for example, a people of quite simple morals, who nevertheless were quite capable of naming an infinity of animals and minerals which we encounter only quite rarely and with great difficulty. "This is obvious proof," adds M. de Buffon, "that these objects of natural history were known to them, and that the Greeks were not only familiar with them but even had precise ideas of them, ideas which they could not have come to except by a study of these objects, a study which necessarily presupposes observations and comments thereon."

These reflections prepare the way for a eulogy of Aristotle and his history of animals, which is perhaps until this day the best we have in this genre. Such praise caused us to take up our Aristotle again, and we recognized in that work all the traits which characterize M. de Buffon. We were naturally led to note that he gives more of an idea of the method of the Philosopher and of his erudition and researches, than do either Theodore of Gaza or William Duval, his two most learned editors. Four pages of our author make us justly appreciate the greatness, nobility, naturalness, and worth of this method, those truly royal characteristics in a word, which brought Alexander to place Aristotle in a financial situation wherein he could make his observations.[2] And this passage ends with the following declaration, which is one of the very best things that has ever been said in praise of a writer.

> This work of Aristotle's appears to me like a table of materials which might have been extended with the greatest care for many thousands of volumes filled with descriptions and observations of all kinds. It is the most learned abridgment that has ever been made, if science is, indeed, the history of facts. And even if one were to suppose that Aristotle had drawn from the books of his time that which he put into his own, the plan of the work, its distribution, the choice of examples, the exactness of the comparisons, a certain form in the ideas, which I shall gladly describe as philosophic in character, all this does not leave one in doubt for even an instant that he was himself far richer than those from whom he supposedly borrowed.

In the sequel there is an excellent characterization of Pliny which we must omit here in order to devote a few moments to the conclusion of the discourse. This is indeed the most exquisite part of the entire doctrine of the learned author [Buffon]. For, rising above the moderns, who often only have fabricated verbose and abstract methods, and bearing his views even beyond those of the ancients who were not as well versed in experimental physics as the moderns, he teaches us how to search out and find the truth in the study of material objects.

Now, this truth consists in the discovery of causes, or, rather, more general effects which take the place of causes. The proof of such discovery lies in repeated, comparable observations, which are capable of certifying the facts and of being used as the consequences which are to be predicted in such cases. And so, to bring together in a few words the complete advice of our author, it is necessary: 1.) to observe natural objects frequently, engrave them in our memory, and familiarize ourselves with them; 2.) to describe them as exactly as possible; 3.) to trace their entire natural history with care; 4.) to compare facts, reason upon them, and connect them in the best way possible with other more general effects. And this is approximately the result of this discourse, all the abundance, details, and advantage of which we are unable to describe. We shall return to this first volume in order to give the plan of the second discourse, which contains the history and theory of the earth, and in order to indicate also, with M. de Buffon, the principal proofs of that theory.

On p. 12 of the discourse of which we are giving an account, these words are to be read:

> Then, examining successively and by order the various objects which compose the universe, and placing himself at the head of all created beings, man will see with astonishment that it is possible to descend by almost imperceptible gradations from the most perfect of creatures to the most formless matter. . . .

Now, this ought to be taken solely in the order and sphere of material beings, for it could not be true that by almost insensible degrees one can descend from a creature purely spiritual even to the most unformed matter. Everyone knows that spirit, which is immortal, and matter, which contains the seeds of its own dissolution, are two incommensurable orders of beings. Thus there can be no gradation so happily placed as to serve as the imperceptible passage from one to the other. M. de Buffon himself furnishes us with the principle of a solution when he says, two lines earlier that man resembles the animals in all material aspects. These last words are a sort of restriction which directs the proposition that follows solely to sensible and corporeal beings.[3]

NOTES

*Memories pour l'histoire des sciences & des beaux arts [Journal de Trévoux], Sept. 1749, pp. 1853-1872. Review of Histoire naturelle, generale et particulière, avec la description du cabinet du Roi, Tome I., pag. 612. De l'Imprimerie Royale. M.DCC.XLIX.

The Journal de Trévoux Reviews 221

[1] This esteem appeared in Articles XCIX and C of this present volume of our Memoirs. We affirm without reservation that we had not yet read M. de Buffon's discourse when we wrote those two pieces; and we give general notice that the first seven articles of this Journal for September had been printed for almost two months [at that time].

[2] [He gave him 800 talents, that is to say 1,440,000 livres of our money.]

[3] [For Buffon's more extended remarks on this transition, see above, p. 102].

REVIEW OF BUFFON'S HISTOIRE NATURELLE

October, 1749

THE JOURNAL DE TRÉVOUX*

Article CXXV

Translated by John Lyon

    After the discourse which stands at the head of the first volume of this great history, we find in the same volume: 1.) a second discourse containing the history and the theory of the earth; 2) Nineteen articles which serve as proofs of that theory. Thus, our extract will be composed of two pieces. In the first we shall give an account of the discourse, and in the second we shall make several comments on the proofs. All this we shall do in a most cursory fashion, in order to instill in our readers a desire to turn to the work itself.[1]

. . . .

    [M. de Buffon] concludes in this regard that the waters of the sea have covered all parts of the terrestrial continents. But let us be careful to distinguish this opinion from that absurd hypothesis of a self-styled philosopher about whom we spoke several months ago in these Memoires. Telliamed (this is the name of the philosopher)[2] clearly has held that matter is absolutely eternal. He has taught that all men owe their existence to the sea, and that originally we were all marine animals. He claims that the changes which have taken place in the terrestrial globe are capable of being used to show its duration, be it for the ages which have preceded us or for future times. In a word, the opinion of Telliamed comprises a system equally contrary to reason and to revelation. The doctrine which M. de Buffon sets forth here contains nothing of the sort. His observations readily lead one to believe that <u>the earth which serves as our abode has been sea-bottom</u>. But he simply suggests by these words that successive partial inundations have produced certain changes in our globe, and have covered with sea shells and marine productions places quite distant from the sea. He acknowledges the authority of the Sacred Books, the history of Creation, and that of the universal deluge. And he is convinced that this general inundation was, in the hands of God, an instrument of vengeance, a miracle of terror.

. . . .

Let us move on to the nineteen articles which contain the proofs.³ But, since the matter is yet extremely vast, we can only touch lightly on the first five of these articles. The rest shall occupy us in another issue.

II. One should not understand by the word proofs the sort of uninteresting records such as one sometimes finds at the end of grand histories. In this case the proofs are as well-wrought and in most respects more interesting than the history itself. There is in each article a circumstantial account of phenomena which the preceding discourse has marked out in general and in historical fashion. A glance at the list [of articles] can give one an idea of these various objects and of the method according to which they are treated.

But before entering into these matters M. de Buffon speaks of the formation of planets, and, in general, of all celestial physics. This first article ["1. De la formation des Planetes"] presents an hypothesis which the author considers more probable than an infinity of others which have been made concerning the same subject. The foundation of the system is that force which we know under the name of gravity [pesanteur]. It is spread about through all matter. Everything obeys its laws: planets, comets, the Sun, the earth. But beyond that force, which is doubtless the effect of the power and will of the Creator, it is necessary to recognize another force, namely, that of impulsion. For, according to the law of gravity, each planet would fall into the Sun if it were not retained or rather withdrawn at every instant by a force which tends to impel bodies in a straight line, and which, in conjunction with gravity, forms a curved line.

"This force of impulsion," our author says, "had certainly been communicated to the stars in general by the hand of God, when he set the Universe in motion. But since, in physics, one ought to abstain as much as possible from having recourse to causes which are beyond nature, it seems to me that in the solar system one ought to account for this force of impulsion in a sufficiently probable fashion, one which allows one to find a cause the effect of which accords with the rules of mechanics."

Here, one sees, the hypothesis is announced. The author prepares the way for it by a sustained comparison to the movement of comets and planets. He then sets forth his position thus: "Could one not imagine with some sort of probability that a comet, falling on to the surface of the sun, could have displaced that star, and that several parts of it would have

The Journal de Trévoux Reviews

been separated, to which would have been communicated a movement of impulsion in the same direction and by the same impact [dans le meme sens, par un meme choc], so that the planets would have belonged formerly to the body of the sun, and been detached from it by a force of impulsion common to them all, a force which they conserve to this day?"[4]

. . . .

Now, according to our author, it is possible to believe that all this took place in the time about which Moses said that God separated the light from the darkness. About this there still would be two difficulties: 1.) Before the separation of the light from the darkness, following the text of Genesis, God had already created the earth, whereas in our author's system the earth had not commenced to exist until the moment of the separation of light from darkness, when the tail of the comet had already been set in action. 2.) Genesis says that the sun, or the great light, was only created on the fourth day, and the separation of the light from the darkness (times which the hypothesis assigns to the action of the comet) had prevailed since the first day. How, then, could the comet have been able, on this first day to form the earth and the other planets by detaching several parts from the sun, which according to Genesis, had not yet been created? We only see a slightly metaphysical way of replying to these objections. That would be to say in the first place that the earth, being material, would have been created before the separation of light from darkness, though it had only begun to exist as a planet at the time of its separation. In the second place, it is possible to conceive of the sun as having existed from the first day in this sense, that the light was created then. For most interpreters think that the luminous matter formed as a result of these words of the Creator: Let there be light served on the fourth day to compose the body of the sun itself.

But however the matter may be, the entire structure of this hypothesis may be considered as a mental construct [jeu d'ésprit] wherein M. de Buffon has inserted much knowledge and sagacity. It appears at first a bit extraordinary that this academician who is the self-styled enemy of speculative systems and unsupported conjectures, should yet permit himself an opinion such as the one set forth in this first article. However, here is the corrective or the justification which he himself presents to us before explicating his thoughts: "We hope," he says, "to put the reader [in a position wherein he can] more readily specify the great difference that exists between an hypothesis, in which there is nothing but possibilities, and a theory based on facts; between a system such as we are going to give concerning the formation of the earth, & a physical history of its actual state, such as we gave in the preceding discourse."

It thus appears that the system in question is only a pretext designed to set off more brilliantly the merit of a theory separated from all conjectures. So the author easily returns to take up arms against the makers of hypotheses, and Articles II, III, IV, and V are used to combat Whiston, Burnet, and Woodward, whose opinions spare neither reason, physics, nor the authority of the Holy Books. We might note in particular how M. de Buffon concludes by finishing off the fantastic system of Whiston: "Every time that one shall have the temerity to wish to explain by physical reasons the truths of theology, or shall allow oneself to interpret in purely human perspectives the text of Sacred Scripture, or shall wish to reason about the will of the Most High and upon the execution of His Decrees -- just so often will one fall necessarily into darkness and chaos wherein the author of this [Whiston] has fallen."

These words deserve to be written in letters of gold. They conclude this extract, which only contains the hundredth part of that which has occupied us in the great work which we are concerned with analyzing.

NOTES

*Memoires pour l'historie des sciences & des beaux arts [Journal de Trévoux], Oct. 1749, pp. 2226-2245.

1[We omit a simple recapitulation of Buffon's text. However, it is significant to note the distinction which the author of the review makes in the paragraph which follows (in the text above, p. 223) concerning maritime creatures whose remains have been scattered about in inland regions.]

2[Telliamed: the name of a fictitious Indian philosopher in the book by that name composed by Benoit de Maillet (1656-1738), French traveler and amateur natural philosopher. "Telliamed" (de Maillet spelled backward) came to a synonym, as used in the text above, for de Maillet.]

3[A list of the titles of the articles is appended in a footnote in the text, and has been omitted here.]

4[See text above, p. 154. There then follows a description of the plate which graphically depicts God moving a comet into the sun, etc. The plate is given above on p. 150.]

REVIEW OF BUFFON'S HISTOIRE NATURELLE

March, 1750

THE JOURNAL DE TRÉVOUX*

Article XXXII

Translated by John Lyon

    This volume contains the history of animals, that is to say, the most perfect works of the Creator. What wonders, indeed, does one not discover in that part of matter which forms the body of an animal! These wonders escape the notice of inattentive minds. The thoughtful and reflective man examines and admires them; but what appears to be most admirable to the author [Buffon] is the succession, the renewal, and the duration of species among animals. <u>This power of procreation is for us</u>, he says, <u>a mystery whose depth, it appears, is not given to us to measure</u>. He undertakes to measure these depths nevertheless, and employs more than 2/3 of the present volume in doing so.
    We shall not give an account of the ingenious details into which the celebrated Academician enters concerning the subject of generation, and the reproduction of species. These kinds of matters do not allow of being rendered accurately in an extract in which brevity is an essential quality. They please few readers, and may perhaps offend some. It suffices that the public should know that the question of generation is treated in this work with a clarity, a breadth, and an ability which leaves nothing to be desired. We shall only observe that the author does not admit of the system of preexistent germs, germs within germs contained into infinity. He shows that it is necessary to have recourse to <u>an ever-active organic matter, always given over to shaping itself</u>, to making itself, and to producing beings similar to those which house it.
    However that may work itself our in the formation and development of man, we shall consider him at the moment of his birth, and follow through the different ages of his life. Before entering upon the natural history of man, M. de Buffon examines the nature of man. He recognizes that man is composed of two substances: the thinking substance is ourselves, and its existence is demonstrated to us; and extended substance, which is as it were outside ourselves. The latter's existence is not very certain. It even appears doubtful to the author, which one must understand as a doubt improperly so-called, or, what amounts to the same thing, a less intimate acquaintance than that of which the knowledge of the soul is the object.

227

But, finally, what is in no way doubtful is that, in comparing soul with matter, one finds such great differences, and such marked oppositions that it is evident that the soul is of a totally different and infinitely superior order. M. de Buffon sets forth these differences, and concludes on the basis of them that the soul is indivisible and immaterial. He concludes, further, that man is of a very different and very distinguished nature, one so superior to those of beasts that it would be necessary to be as poorly informed as they are about it in order to confound the two.

There speaks and judges a wise and enlightened philosophy. The self disappears, one knows not on what to fasten, when one assumes thought and reasoning to be only the agitation of corpuscles. The same result ensues when the vast idea of the infinite and its proportions, the will and the process of deliberation, are assumed to result from the circular or linear movement [du mouvèment, en cercle ou en quarré] of particles of matter. It is impossible to find the least appearance of reason in such thought processes. The foolhardy author of the Histoire de l'Ame[1] has dared to maintain similar extravagances. But what is perhaps more extravagant than his system is that he does not conclude to the materialism of the soul, but only that it depends on the various organs of the body for its operations. This is an unwarranted conclusion, and this new Doctor of Materialism ought to be ashamed to stammer out such impious lessons. Proceeding with rigorous logic, what ought one to conclude the soul depends on in the body for its functions? One ought to conclude that the soul and the body are two substances closely united and mutually dependent; but can one conclude that body and soul are the same substance, a single substance? That is not obvious. How comes it to be obvious to the author?

But let us press further the demonstration of this truth. Let us suppose that God had wished to make an individual, called man, of two substances, the one substance thinking, the other extended, and that he had established laws in consequence of which the operation of the thinking substance depended upon the movement of the extended substance. The operations of the soul would be explained by chance [par le jeu], and the diversity of organs in the same manner as the historian of the soul has explained them. However, under this supposition, of which we do not think he would dare to deny the possibility, soul and body would not be a single substance, but two substances mutually distinguished from each other. It is thus evident that the work of this new materialist is a continual paralogism and that it ought to lead astray only those persons who do not know how to evaluate the force of an argument and who let themselves be led astray by a grand impudence, a vain display of erudition, and by the deviations of perfervid imagination. But let us return to M. de Buffon and follow him in the curious details which he shows us.

The Journal de Trévoux Review

The second chapter speaks of childhood. To speak of this state of misery, tears, and weakness which immediately follows our birth would appear to be appropriate only to humiliate us. We return with M. de Buffon to this first moment of our existence and, without revulsion, and perhaps with a delicate philosophical pleasure, we see ourselves once more in a cradle. The tears and cries of the newly-born infant result from the pain which the action of the air causes him. The impression of this active fluid set in motion the fibres of this delicate body and thus bring about a sensation of pain. Another effect of the action of the air is to enter the lungs and dilate and distend them. This is the cause of respiration, which begins with birth and ends only with death. M. de Buffon reports some experiments which appear to prove that respiration "is not as absolutely necessary to the new-born infant as it is to the adult, and that it might perhaps be possible, going about it with caution, to prevent in this fashion the foramen ovale [between the auricles of the heart] from closing and develop by this means excellent divers, and species of amphibious animals which could live in air as well as in water."

The infant's eyes are open from birth, but the operation of this organ, as well as others, appears still imperfect. "The senses," M. de Buffon ingeniously notes, "are kinds of instruments of which it is necessary to be aware to make use of them."

The infant who cries and wails at birth only begins to laugh at the end of 40 days. It is true that it is also not until this time that he begins to shed tears, which causes one to suspect that its cries and groans are only, as with animals, a mechanical response, and that the soul only develops and acts at the end of 40 days.[2]

. . . .

M. de Buffon ends this Chapter [V] with a table of probabilities concerning the length of life. This table is ingeniously thought out and exactly put together. "It shows," says the author, "that one can reasonably hope --that is, one can bet even - that an infant who is just born will live to be eight years old; that an infant who has lived to be one year old, or has aged one year, will live 33 years more [vivra encore 33 ans]; that an infant who has completed two years of life will live 38 more; that a man of 20 will live yet 33 years and 5 months more; that a man of 30 will live 28 years more, and so on for all other ages."

NOTES

*Memoires pour l'histoire des sciences & des beaux arts [Journal de Trévoux], March, 1750, pp. 581-604.

[1]The reference is probably to Julien Offray de La Mettrie (1709-1751), French surgeon and materialist philosopher, whose Histoire naturelle de l'ame appeared in 1745.

[2]["Ce qui fait soupçonner que ses cris & ses gémissements ne sont, comme dans les animaux, qu'une impression machinale, & que l'amê ne se developpe, & n'agit qu'au bout de 40 jours." The next sections follow Buffon's observations on cultural relativity in exposing infants to climatic rigors; on child prodigies and education: on puberty (Chapter 3). Here again the reviewer thinks it out of place in a periodical to go into these matters, and congratulates Buffon on his circumspection and finesse in this; the reviewer praises Buffon's style; continues with Chapter 4 (l'age viril) wherein he notes that Buffon thinks it wrong to judge the soul or beauty of character by physiognomic size and shape; moves on to Ch. V., Old Age & Death - adding to Buffon's reflexions on the painlessness of most deaths the advice that a virtuous and Christian life is the best preparation for death; and concludes this (fourth) article as continued.]

10. The *Journal des savants* Reviews (1749) (selected)
John Lyon

The Journal des Savants (Journal des Sçavans) was founded in 1665 by Denis de Sallo, a councilor of the Parlement of Paris. It was the first literary journal in France, and remained the "premier" journal, in another sense of that word, throughout much of the Eighteenth Century.[1] Voltaire called it "the father of all works of its sort."[2] Virtually every French writer of any repute in the Eighteenth Century contributed to it.[3] Its review of Buffon's critical volumes appeared in October, 1749 and March, 1754. The initial review was blandly appreciative, simply recapitulating the argument and structure of the work. The first paragraphs, given here, allow one to become acquainted with the tone of the review, a tone quite like that of the Journal de Trévoux in its irenicism, when contrasted with the polemic stridency of the Nouvelles ecclésiastiques review, which follows it.

NOTES

[1]Eugene Hatin, Histoire politique et littéraire de la presse en France (Paris: Poulet-Malassis et de Broise; 1859) T. 2. 152

[2]Siècle de Louis XIV, cited in Ibid.

[3]Ibid., T.2. 208.

REVIEW OF BUFFON'S HISTOIRE NATURELLE

October, 1749

THE JOURNAL DES SAVANTS*

Translated by John Lyon

    Many learned naturalists have indeed felicitiously treated many parts of natural history, but up to the present no one has had the courage, or, to put it better, none have felt strong enough to embrace this history in its complete extent. Nevertheless, grand as such a design may be, views even more exalted are set forth in the work which we announce today, a work which is in part that of M. de Buffon, and in part that of M. Daubenton. The work does not limit itself to giving us exact descriptions and of ascertaining facts. Above all, this work opens new routes for perfecting the various parts of natural philosophy and teaches us to discover such connections as particular facts might have with the phenomena of Nature.
    The first volume, which is entirely the work of M. de Buffon, serves as an introduction and foundation to the complete work. It begins with a discourse on the manner of studying and expounding natural history. There this illustrious Academician gives several preliminary notions of the methods which have been imagined to try to make natural history intelligible. He shows the advantages and disadvantages of them, and he does this as a man accustomed to see things in their first principles and, consequently, at quite a distance from the ordinary fault of most of the inventors of methods, "which is the desire to judge all by a part, to reduce nature to petty systems which are foreign to her, and to form arbitrarily many loose assemblies of her immense works, and finally, by multiplying names and representations, to render the language of science more difficult than science itself."
    He shows that it is impossible to give a general system, a perfect method, not only for natural history in its entirety, but even for one single branch of it, which he proves in particular for botany. Though the terms are possibly useful, the method involved may have been, for those who have distinguished themselves by it, a sort of philosopher's stone for which they have searched with infinite pains and travail. In doing a critique of most of these methods, he demonstrates their inadequacy. This is, above all, the case with M. Linnaeus' method. But at the same time he admits that none of them are more ingenious or more complete than that of M. de Tournefort.

All that M. de Buffon says on this subject results in the conclusion that the best method that can be chosen is only useful insofar as it leads us to the one thing he recognizes as true, that is, the description and history of each thing in particular.

Coming next to the method of distribution of the various subjects of natural history, he takes up an examination of whatever corresponds to reality in such divisions as have been devised up to the present.

In order to scout out the subject, Buffon advises that it is necessary to rid ourselves of our prejudices, and even deprive ourselves of our ideas, for a moment. He wishes that we could put ourselves in that imaginary situation which a man would be in who should have forgotten everything and who wakes up completely anew to all that surrounds him. He places that man in a field where animals, birds, fishes, plants, and stones successively present themselves to his eyes.

NOTE

*Journal des Sçavans, October, 1749 (pp. 648-657): Review of Buffon's Histoire naturelle générale et particulière, avec la description du cabinet du Roy . . . A Paris, de l'Imprimerie Royale, 1749. See also ibid., March, 1754: review of Tomes 2,3,4,5,6 (Paris: 1749 & 1750). The rest of this review simply recapitulates Buffon's argument and need not be repeated here. There is a second installment of this review (Journal des Sçavans, March, 1754) which, similarly, presents no serious critique. The usefulness of the material here consists largely in the contrast in tone and style which it allows one to see between this review and that in, e.g., the Nouvelles ecclésiastiques.

11. The *Nouvelles ecclésiastiques* Reviews (1750)
John Lyon

The weekly Nouvelles ecclésiastiques first appeared in 1728, and remained for thirty years under the direction of a cure of Touraine, Fontaine de la Roche. It was printed secretly, and lived a clandestine existence until 1803.[1] Condemned by the Parliament of Paris in 1731, this Jansenist, anti-Jesuit periodical circulated surreptitiously through a network of monasteries, circulating apparently among clergy of the second rank, in particular, where a defense of the Gallican Church and an opposition to the Bull Unigenitus were most noticeable.[2] The Nouvelles ecclésiastiques was printed on a single sheet, folded to four pages. "Vast and formless, without literary grace, a jungle of theology almost impossible to explore, it is as forbidding as the grim readers to whom it is addressed," wrote one modern historian.[3] Yet it would be unkind to characterize the present review solely in such terms.

Nevertheless, the strident and contentious note is apparent from the first line of the review's article on Buffon's Histoire naturelle, as is the author's disgust with the time- and place-serving of the Jesuits and their characteristically laudatory review of Buffon. Presumably, the author of the review was Fontaine de la Roche; but the necessarily conspiratorial secrecy which surrounded the composition, printing, and distribution of the Nouvelles ecclésiastiques make the positive ascertainment of authorship most difficult.[4]

NOTES

[1]There had been an earlier series under the same title, 1672-1698, under the editorship of Louis Fouquet, Bishop of Agde. See Louis Trenard, "La Presse Française des origines à 1788," in Claude Bellanger, Jacques Godechot, Pierre Guiral, and Fernand Terron, eds., Histoire generale de la presse Française. (Paris: Presses Universitaires de France; 1969). Tome I, p. 281.

²Ibid; 281-283. See also Charles Ledre, Histoire de la Presse (Paris: Librairie Artheme Fayard; 1958), p. 87. The Bull, issued in 1713 by Clement XI, condemned a series of so-called Jansenist propositions as heretical.

³R. R. Palmer, Catholics and Unbelievers in Eighteenth Century France (Princeton Univ. Press; 1939), p. 27.

⁴Trenard, loc. cit., pp. 281-82. Guenin de Saint-Marc is given as Fontaine de la Roche's successor, and "les freres des Essarts" and the Abbé d'Etemare as his collaborators.

REVIEW OF BUFFON'S HISTOIRE NATURELLE

February 6, 1750

THE NOUVELLES ECCLÉSIASTIQUES*

Translated by John Lyon

The book whose venom we believe ourselves obliged to expose today has as its title: L'Historie naturelle, générale et particulière, avec la description du cabinet du Roi. [Paris. De l'Imprimerie Royale, 1749] This book appears with all the externals which could give it a reputation. Its author is a celebrated Academician. It is dedicated to the King. It is printed at the Louvre. The Journal des Savans [sic] gave it the highest praise. The journalists of Trévoux have an exalted idea of it, and if they found some flaws in it they immediately hastened to cover them up. They carefully brought together places where the author shows some respect for the divine Scriptures; and the joy of finding in the book of a good mind marks of veneration for religion leads them to ascribe to him things which he never said. But is not necessary to be taken in by this. There is a language which an author assumes in case of necessity. A man of position, who writes in a Catholic realm, takes care to speak guardedly. Those who think as he does know how to reconcile this with his position, and regard as a bit out of line Catholic characteristics which the entire system belies. Those who think differently but who study nothing thoroughly, content themselves with these equivocal marks of Catholicity, and pass on without dwelling on the matter. In a time when the spirit of libertinage makes such rapid progress, we ought not allow ourselves to be misled.
"The first truth which issues from this serious examination of nature," says our Academician, "is a truth which perhaps humbles man."

> This truth is that he ought to classify himself with the animals, to whom his whole natural being connects him. The instinct of animals will perhaps appear to man even more certain than his own reason, and their industry more admirable than his arts. Then, examining successively and by order the various objects which compose the universe, and placing himself at the head of all created beings, man will see with astonishment that it is possible to descend by almost imperceptible degrees from the

> most perfect of creatures to the most
> formless matter, from the most perfectly
> formed animal to the most amorphous
> mineral. He will recognize that these
> imperceptible nuances are the great work of
> nature, and will find them not only in the
> size and shape of things, but in changes,
> productions, and successions of every sort.
> (Volume I, page 12)¹

That which is put forth here is not drawn from the Holy Books. Those who have read the three letters against Pope's "Essay on Man" will find here also that pernicious system which was refuted there. Pope puts man in the class of animals, and is content to assign him the first rank. It is agreed that man is reasonable; but he undertakes to show that the instinct of animals is more certain than reason, and their industry more admirable than man's arts. We read of almost imperceptible degrees by which to descend from the most perfect creature to the most formless matter. What does this language mean? What are these imperceptible nuances which are the great work of nature and which lead from the best organized animal to the most crude mineral? Is it by imperceptible nuances that we descend from intelligent beings to corporeal beings? Do man and the brute differ only in the structure and refinement of organs? Is one far from the materialists, when, after having degraded man by allowing him to wonder whether the animals do not have a better position than he, one is content to regard himself as the best organized animal? Can M. de Buffon (the author in question, of whose work we examine only the first volume) ignore the fact that when God created the beings over which man should preside, He only used the fiat; but that when He created man He aroused Himself so as to form a work worthy of Himself? "Let us make man," He said, "in our image and likeness." Words (our best claim to nobility and the best sign of our origin) which distance us from the animals, and draw us near to God! A Christian can forget these?

We are going to transcribe here a rather long passage. But what it contains is so astonishing that none of it can be omitted. It is necessary that the reader be put in a position to judge for himself of the justice of our critique. Furthermore, few persons are able to procure the work of M. de Buffon. "Truth," he says, "that metaphysical entity of which everyone

> believes himself to have a clear idea,
> seems to me to be confounded with such a
> great number of strange objects to which
> its name is applied that I am not at all
> surprised that it is hard to recognize...
> The word truth gives rise to only a
> vague idea: <u>it never has had a precise
> definition</u>. And the definition itself,

taken in a general and absolute sense, is
but an abstraction which exists only by
virtue of some supposition. Instead of
trying to form a definition of truth, let
us rather try to make an enumeration of
make an enumeration of truths. Let us look
closely at what are commonly called truths
and try to form clear ideas of them.

There are many kinds of truth, and
customarily placed in the first order are
those of mathematics, which are, however,
only truths of definition. These definitions are concerned with simple but
abstract suppositions, and all the truths
of this sort are nothing more than the
worked-out and always abstract consequences
of these definitions. We have made the
suppositions, and we have combined them in
all sorts of ways. The body of combinations that results is the science of mathematics. There is, then, no more in that
science than what we have put into it, and
the truths which are drawn from it can only
be different expressions under which the
suppositions which we have used are presented. Thus, the mathematical truths are
only the exact repetitions of definitions
or suppositions. The last consequence is
true only because it is identical with that
which preceded it, and this latter in its
turn with its antecedent. Thus one may
proceed backward right to the first presupposition. And since definitions are the
sole principles upon which everything is
established, and since they are arbitrary
and relative, all the consequences which
can be deduced from them are equally arbitrary and relative. Hence, that which we
call mathematical truth is thus reduced to
the identity of ideas, and HAS NOTHING OF
THE REAL ABOUT IT. We make suppositions,
we reason on the basis of our suppositions,
we draw the consequences of them, we come
to conclusions. The conclusion, or the
last consequence, is a proposition which is
true in proportion as our supposition was
true. But the truth of this proposition
cannot exceed that of the supposition itself. This is hardly the place to discourse on the methods of the science of
mathematics, or on the abuse of such
methods. It is sufficient for our purposes

to have proved (has that been done?) that mathematical truths are only truths of definition or, if you prefer, different expressions of the same thing, and that they are only truths <u>in relation to</u> the very definitions with which we started. It is <u>for this reason</u> that they have the advantage of always being precise and conclusive, but abstract, intellectual, and <u>arbitrary</u>.

Physical truths, on the other hand, are in no way arbitrary, and in no way depend on us. Instead of being founded on suppositions which we have made, they depend only on facts. A sequence of similar facts or, if you prefer, a frequent repetition and an uninterrupted succession of the same occurrences constitute the essence of this sort of truth. What is called physical truth is thus <u>ONLY A PROBABILITY</u>, but a probability so great that it is equivalent to certitude. In mathematics, one supposes. In the physical sciences, one sets down a claim and establishes it. There, one has definitions; here, there are facts. One goes from definition to definition in the abstract sciences, but one proceeds from observation to observation in the real sciences. In the first case one arrives at evidence, while in the latter the result is certitude. The word "truth" is used for both, and consequently corresponds to two different ideas. Its signification is vague and complicated, and it thus has not been <u>possible</u> to define the term in a general way. It has been necessary, as we have just seen, to distinguish the kinds of truth in order to form a clear idea of it.

I shall not speak of other orders of truths: those of the moral order, for example, which are in part real and in part arbitrary, would demand a lengthy discussion which would take us away from our goal, and that more especially because they have as object and end <u>ONLY DECORUM AND PROBABILITIES</u>.

Mathematical evidence and physical certitude are thus the only two aspects under which we ought to consider truth. As soon as it withdraws from one or another of these, it is no more than appearance and probability.[2].

These principles posed, the author examines what can be known of science with evidence or certitude, and he says:

> We know, or we can know, in evident science, all the characteristics or, rather, all the relationships of numbers, lines, surfaces and of all the other abstract quantities... Since we are the creators of this sort of knowledge, and since it takes under consideration <u>absolutely nothing</u> except what we ourselves have already <u>IMAGINED</u>, it is impossible to have therein <u>EITHER OBSCURITIES OR PARADOXES WHICH MAY BE ACTUAL OR IMPOSSIBLE OF RESOLUTION</u>.[3].

Either the terms signify nothing, or M. de Buffon is a perfect Pyrrhonian. To say of Truth that it is a metaphysical entity or, if one will, a being of reason; to say that the word "truth" brings to birth only vague notions and that it has never had any precise definition; that the definition of truth is but an abstraction, which only exists by virtue of some supposition; that mathematical truths have no reality; that they are only truths relative to definitions which are made of them, these definitions in turn being arbitrary; that we are the creators of mathematical truths; that the science of mathematics contains absolutely nothing that we have not created in imagination, and that it is for this reason that it does not contain either obscurities or paradoxes; to make the essence of physical truths to consist in the uninterrupted succession of the same events; to add that what is called physical truth is but probability; and, finally, to dare to teach that moral truths have as their object and end only convention and probability: Is this not to introduce Pyrrhonism into religion and all the human sciences? If mathematical truths are only arbitrary truths and suppositions; if the truths of natural philosophy, which rest upon facts, only result in probabilities; if moral truths have as their object only matters of decorum; then there is an end of certitude. We have then no more principles which are capable of guiding us as we work toward a knowledge of that which it is most important to know. We shall not be able to demonstrate the existence of God. The observation of creatures may no longer lead us in an infallible manner to the knowledge of the Creator. From the first Being to the last, all is surrendered to incertitude. If I say that the whole is greater than a part; or that the whole is equal to all its parts taken together; or that quantities equal to the same quantity are equal to each other; or that, if to equal quantities the same quantity is added they remain equal; or if

I say that two plus two equals four; or that four multiplied by itself gives sixteen; or that thus the number four is found four times in sixteen; all that, according to M. de Buffon, has nothing actual about it. These are but arbitrary truths. M. de Buffon nevertheless allows for evidence in mathematics, and for certitude in natural philosophy. But one must be wary here, for the evidence is not absolute but only <u>relative</u>. You say: "The whole is greater than a part of it." You draw from this principle an appropriate consequence. The consequence will be evident, according to M. de Buffon; but the principle being arbitrary, the consequence can have no more reality than the principle. Who would believe that a man of intellect seriously pronounced parallel extravagances? But God always takes pleasure in convicting them of folly who would be wise without Him. It is to add ingratitude and impiety to folly to reduce moral truths to matters of decorum and probability. What? To love God with all your heart, with all your soul, and all its powers, and your neighbor as yourself - that is not an essential obligation? Not to commit perjury, theft, adultery, homicide, - all that is only proper manners and probability? Believe in God, when one deliberately admits the principles which directly result in these consequences? If the first principles of the law of nature only have as their object and purpose decorum and probabilities, there is no more obligation for man to worship God and render homage to Him; there is no more obligation for subjects to submit to their princes. That which is only a matter of decorum is not at all a strict and indispensible obligation. These principles, as we have seen, tend toward the destruction of all religion and all states. We do not pause to make known the contradictions which prevail in the extract which we come to relate. But we are not able to refrain from pointing out that which M. de Buffon said in speaking of mathematical truths, namely, that the reason why the science which comprehends them contains neither impossible obscurities nor real paradoxes is that we are the <u>creators</u> of this science, and that it takes in absolutely nothing but that which we ourselves have imagined. Doesn't that teach man to regard as <u>real paradoxes</u> and <u>impossible things</u> the mysteries of religion, which certainly are not the inventions of man?

M. de Buffon next undertakes to explain the manner in which the earth was formed. He conjectures that the earth owes its birth to the passage of a comet which fell on the sun, shifted it, and caused our globe and other planets to come forth out of this star. M. de Buffon says that the earth, in leaving the sun, was in a state of liquefaction. It was as a torrent of fire, which little by little cooled down for lack of combustible materials: "<u>so the sun will die out probably for the same reason, but in some future age, and proportionately as far from the times that the earth and the other planets become extinct as its greatness exceeds that of the planets.</u>" <u>These are the</u> very words of our Academician [....] How is it possible to reconcile this hypothesis with the account of Moses? It would be tried in vain. What is singular here is that M. de Buffon puts Holy Scripture over against several Englishmen who have written on this matter. He says of one of them in particular (i.e. Whiston) that every time that one has the temerity to wish to explain theological truths by physical causes or allows oneself to explain in purely human terms the divine text of the Holy Books, and wishes to reason on the will of the Most High and upon the execution of His decrees, then one necessarily falls into the darkness and chaos where Whiston fell [....] The latter, however, is not as opposed to the Sacred Books as is his censor. M. de Buffon reproves Whiston with having said that the earth shall perish by fire; that the destruction will be preceded by dreadful tremors, thunder and fearful meteors; that the sun and the moon will take on hideous aspects; that the heavens will appear to collapse; and that fire will be general upon the earth [....] M. de Buffon calls these <u>reckless, not to say extravagant, assertions</u>. At least (he adds) that is what they seem to be at first glance. But here Whiston only echoes the Apostle, St. Peter, who says that the heavens and the earth are at present reserved for a future conflagration, and that in the noise of a terrible tempest the heavens shall pass away, the elements dissolve, and the earth be burnt with all that it contains.[5] So far as the tremors in the earth are concerned, and the darkening of the sun and moon which are to precede the last judgment, it is Jesus Christ himself who said this: and who is M. de Buffon that he dare contradict Him?

God declared to Noah as he left the Ark that he would send no more deluges upon the earth. M. de Buffon's ideas are quite different. If one gives credence to them, the earth we now inhabit was for a long time at the bottom of the sea. "It is," (he says) "the waters gathered together in the vast extent of the seas which, by the continuous movement of flux and reflux, have produced the mountains, valleys, and other irregularities on the surface of the earth. It is currents in the sea which have hallowed out the glens and raised up the hills, while giving them their corresponding directions; it is these same waters of the sea which, by carrying the earth as sediment, have deposited it here and there in horizontal beds, while it is the waters of the heavens which have, little by little, destroyed the work of the sea, continually wearing away the heights of the mountains, filling in the valleys, the mouths of rivers, and gulfs, bringing everything back to level once more. These forces shall one day give back the earth to the sea, which shall repossess the land bit by bit, while bringing to light new continents intersected by glens and mountains and completely like those which we inhabit today."

Thus we have a world far older than Moses made it out to be. Who shall tell us even when it began? How many centuries were necessary in order that the flux and reflux of the waters should form the mountains which are on the earth? But whereas the sea covered all the earth which is inhabited today, the vast extent which the sea presently occupies will come to be dry and filled with mountains, which must in turn be worn away by rain, until the entire surface of this former continent being levelled, the sea may have taken it and given ours once more to discovery. Those who make the world eternal, and who see only a continual recurrence of the same events, do they think differently from M. de Buffon? However that may be, this Academician quite clearly contradicts the account of Moses which says that God, on the third day of creation, brought together in one place the waters which were under the heavens to form the sea, and caused to be brought to the land all the plants and trees which are its ornament. If the sea should cover the land once more, what will become of the promise God made not to inundate it again as it had been at the time of the Deluge? What will become of the promise of the Apostle Peter which assures us that the earth which we inhabit must perish by fire? Will M. de Buffon say to us what he says to Whiston, that we reason about this like disputatious theologians rather than like enlightened philosophers? As if the philosopher, in order to avoid bewilderment, need not take Revelation as his guide! Who is the philosopher who has reasoned justly in the absence of the light of faith? However, our Academician is so enamoured of his system that he thinks it impossible to doubt it. He says:

> I have often examined quarries, the stones of which were full of shells. I have seen whole hills composed of shells, and chains of rocks intermixed with shells through their whole extent. The quantity of shells, and other productions of the sea, is, in many places, so prodigious, that it is difficult to believe any more of them existed in their natural element. It is from this enormous quantity that no doubt remains of the earth's having continued for a very long time under the waters of the sea. The number of sea shells found in a fossil or petrified state is so amazing, that, were it not for this circumstance, we never should have had a proper idea of the surprising quantities of those animals to which the ocean gives birth. We must not, therefore, imagine, like those who talk and reason concerning things they never saw, that shells are only to be found scattered here and there by chance, or in small heaps, like those of oysters thrown from our doors. They appear, on the contrary, in masses like mountains, in banks of 100 or 200 leagues in length. They may often be traced

through whole provinces, and in masses 50 or 60 feet thick. It is only after having learned these facts that a man is entitled to reason on this subject.[6]

What M. de Buffon has just said he repeats in another place further on (p. 581):

> It was no sooner suspected that our continent might formerly have been the bottom of the sea, than the fact became incontestible. The spoils of the ocean found in every place, the horizontal position of the strata, and the corresponding angles of the hills and mountains, appeared to be convincing proofs; for, when we examine the plains, the valleys, and the hills, it is apparent, that the surface of the earth has been figured by the waters. When we descend into the bowels of the earth, it is equally evident, that those stones which include sea shells have been formed by sediments deposited by the waters, since the sea shells themselves are impregnated with the same matter that surrounds them. And in fine, if we consider the corresponding angles of the hills and mountains, we cannot hestitate in pronouncing, that they received their configuration and direction from currents of the ocean. It is true, that, since the earth was first left uncovered with water, the original figure of its surface has been gradually changing: the mountains have diminished in height; the plains have been elevated; the angles of the hills have become more obtuse.... But every thing has remained esentially the same. The ancient form is still recognisable; and I am persuaded that every man may be convinced, by his own eyes, of the truth of all that has been advanced on this subject; and that, whoever has attended to the proofs I have given, must be fully satisfied, that the earth was formerly under the waters of the ocean, and that the surface, which we now behold, received its configuration from the currents and movements of the sea.[7]

We shall conclude this article in the next issue.

REVIEW OF BUFFON'S HISTOIRE NATURELLE

February 13, 1750

THE NOUVELLES ECCLÉSIASTIQUES*

Translated by John Lyon

In order to give authority to this system which criticizes the Mosaic narrative, M. de Buffon calls in witness M. de Fontenelle who, in his History of the Academy of Sciences (1716, p. 14) says: "It has been well proved that all rocks once were a soft paste, and as there are quarries almost everywhere, the surface of the earth has thus been in all these places, at least to a certain depth, a vessel of mud. The seashells which are found in almost every quarry prove that this receptacle [the earth] was beneath the sea, and that, consequently, the sea covered all these places. It could not of course cover them without covering all places level with them or lower than they are. And it could not do this without covering the entire surface of the globe [to such a depth]... (sic). The sea, therefore covered all the earth, and from this it follows that all the banks and beds of stone which are on the plains are horizontal and parallel to each other. Fishes, then, will have been the oldest inhabitants of the globe, which could not yet support either terrestrial animals or birds."
The creation of the fishes only preceded that of terrestrial animals by one day. But birds were created the same day as the fishes. Here, then, Holy Scripture is contradicted, and what Moses said is counted for nothing. However, M. de Buffon, who cites Fontenelle in his favor for placing the earth at the bottom of the sea, reprimands Leibniz for having thought the same thing. "To say (as Leibniz does) that the sea once covered the whole earth, that it enveloped the entire globe, and that it is for this reason that one finds sea shells everywhere - this is [says M. de Buffon] to pay no attention to a most essential thing."

> The instantaneous creation of the world destroys the notion of the globe's being covered with the ocean, and of that being the reason why sea shells are so much diffused through different parts of the earth; for, if that had been the case, it must of necessity be allowed, that shells, and other productions of the ocean, which are still found in the bowels of the earth, were created long prior to man, and other

> land animals. Now, independent of scripture authority, is it not reasonable to think that the origin of all kinds of animals and vegetables is equally ancient?[8]

M. de Buffon seems to condemn in Leibniz what he approves in Fontenelle. He appears to recognize the authority of the Sacred Books, whereas he expressly contradicts them. What are we to make of this double-talk?

Catholics, Protestants, the very Fathers of the Church such as Tertullian, have thought of all these seashells and other marine productions which are found in mountains and the interior of the earth, as proofs of the universal Deluge. M. de Buffon opposes this opinion with all his strength.

> The notion that the shells were transported and left upon the land by the deluge, is the general opinion, or rather superstition, of naturalists. Woodward, Scheutzer, and others, call petrified shells the remains of the deluge; they regard them as medals or monuments left us by God of this dreadful catastrophe, that the memorial of it might never be effaced among men. Lastly, they have embraced this hypothesis with so blind a veneration, that their only anxiety is to reconcile it with holy writ; and, in place of deriving any light from observation and experience, they wrap themselves up in the dark clouds of physical theology, the obscurity and littleness of which derogate from the simplicity and dignity of religion, and present to the sceptic nothing but a ridiculous medley of human conceits and divine truths.[9]

But which is the more religious, the view which sees nowhere on earth marks of the Deluge, or that which sees them everywhere? Which of the two tips the scale toward unbelief? When the incredulous shall have been released from the mute testimony given to Scripture by so many fishes, seashells, and petrified marine plants; when they shall have been persuaded that all these productions owe their birth to a time when the earth was beneath the sea, although this system be diametrically opposed to the mosaic account: How, then, will M. de Buffon begin to lead these unbelievers back to the salutory yoke of faith? Our Academician thinks it useless either to guess what the results of the Deluge have been or to add facts to the account in Holy Scripture. He asks if it is said in Scripture that the waters were agitated enough to raise up seashells from the bottom of the sea and scatter them all over the earth. He replies: No. The Ark sailed tranquilly upon the waves (p. 203). But doesn't Scripture say that God caused a wind to arise on the earth, and that the waters being agitated from side to side, they retired, and began to diminish after 150 days? During this agitation would it have been so hard for the fishes and seashells to change their habitation? If God preserves in the earth until this day the marks of his wrath against the abominations of Sodom and Gomorrah, is it astonishing that He has willed likewise to leave us some imperishable traces of the Deluge to serve for the instruction of all men, to make them behold in fear the terrible effects of His justice?

M. de Buffon reproaches his adversaries with having attributed to the Deluge the formation of mountains by the changes which the waters made on the surface of the earth. There is reason to maintain that the mountains existed before the Deluge. It appears in various places in Scripture, and in particular in Psalm 103, that the mountains existed from the beginning. They were not formed bit by bit by the flux and reflux of the sea and from sediment in the waters, as M. de Buffon falsely claims. Rather, they were like the earth, created from the beginning. "[Thou] hast," David says,

> Founded the earth upon its own bases; it shall
>   not be moved for ever and ever.
> The deep like a garment is its clothing: about
>   the mountains shall the waters stand.
> At thy rebuke they shall flee: at the voice
>   of thy thunder they shall fear.
> The mountains ascend, and the plains descend
>   into the place which thou hast founded for them.
> Thou hast set a bound which they shall not
>   pass over: neither shall they return to cover
>   the earth.[10]

The mountains ascend and the plains descend into the place which Thou hast founded for them. It is possible to set this text and others like it against those in which the mountains are posterior to the Deluge. As to M. de Buffon, it is necessary to refer him to what God said to Job:

> Who is this that wrappeth up sentences in unskillful
>   words?
> Gird up thy loins like a man. I will ask thee, and
>   answer thou me.
> Where wast thou when I laid the foundations of the
>   earth? Tell me of thou hast understanding.
> Who hath laid the measures thereof, if thou
>   knowest? Or who hath stretched the line upon
>   it?
> Upon what are its bases grounded?....

> Who shut up the seas with doors, when it broke
> forth as issuing out of the womb:
> When I made a cloud the garment thereof, and wrapped
> it in a mist as in swaddling bands?
> I set bounds around it, and made it bars and doors.
> And I said: Hitherto thou shalt come, and shalt
> go no further. And here thou shalt break thy
> swelling waves....
> Where is the way where light dwelleth; and
> where is the place of darkness?....
> Shalt thou be able to join together the shining
> stars of the Pleiades, or canst thous stop the
> turning about of Arcturus?
> Canst thou bring forth the day star in its time,
> and make the evening star to rise upon the
> children of the earth?[11]

Job replied to the Lord:

> Who is this that hideth counsel without knowledge?
> Therefore I have spoken unwisely, and things
> that above measure exceeded my knowledge....
> With the hearing of the ear, I have heard thee: but
> now my eye seeth thee.
> Therefore I reprehend myself, and do penance in
> dust and ashes.[12]

How happy M. de Buffon would be if he took the holy Job as a model? If he does not do so, can one without blame leave uncriticized a work as pernicious as his? Beyond the injury which this book does to God, it dishonors the name of the king, to whom it is dedicated. That realm is discredited in which works which destroy religion are allowed to be printed and sold with official knowledge. The evil is more extensive than one thinks. The extract from the <u>History of the Academy</u>, which we have cited after M. de Buffon, shows that since 1716 there have been set forth, under the name of the Academy, systems which derogate from the truth and authority of the Sacred Books. The author of the <u>Journal des Savans</u>, review who knew about M. de Buffon's work beforehand, isn't he also suspect of endorsing the irreligious system of this Academician? We say it without hesitation. There creeps about the Academies a spirit which ought to alarm those who have religion. Pope's "Essay on Man" continues to be sold under privilege from the Academy of Belles-Lettres. M. l'Abbé de Resnel, author of the versification in our language, was received by the French Academy, without anyone demanding of him any retraction. Voltaire himself, yes, Voltaire, has also become a member of that Academy. To have given its suffrage to him is to become almost as guilty as he. It will be said that the author of NN. [sic] [Eccl." ?][13]. does not understand his own interests, in taking so little caution with the men who set the tone of society. But the author of NN [sic] may not be the servant of Jesus Christ if he attempts to please men. (Should I seek then to please man? If I should seek to please men, I would not be a servant of Christ.)[14] The author of NN [sic] esteems the talents involved; but when such talents are used against religion - is it possible to see this, and say nothing? Those whom we attack are men of intellect, we know. But we also know that God detests the pride of mind of those who use the intellect against Him.

In conclusion, let us refer these persons, and M. de Buffon in particular, to a man whom they do not apparently regard as a man of little ingenuity, nor as a superstitious intellect. The great Bossuet (after having said in the second volume of the Élévations) that the Magi were the savants of their time, observers of the heavens, to whom God made himself known, and who had renounced the cult of their land, adds:

> "this is what the high sciences ought to lead to. Philosophers of our day, of whatsoever rank you may be, be it observers of the stars, or contemplators of nature below, devotees of that which is called natural philosophy, or busied with the abstract sciences which are called mathematics, where truth appears to preside more often than in the others: I do not wish to say that the objects of your study are ignoble, for from truth to truth you can proceed even to God, who is the truth of truths, the source the truth, Truth itself, in whom subsist those truths which you call eternal, truths immutable and invariable, truths which cannot but be truths. Those who open their eyes see these in themselves, and nevertheless above themselves, for they rule over their reasoning and preside in the consciousness of all who see and hear, be they men or angels. It is this truth that you ought to search for in your sciences. Therefore, cultivate these sciences, but do not be absorbed by them. Do not presume, and do not believe yourselves to be something more than others because you know the properties and the reasons of things great and small. Such sciences are the vain ornament of curious and weak minds, which after all leads to nothing that is real and has nothing substantial of its own, except insofar as through the love of truth and the habit of ascertaining it in certain objects, practitioners of these sciences are led to

search for true and useful certitude in God alone!

NOTES

*Nouvelles ecclésiastiques, ou memoires, pour servir a l'histoire de la constitution unigenitus, Pour L'annee M.DCC.L. Feb. 6, 1750, pp. 21-24, and February 13, 1750, pp. 25-27.

[1][See "The 'Initial Discourse', translated by John Lyon, above p. 102.]

[2][Ibid., pp. 122-124. Passages underlined or capitalized are in italics or capitals in the review. (The reviewer is working from the first French edition; the translator from the last edition printed during Buffon's lifetime.)]

[3][Ibid., p. 124.]

[4][For an extended and more temperate commentary on many of the issues raised by the reviewer, see Lamoignon-Malesherbes' Observations, below, pp. 297-314.]

[5][See 2 Peter, Ch. 3.]

[6][Buffon, "Preuves de la theorie de la terre," Art. viii, Histoire naturelle Vol. I (1749), as translated in Buffon, Natural History, General and Particular, translated with notes and observations by William Smellie, in twenty volumes (London: Cadell and Davies, 1812), I, 208-9.]

[7][Ibid., II, 176-77.]

[8][Ibid., I, 126.]

[9][Ibid., I, 131.]

[10][Psalm 103, verses 5-9 (Douay-Rheims ed.).]

[11]["Job," Ch. 38, verses 2-6, 8-11, 19, 31-32.]

[12]["Job," Ch. 42, verses 3,5-6.]

[13][The context suggests that the author of the review in the Nouvelles ecclésiastiques is referring to himself.]

[14]An quaro hominibus placere? Si adhuc hominibus placerem, Christi servus non essem.

12. The *Bibliothèque raisonée* Reviews (1750–51)
John Lyon

The Bibliothèque raisonnée des ouvrages des savants de l'Europe appeared in 1728, the same year as the Nouvelles ecclésiastiques. It was founded in Amsterdam by Armand de La Chapelle, and came to be edited by Pierre Maussuet, a former Benedictine from the Ardennes.[1] Among the founders and collaborators were Jean Barbeyrac, Pierre Desmaizeaux, and [Willem Jakob] s'Gravesande. Altogether the journal went through 52 volumes between 1728 and 1753.[2]

In the interests of impartiality and competency, the work of the review was parcelled out to contributors who remained anonymous and unknown even to each other, each working on materials particularly appropriate to his professional interests and capabilities.[3] The review pledged neither to assert anything nor to censure anything in the absence of knowledge and reason, to allow authors to reply freely to their critics, to respect the rights of morals, religion, and the laws, and to be a genuinely European, as opposed to a national, journal.[4]

The two reviews included here proceed in a measured fashion to take the dimensions of Buffon's first and second volumes, the first focusing in particular on the taxonomic and geological features of Buffon's work and the second analyzing his theory of generation. They provide a more detailed critique of the latter than any other early review with which we are familiar, except for the critiques by Haller translated below.

NOTES

[1] Louis Trenard, "La Presse Française des origines à 1788," in Claude Bellanger, Jacques Godechot, Pierre Guiral, and Fernand Terrou, eds., Historie générale de la presse française (Paris: Presses Universitaires de France; 1969), Tome I, 292. See also Eugene Hatin, Histoire politique et litteraire de la

presse en France (Paris: Poulet-Malassis et De Broise; 1859), Volume 2., pp. 303-306, on which Trenard's coverage is based.

   ²Ibid.

   ³Hatin, T.2, 303-04.

   ⁴Ibid. p, 304.

REVIEW OF BUFFON'S HISTOIRE NATURELLE

October-December, 1750

THE BIBLIOTHÈQUE RAISONNÉE*

Translated by John Lyon

This volume, which is the first of a vast work of which three volumes are actually printed and of which the fourth is promised imminently is a bit different than its title suggests. It has no connection with the Cabinet du Roi, and is instead on the formation of the earth. It is the work of M. de Buffon, and on many counts it deserves to be introduced to the reader.

The "Initial Discourse" is concerned with the manner of studying and expounding natural history. It is not on the basis of the "Discourse" " that the entire work should be judged, for much would be lost thereby. I do not know if it is passion or prejudice, or both, which has driven our author to maltreat as he has Mr. Linnaeus and authors of [taxonomic] systems. Indeed, we hold with him that a method based on a number of characteristics of classified entities is always infinitely superior to others which set apart and bracket together entities by means of a single characteristic, whatever it might be. Furthermore, it is true that Aldrovandi and other compilers of collections have gone to excessive lengths in their compilations. They have gathered together in their compilations both that which is worth the trouble of conserving and that which is not. But as to the critique which Mr. (sic) de Buffon performs on the new denominations of Mr. Linnaeus, I must be allowed to dissent from its harshness. He completely rejects the use of characteristics derived from the feet of animals, and holds that it is better to classify the horse, which is soliped, with the dog, which is fissiped, since in our experience the dog trots after the horse. In a word, Mr. (sic) de Buffon absolutely rejects classes based on feet, teeth, or mammary glands, such as Ray sketched out and Mr. Klein retained. Buffon disapproves of scientific names which are made up of the genus and the specific difference. That an ass should be called a horse (<u>Equus cauda strigosa</u>), [horse with a thin tail] for example, is not to his liking.

I am surprised, as much as one can be, by this critique. Taken rigorously, it leads to no less than a relapse of all natural history into the confusion from which the industry of modern men has drawn it. We have pointed out elsewhere [T.XXXIV, p. 11, article "Klein"] that the worst method is preferable to absolute confusion, and have drawn together there, further, several instances of this.

Mr. de Buffon wishes that the name "ass" be retained, which is different from that of the "horse." In the natural system of grammar [la caracteristique naturelle], which ought to have guided all the languages of the world, similar words ought to denote similar beings, while different names denote different things. The public lacked nothing when it instructed itself in these resemblances and differences. It calls tulips all that vast assemblage of flowers which abound daily, and all the dogs of the world are dogs in French, despite their accidental differences. But it has happened that the public, prepossessed by whatever mark of distinction strikes the imagination, has separated by different names beings which are in fact of the same kind. This happens in particular when the animals in question are domestic ones. Thus it happens that the ox and the cow are distinguished by different names, and calves likewise from each of these. This is scarcely a philosophical use of language, and is internally inconsistent even in general usage, for the public has no separate word for female pigeons nor for their young, and it generally lumps together under the same name both sexes and all ages of almost all animals.

It is for this same reason that people separate the ass from the horse. As a result of the slowness of the former, and the contempt people have for it, they have been reluctant to classify it with the horse, because of the swiftness and exalted character of the latter. But it is not proper for a philosopher to follow the prejudices of the vulgar. It is necessary that one choose: either there is no system whatsoever, and as many names must be given as there are individual bodies; or there is a system of genera, and natural classes, which God created, and in which the eye of the most vulgar can detect the common character. Such is the genus of the horse, whose unique feet, croup, head, and teeth are such evident and recognizable characteristics that the least educated explorers have recognized the zebra of the Cape of Good Hope as belonging to this genus. But if animals are the same within a genus, is it not more philosophical to give them the same generic name, and to further distinguish them more particularly by means of characteristics drawn from their nature itself? Wouldn't it be a supreme blessing for all natural history if we were to have for all beings a nomenclature as rational as Equus cauda promissa and Equus cauda strigosa - names which Aristotle, whom Mr. de Buffon admires, would have himself admired, because they express the genus and the specific difference, and yield a definition?

That which appears to have prejudiced Mr. de Buffon is the fact that the ass is a common, plebian, animal, an animal with which all are familiar from our youth, an animal which has no need for a philosophical name in order to be recognized. But this is not the case as soon as we are concerned with strange or little known animals or plants. Seeing that the straightforward similarity of names is of use in signifying to us that the animals thus named are similar, and that the detailed description completes the clarification, Mr. Linnaeus calls the Coatimundi; <u>Ursus cauda promissa</u>. This name arouses in us an idea which the barbarous name that I have just cited would never arouse. It notifies us that the animal in question should look like a bear, that it has feet, teeth, and mammary glands, if you will, like bears, and that it is different from the bears by its long tail. Mr. Linnaeus' nomenclature gives us a description, while the plebian name which Mr. de Buffon defends is only a sound, giving rise to no idea whatsoever.

This truth is even more apparent as soon as one is concerned with plants, for the genera of plants have infinitely more species than do the animal genera. It would be necessary to abandon botany, which would have succumbed under the weight of the 20,000 different names which it would have been necessary to retain, without the methodical arrangement which has come to the aid of humankind in the century in which we have the greatest need of it. One need learn no more than 800 names; and, moreover, these names inscribe themselves in the memory by the classical characters which join them together. I would recognize a new plant as a Veronica anywhere in the universe, as soon as its characteristic is given. And once I know its genus, a quarter-hour's work will lead me to the species, if it is known, or would convince me that my specimen is of a new species. That which would have been an immense labor for Bauhin is only an amusement for me.

It also appears that the imperfections of systems have offended Mr. de Buffon. All classes are not yet based on natural distinctions, although there are many which appear to be so. This is a defect; but artificial classes take nothing away from the value of natural ones, and ought to be tolerated, because they fill out a system, and they are always worth infinitely more than absolute disorder.

All the praises of the Ancients that Mr. de Buffon sings with such emphasis will never disenchant me with the moderns. The Ancients knew neither number nor measure. Their style of life removed them from Nature. They lacked instruments, illustrations and patience. Their descriptions of plants are next to useless, for they indicate nothing, and they did not even go so far as to perceive the rudiments of systematization. The color and shape of flowers; a slightly longer or shorter life span, a stem a bit higher or lower, were as essential in their eyes as the petals or fruits of a plant. Aristotle had a more exact and philosophical genius than his predecessors. His treatises on animals have much of value in them, and, furthermore, much that is beautiful. But would it be appropriate to compare his scraps of imperfect dissections to our designs, our descriptions, our anatomies?

Let us pass on to the complaints which Mr. de Buffon makes in an abusive manner on the methods of mathematics and calculus. This latter is the perfection of human knowledge, and often apprises us of truths which subsequent labor only belatedly and slowly confirms. Newton sensed the true figure of the earth, Mr. Leibnitz foresaw the existence of polyps; ought we hold in contempt those luminaries who inform us of the sight of things yet unseen?

The <u>Theory</u> <u>of</u> <u>the</u> <u>Earth</u> <u>and</u> <u>its</u> <u>History</u> (sic) provide the material of the second <u>memoir.</u> Let us pursue Mr. de Buffon here. He adopts an old system which has become fashionable once more in our day. Mr. Leibnitz proposed it in his <u>Protogée</u> (sic), which we reviewed for the reader in Part I of Tome XLIII of this [Journal]. Linnaeus and Celsius have followed him. Mr. de Maillet gave us a travesty of it in a sort of philosophical novel, and Mr. Le Cat opposes its embellishments by Buffon himself. The earth, after having been fused through the power of fire, cast off vapors. These vapors condensed, fell back on the earth, and created a universal sea which covered the entire surface of it. The flux and reflux of the waters has been the principal agent which drew forth our globe from this state. As the flux and reflux acted with most force beneath the equator, earth and silt slowly built up there the oldest and first of the continents. The reflux did not carry away that which the flux brought in, since we see banks elevated in the same manner every day on our coasts without being destroyed in the process. Entire centuries' have piled up new banks of earth, forming mountains and glens. The inclinations and declinations of these mountains and glens correspond with each other, for the force which created a glen on one side also made a mountain opposite it. This observation, which is that of Mr. Bourguet, holds good for the whole earth, and I have not seen it contradicted. But Mr. de Buffon offers another condition here which does not seem to be equally incontestable. He would have it that the mountains which form the opposite walls of valleys should be almost equal. The highest mountains then will be those under the Equator: the Andes in the new world and those of Africa in the old. They were formed first and have been longer a-building than the others, and the flux and reflux has worked on them more significantly. Other, lower mountains are the work of wind, water currents, and other irregular movements. Such is the system of Mr. de Buffon.

The movement of the sea, which is generally from East to West, began to dry out the land, sweep out the sea, and form the eastern continent of Asia from a mass of silt and earth which should thus be the oldest of the continents. Europe, and the lands about the Poles, were the last to be formed. If this process formed solid land, it also formed the seas at the cost of the original sea.

The ocean broke into valleys, which had been separated from it only by low banks, which time and earthquakes had

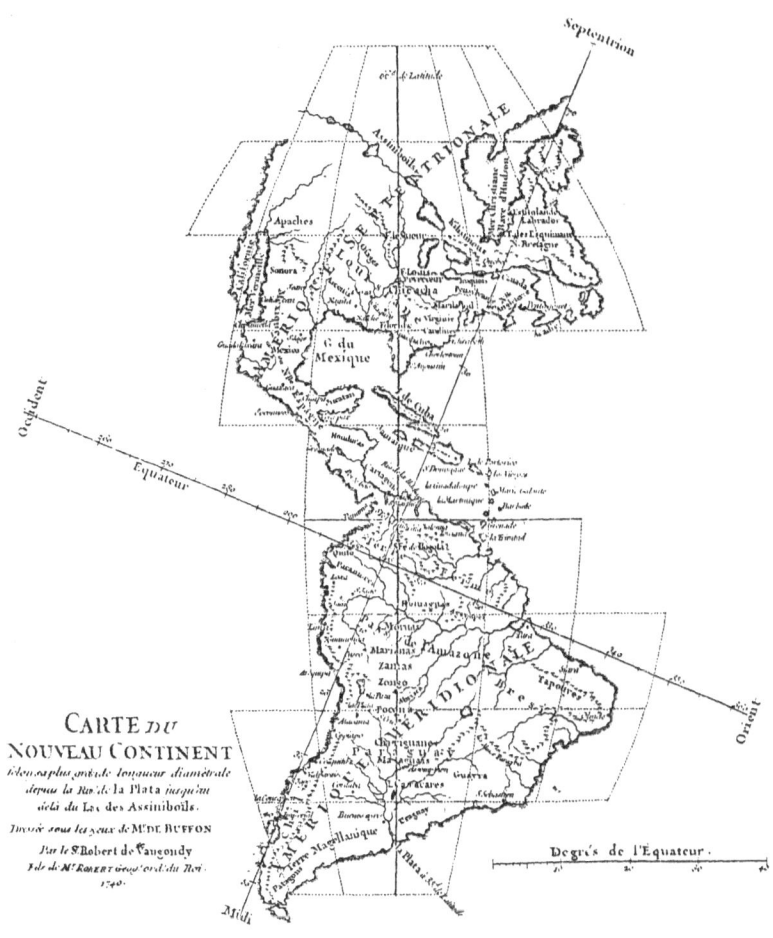

Figure 16 - Map of the New World from first edition of *Histoire naturelle*. Uncertainty over the orientation of the New World and the geography of the North Pacific results in the common appellation of the Atlantic as the "North Sea," and the Pacific as the "South Sea."

managed to wear down. It flooded a part of the old world, and formed the Mediterranean Sea. The Pacific Ocean strained against the east coast of China, and made an immense irruption there which separated Asia from New Guinea and produced a sort of Mediterranean Sea, of which the [Indonesian] Archipelago is all that remains of the old mountains. The Kamchatka Sea is of the same sort. The Black Sea came about through other means. It is only a vast lake produced by the great rivers which flow from the high plains of Poland, Hungary, and Russia. The Caspian Sea was likewise formed by the waters of the Volga. Other lands grow daily. The Nile, the Rhine, and all the great rivers carry silt and provide the basis for the extension of costal areas. The Red Sea is higher than the Mediterranean. It would flood Egypt, if it were able to get over the Isthmus of Suez. That will take place. One daily sees land slip beneath the sea, and sea change into arable land. Our globe has not yet lost its habit of changing, and never will.

The waters, running off, left the first beds of earth in a dried-out state. These are cracked, and have formed perpendicular fissures, which are still found in most parts of mountains. It is in these fissures that vitreous rocks are found, along with metals, boulders, crystals, gravel, sand, and limestone.

Volcanoes filled the face of the Earth after their fashion; rivers add to their coasts. In opposition to the great principle by which the mountains were formed, they lower the height of them, raise up the rocks, raise the sea level, hollow-out the earth, and get it ready to be inundated once more and returned to its initial state of submersion.

Seashells and imprints of fishes which are found in the rocks are the relics of the universal empire of the Ocean. It left them in the banks from which it withdrew, and covered them with new beds of silt. The beds turned to stone, and the seashells found themselves embedded in masses of rocks or slate. Why do we not find in the earth seashells which have been at the bottom of the sea? They are so found, without number, without measure, since limestone, chalk, and marl are nothing but seashells dissolved and sedimented together. Red porphry is only formed, for example, from the spines of sea-urchins.[1]

There is a <u>precis</u> of the discourse, upon which we intend to make several remarks, with all the deference due to the merit and to the great vision of Mr. de Buffon.

Is it certain that the ebb and flow are capable of forming the Alps, mountains 12,000 to 18,000 feet in height? Has anyone ever seen it even raise land beneath the Equator higher than 100 feet? The earth rises only to 10 1/2 feet at Para in the vicinity of the Equator. Has Mr. de Buffon posited a cause which gives to the original ocean a flux infinitely greater than that movement which we observe in it today? It was indeed more extensive then, since the ocean covered the globe, but it was only a matter of degree. In our own day the sea still covers most of the earth.

Mountains which form the opposite sides of valleys are not equal. I have viewed the contrary state thousands of times: mountains of a prodigious height on one side, and those of quite mediocre elevation on the other. Switzerland is in general a large valley of 10 leagues or more, taking its measure from Stockhorn to Mount Jura. But the Stockhorn and the range of which it is a part, is twice as high as Mount Jura. In the narrower valleys, wider than a league, as for example the Hasli in Switzerland, there are mountains 10,000 feet high on the east, and others on the West of less than half that height. From Stockhorn to the other end of the valley through which the river Aare flows, it is about two leagues. The mountains situated in the North-East are only 1,000 feet high, while those in the South-West of it are 3,000 feet. That is extremely apparent in valley lands. The valley of Neufchatel has the Jura to the North-West which may be 2,000 feet high, and hills perhaps several hundred feet high to the South-East. One sees every day that the descent into a valley is longer or shorter than that on the other side.

Is it the case that the highest mountains are found beneath the Equator? Isn't most of the earth there really sea, an immense sea, the great valley of this globe? True, the Andes are there. But the equator cuts across them in a very short line, a line not more than three or four degrees wide. These same Andes extend an immense distance toward the [South] Pole rather than toward Guiana, and the part of equatorial America to the East of the Andes is a flat land, in fact the greatest plain in the universe.

As to Africa, I do not know if Mr. de Buffon has had correspondence which would lead him to believe that that part of the continent which is below the Line is elevated. The Atlas Mountains are a way off from it. The Senega (sic) glides through vast plains and between quite modest hills at a short distance from the Equator, and cuts through almost the entire African continent.

So far as Asia is concerned, I do not know that it has any high mountains in that small part of it which falls beneath the Equator, whereas the Alps are a great distance from it, and as close to the Pole as to the Equator. Other great mountains run almost in the same parallels of latitude as the Alps, from one end of Asia to the other. They separate Tartary and Siberia from the Indies. This is the longest chain of mountains that exist on the earth, and they appear to be extremely high, judged by the greatness of the rivers which flow from them, and which need not yield anything to the Amazon, since they have a course almost equal to 30 degrees. They appear to be so, further, by reason of the coldness of the valleys which separate this complex chain of mountains from Tartary.

The ebb and flow added to the movement of the Sea from East to West, formed the greatest mountains: Such is the system of Mr. de Buffon. That works alright for the principle chain of American mountains, which extends from South to North.

But how did this movement form the immense mountains of Asia, whose axes parallel the general current of the Sea? Why did silt come to a stop at a length of 100 degrees upon a mediocre breadth? Why didn't it build up the Alps in the Southern part of Asia (the Alps, which ought to have raised themselves up from North to South), to become higher as they approach the Equator, and to join China to New Guinea?

Is it certain that Eastern lands are the oldest? They should, then, be the highest, and they are not. The Andes are near the western shore of America, and are extremely far from the Eastern shore. That Eastern shore, which has withstood the effort of the North Atlantic [mer du nord]², is nothing but a plain the elevation of which never exceeds 10 1/2 feet for more than 200 leagues, according to Mr. de la Condamine. America is the reverse of Mr. de Buffon's system. It is high and mountainous (and thus old) in the West, and flat in the East. And by another hypothesis of the same author which we have already related, it ought to be old in the East and new in the West.

China, which is at the extremity of the Orient, is quite flat for being so old, its mountains are quite small, and its greatest rivers enter the sea by the very shore which faces East, and which ought to be the oldest and the highest, according to Mr. de Buffon. All histories and all fables agree that man did not migrate from the eastern part of Asia, but from that part of it closest to Europe, from a low land, the plains of Assyria. Our Europe, which Mr. de Buffon claims to be of recent origin, might claim the greatest antiquity by virtue of its Alps. They were the highest mountains known until the discovery of the Andes, and they are far higher than the mountains of China, which are completely arable and cultivated.

Perpendicular fissures are in fact found in mountains of medium height. There they appear to be the result of waters flowing out of the interior of a still soft earth. But they seem to me to be too rare an occurrence to support a sytem as extensive as that for which Mr. de Buffon uses them. I have seen the Harz mines, the richest in Europe. There the metals are not found in the fissures: I saw none of them there. They are in the rock itself, rock of an immense extent, which is a block formed of black stone, a little silver, and much lead.

I am not completely persuaded either of the fashion in which the Mediterranean Sea, or that of Kamchatka, was formed, according to the ideas of Mr. de Buffon. If these seas were created by an inundation of the ocean, and if they were solid ground before being seas, it is necessary to suppose that there was a basin, an immense valley, hollowed out of the surface of the earth, deeper than 1000 or 2000 feet below sea level in certain spots, and everywhere lower than the surface of the sea, ready to receive it as soon as it had broken through its barriers. Now, I do not believe that an example of such lands below sea level can be found. There are people who claim that the flat lands of Holland are below sea level, but that is not exactly true. Dutch rivers and canals in the towns are lower than the streets; yet these rivers empty into the sea. If the sea were higher than the rivers, it would flow up their channels and inundate all the villages, which are never flooded by the highest seas. But this would be something still quite different from depressions of 2,000 or 3,000 feet, deeper than the deepest mines, where the mercury rises to 33 and 34 inches [pouces]. There is nothing of this sort on the face of the earth, and nothing gives Mr. de Buffon warrant for such a supposition.

As far as my views extend, it seems obvious to me that the mountains, rivers, and the sea are to be found on the earth just as they came from the Creator's hand. There is an exact and perfect interconnection between these parts of the globe, an interconnection which would not prevail in the system of chance which Mr. de Buffon adopts. The rivers which empty into the Mediterranean are proportionate to this basin, which was hollowed out for them. The evaporation from its immense surface takes up the waters which they bring there, without there being either too much or too little for the Sea. The mountains which produce these rivers are in turn proportioned to the rivers and to the sea which receives them. They supply a regular quantity of water, which hardly changes in volume at all, and which keeps the sea level constant. Given the irruption of the Ocean posited by Mr. de Buffon, these mountains would be higher and still more immense. Consequently, they would have more snow, being higher and colder, and they would supply more water than at present. Instead of the Mediterranean Sea, this immense basin in which today the waters of the Alps, the Pyrenees, the Taurus, the mountains of Ethiopia and Libya are swallowed up, there would have been nothing but a plain, strewn with several lakes, perhaps, which would never have sufficed to evaporate these waters from the Alps, Pyrenees, Taurus, and the Ethiopian mountains, since the modern sea which Mr. de Buffon posits, the Mediterranean Sea, alone would suffice for this evaporation. Our reasoning is even more telling, since it is certain that the ancient earth [globe] was more wooded and less cultivated than is ours, and that the vast woods augment the exhalations of the earth and rain, and that all lands of the world become more dry and arid in proportion as they are deforested.[3] Following this line of reasoning, the East Indian Ocean [Mer des Indes], that of Siberia, and the North Atlantic evaporate exactly the amount of water brought them by the Ganges, the St. Lawrence, and the Amazon; and the quantity of water these rivers bring here is determined by the extent of the lands they traverse and the mountains which give rise to them. This reflection makes it more than probable that the earth has been just about the same since creation. A more constricted Ocean would not have sufficed to receive these rivers. What would have become of their waters? And such would have been the Ocean before the prodigious conquests which Mr. de Buffon attributes to it, before that irruption which flooded the Gulf of Mexico and the terrain which to this day forms the West Indian Sea [Mer des Indes].[4]

One can get an idea of the changes which may have taken place over 4,000 years on the earth by judging by the sorts of changes which have taken place in our times. Nothing gives us a warrant to posit causes different from those which exist in our day. These changes are small, and almost imperceptible. For whatever lowlands the sea may flood, it retires from other equally low areas. The most ancient geography we know shows the position of rivers, lakes, and seas to be today almost the same as they were at the beginning of history. If the sea had been higher Egypt would not have existed. If it had been lower, Jaffa, Hezlongueber, Tyre, Sidon, and Carthage would not have been seaports.

But of all the doubts which present themselves against Mr. de Buffon's system, the strongest is that which arises against his manner of examining figured stones [fossils] or the relics and imprints of plants, animals, and shellfish which are found in so great number in all the mountains of the earth, with the sole exception of the Alps, the highest mountains of Europe or America, which in effect have no such stones at all. Mr. de Buffon attributes the presence of these figured stones to the retreat of the waters of an universal ocean. Its retreat left bare that part of the globe which became solid ground, and which was once the bottom of the sea. The seashells left thus had been covered with silt, etc. This supposition might have weight if, in the rocks of the Alps, one only found plants and animals of the sea, and above all animals such as the climate of the Alps was able to produce. But such is far from the case.

There have been found in Swiss quarries [the remains of] crocodiles, men, wheat, ears of barley, and ferns of the Indies. Almost similar ferns have been found in France, England, and Germany. Mr. Scheuchzer found enough of them to form a herbal of them. Consequently, these quarries were not formed out of the silt of a universal sea, before which there were neither men nor ferns nor crocodiles, nor crayfish, nor fresh water fishes. It is thus necessary that there had been an habitable earth before that immense inundation, which left its seashells on all the mountains of the globe, an earth which was cultivated. Mr. de Buffon will respond to this objection that the presence of seashells in lands in which terrestrial animals have been found were formed by particular inundations and by catastrophes posterior to that of the retreat of the primeval ocean. That does not speak too well for his system. The Alps, because of their height, ought to be the oldest parts of earth, and thus overlain with less dramatic changes, which, according to Buffon's avowal, took place after their formation. But these American ferns, and seashells of the Indies - aren't there difficulties here? Doesn't he see that the climate of Switzerland was incapable of producing ferns and animals which require continuous heat and no winter? He replies that the Sea brought them there. It brings to Ireland American beans. It is sufficient to reply to him that beans are capable of floating with the currents but this is not so with the skeletons of crocodiles. Did the ebb and flow ever bring to Europe animals and seashells of the Indies? And whence come it that the ancient Ocean might transport things which the modern ones cannot? How could it transport ponderous bodies 4000 leagues which today it cannot move one league?

Furthermore, if this all powerful flow of the sea brought to the Alps ferns of the Indies, and left them on these beaches, they ought to be the only ones, that is, one ought to find there only those things brought there by a strong South-West [current]. But one finds there all mixed together European animals, barley (which has been through all times the staple of the people in the vicinity of the Alps), and even ferns and Equisetum which are native to the region. One finds in the slates of Marsfeld herrings and carp mixed together, and in the white stone of Pappenberg terrestrial and marine forms of crayfish. This ebb, so far-reaching, which brought remains from flooded America, brought also those from flooded Europe at the same time. There has been, therefore, a universal Deluge, a general inundation of a world already inhabited. And it is precisely this Deluge that Mr. de Buffon, along with the greatest part of modern philosophers, adapt to their fashion or disbelieve in or exclude from the formation of the imprints of animals and the transport of seashells. As for myself, I choose to believe consequently in something which convinces me - that is, the testimony of Moses - which is not objectionable proof for me. It is so obvious, so evident, that the habitable earth, cultivated, peopled with men and animals, covered with their works, was inundated everywhere and at the same time; that the peaks of the highest mountains remained dry, and, for this reason, have no imprints of seashells or marine plants, while mountains of moderate height - those of up to 6,000 feet above sea level or thereabout - have been covered by the waters of a Sea which had brought together the remains of two continents and all climates.

Moving on to the second memoir of Mr. de Buffon, it has indeed a title chosen to his liking. "Proofs of the Theory of the Earth," we read, whereas it is an hypothesis concerned with the formation of planets. Mr. de Buffon cuts his pattern out of whole cloth. A comet fell into the sun and caused a portion of matter equal to $1/650^{th}$ of its mass to break off. This was matter fused by the heat of the sun, and leaving it with an accelleration in proportion to its distance from the center of the sun. It broke apart, for it was composed of heterogenous material; its lightest and rarest particles were transported further and formed Saturn. The densest only travelled a short distance, and became Mercury. The orbit of the planets is marked by the line of their departure from the sun, and by the attraction which they retained for it and which tried to recapture them for it. Their matter cooled down once more, congealed into a spherical shape due to the equal attraction of its particles, a small inequality coming about from the compound motion which moves [the planets] and which is in part that of tangential motion and in part that of rotation upon their axes (the reason for this is not explained yet, and will be difficult to come by, for it is not certain that it has been communicated to all the planets). This matter cooled down, and formed a vitreous crust on the earth which mud, sand, and clay have covered. Its exhalations fell back to earth in the form of dew. This dew created a kind of viscid jelly. It is this viscid jelly which became the cultivable earth, with which the surface of the earth is covered. After this hypothesis, should Mr. de Buffon be so cruel toward Whiston, Woodward, and Burnet? Does he alone have the privilege of setting comets in motion?

All the other treatises which Mr. de Buffon continues to call proofs of his theory are of new systems, wherein are to be discovered everywhere curious facts, new remarks and researches, which are in general a sort of commentary on the first thesis of which we have given an extract. One will find there still more facts and hypotheses subject to strong objections. Mr. de Buffon assures us there, for example, that the greatest rivers flow from East to West. I have found as many that flow from South to North, such as the Nile, and those prodigious rivers of Siberia, the Yenesei, the Ob, the Lena; or from North to South, such as the Ganges, the Indus, the Euphrates, the Rhone, and the river of Siam. Further, it is not obvious to me that river channels become tortuous as they approach the sea. Rather, the contrary in the case. They are tortuous for a day's journey from the Mountains, where the valleys are almost always serpentine, and they become more straight when they move freely through the plains which verge on the sea. It is not exactly true, further, that lakes which have no rivers flowing from them are salt, while those which do give rise to rivers are not. There are in the Alps an infinity of lakes, of which a goodly number give rise to no rivers, which are not salty for all that. But we do not know how to follow our author in the extraordinary number of useful things, or the smaller number of curious things, that he says concerning the winds, the inequality of the elevation of rivers, volcanoes, subterranean rivers, the formation of various kinds of stones, etc. And we hope to find more room for important ideas which he has about generation in the second volume of his work.

NOTES

*Bibliothèque raisonnée des ouvrages des savans de l'Europe (Pour les Mois d'Octobre, Novembre, and Decembre, 1750.)

The Bibliothèque raisonnée Reviews   267

Article 1:   [Review of Buffon's Histoire naturelle, générale &
particulière, avec la description du cabinet du Roi.   Paris.
De l'Imprimerie Royale; 1749. Tome 1.] pp. 243-263.

   [1][Porphry is an igneous rock and could not be formed in
the fashion proposed here.]

   [2][See plate P, p. 259. The discussion is clarified by
noting the cartographic orientation of Buffon's day, in which
the Altantic (Mer du Nord) is more obviously North of the
South Pacific (Mer du Sur) than is the case. This reflects
the apparent uncertainty of the location of the true North
Pole.]

   [3]See Bibl. Rais. T. xxxvii, p. 288.

   [4][In this and similar passages, the reviewer does not make
clear distinctions between East and West Indian Oceans. The
only obvious way to make sense of such passages is to supply
the appropriate directional adjectives.]

REVIEW OF BUFFON'S HISTOIRE NATURELLE

January-March, 1751

THE BIBLIOTHÈQUE RAISONNÉE*

Translated by John Lyon

We have given an account in the issue preceding this one of the first part of the work in question, and we found there much that was novel. That section with which we are now concerned is characterized by the same striking novelty. It involves nothing less than a total overthrow of the generally accepted opinion, an overthrow with extensive ramifications which appear to impinge upon religion, since the Sorbonne has found cause here to reflect on the permission to continue the work in question. Truth has rights superior to those of all hypotheses, and we shall not spare any pains to acknowledge before the public the effect that the discoveries of Mr. de Buffon have had upon us. Since this volume is extremely rich, we shall pay particular attention to that part which concerns the generation of animals. In a display [cabinet], wherein one cannot see everything, it is appropriate to fix one's attention on that which is most singular and unique.

The system of the unfolding [développment] [of preformed germs] had been adopted with the consent of all of Europe. Despite the disputes of the ovists, the spermatists, and those who would discern the germs of plants and animals dispersed in the air, and in the elements, it was a matter of common consent to hold that plants and animals had been created in miniature, and that their subsequent growth was nothing more than the opening out of a previously completed mechanism, which came forth from the hands of nature in all its resiliency and which had only to grow in order to pass from being called grain or seed to being called plant and animal. In this fashion was the theory of the formation of animals cut short. A little facile physiology sufficed in order to insert the nutrient humours in the pliant vessels, which only had to be multiplied. To understand the marvellous beauty of the structure of organized bodies, no more trouble than this was necessary, for such bodies issued directly from the hand of God. It was generally believed that this system served to confirm the idea of a Creator, in leaving to nature or to the known forces of matter the sole privilege of being the cause of the growth of animals, and in entirely denying to nature and matter the power of forming them. Certain experiments appeared to attest to the fact that no animal was produced without the coming together of the two sexes. Aristotle was readily dismissed and spontaneous generation treated as barbarous and impious.

Things are quite different with Mr. de Buffon's work. No more primordial germs, no more unfolding. Nature herself enjoys the right of forming herself, of organizing herself, and of passing freely from the inanimate state to that of a plant, or that of an animal. Putrefaction has claimed its privileges once more, and natural philosophy, which had taken them away, now, better instructed, repays by authenticating it with high sanctions. Let us follow Mr. de Buffon in his experiments, and even, at times, in his reasoning.

Salts have unvarying contours. Their molecules are prismatic and cubical. They exhibit some solid form, whose characteristic structure they return to as soon as artificial restraints on them are relaxed, or as soon as the action of water, which dissolves them, ceases. Animals and vegetables themselves approximate the condition of salt. They also have their molecules, of determinate structure, which set their structures, and which are, by and large, of a constant figure. The crystal figures one sees in snow proceed by imperceptible degrees to re-configure themselves in the formation of a Caesar or an Alexander. The tendency of this line of thought is obvious. It is a system of Nature exactly like the homoeomeria of the Ancients.[1] Bodies are composed of small organized bodies. Trees are an accumulation of buds which unfold themselves bit by bit, and then form others once more, and so on to infinity. Let us add that the small particles of plants and animals resemble each other (at least in the greater number of instances) when inspected visually. They are cylindrical or arranged in laminae. The cellular tissue, of which everything vegetable and animal in nature is composed, itself shows no more basic unit of composition. But these laminae and cylinders which so strongly resemble each other--are they alike in everything? That is another question.

In order to understand how these organized molecules take the figure of one animal rather than another, why they form a pear tree rather than a woman, Mr. de Buffon proposes an hypothesis. Nature makes the molds, and introduces in the cavity of the vessels of animals a material appropriate to be molded there. A better craftsman than we are, she does not make this matter take on the hollow figure of these vessels themselves (for that would give us only a preparation after the fashion of Mr. Lieberkuhn), but molds this matter in such a fashion as to penetrate the model and represent it as a solid replica and not merely its superficial appearance.

Throughout the extent of Mr. de Buffon's system which we have set out to follow, the organized particles of the food are molded according to the pattern of each part of man. They receive the imprint of each part, and they conserve it in the seed which has, in the fair sex, all that which is necessary to form a woman, and all the parts which compose, for our sex, the body of a man, including the distinctive feature of it.

The Bibliothèque raisonnée Reviews

For this molding process it is necessary to believe that everything is a vessel in the human body. For the nutritive matter only circulates through the vessels, and it is only through the veins that it is able to return from the brain to the heart and carry in the seminal fluid the imprint of a pineal gland and that of a cerebellum. But the human body is far from being composed solely of vessels. Anatomical experiments demonstrate that it has many more of them than the ancients believed it to have, but many less than Boerhaave maintained. In a word, man is composed either of cellular tissue which often joins and adheres together to form a cylinder, but which also quite often does not do so; or he is made up of a hardened jelly, which has no constant figure and which takes on that of neighboring parts of his body. The cylinder itself, or the vessel, is not formed by the hollow threads [filets creux]; it is a stuffing, a composite of solid and interlaced filaments. The structure of the human body being only partly vascular, the nutrient matter does not come in contact with everything. How is it possible, then, for it to take on the imprint of an entire body? In a femur there are tubules which penetrate through the pores of the bones. But there is infinitely more solid matter, thickened jelly, which fills up the intervals between the fibres, and which is not either hollow or extended to the arteries. The femur is a long cylinder, which small cylinders pierce at right angles here and there, leaving quite considerable intervals in adults, who are the only ones whose [internal] mold proceeds to the seminal fluid. Instead of resembling a long tube, the nutritive matter shaped in these tubules will thus look like a sort of net made of rows of alternating tubes with empty interstices.[2]

But there is another quite considerable difficulty with these molds, and that is that the particles of the seed are infinitely smaller than the parts according to which they are modelled. Take the nose, for example. In order that it should reappear in the infant with a resemblance to that of the mother, it would be necessary that the nose of the infant should be molded by that of the woman. Is it conceivable that a material could possibly be molded according to a model without taking upon itself the dimensions of the model, and that it could possibly show the resemblance to it, at the same time that it is almost infinitely smaller than its model? For I do not believe that anyone has wished to say that in order to form a small nose the elements of this small nose might be different from those of a large one, and that it might be that the seed of the father would vest its contours in elements of the nose. It is not obvious that the elements of the muscles, bones, or viscera differ in man; it is only in great souls [grandes âmes] that the difference is sensible. We know that Leeuwenhoek had not been able to find any difference between the flesh of man, and that of the mouse or elephant.

Furthermore, a one-armed man engenders children with both limbs. Mr. de Buffon would reply that the mother alone furnishes that limb whose matter is missing from the seed of the Father. But the Hottentot, who has had only one testicle since earliest infancy, and the Swiss who has lost one of these organs after an operation, engender quite regularly infants who have their two testicles quite complete. And from whence then comes this second testicle, since the mother is not able to supply it?

Attentively following Mr. de Buffon's system and adopting his ideas about the hypothetical mold [la moulure possible] of nutritive matter, I can only visualize a sort of gelatinous polyp, which might replicate the hollow contour of all of a man's tubules, a complete vascular system, but which could be greatly stretched out to look like a man. I would like to be able to satisfy myself concerning a new property which Mr. de Buffon gives to Nature, that is, the power of forming the internal molds, which might transmit their solid figure to their imprints. That is hard to comprehend; but it would make only such a solid system of vessels as spoken of above, without explaining the resemblance of a microscopic imprint made of the vascular system of an elephant or a whale, to a complete elephant or whale. Is it, furthermore, very likely that two bodies which are in contact can take any other figure than that of their surfaces, or that their exterior surfaces which do not touch are capable of configuring [se mouler] each other? Doesn't Mr. de Buffon fall into inconveniences as great as those which he refutes?

If one were to adopt another hypothesis, and were content to suppose that small organized bodies, such as one sees in prepared infusions [infusions du soin], or in the seed of animals, were molded after some vessels of the parents, and cylindical like them, how would one explain the perfect consistency with which the tubules, [cilindres] so equal to each other, form with infinite regularity complete bodies, the parts of which are so unequal, one to another? Is it possible to conceive the art with which these small cylinders form an eye, or a hand, and ears at an appropriate distance from the eye? What faculty arranges these particles, and what hinders the imprints of the eyes from joining with those of the heart or the limbs? After having granted everything to Mr. de Buffon, do we still not need a directive intelligence, in order to place the organic particles appropriately and in perfect symmetry? For to say that the particles which come from each sex conserve their primordial position and can arrange themselves only according to it, is absolutely opposed to what Mr. de Buffon has himself seen. He has only seen globules swimming in the liquid or heaped up, all alike, without order, without connection, and without system. I do not know if I have convinced the reader, but I do know that I find myself convinced of the truth of these objections. They prevent me from having the pleasure of sharing Mr. de Buffon's ideas, and of believing myself instructed in a true system of generation. This would be a significant pleasure to me, and one that I would hope to savor, after having been informed of the commendation which this system is supposed to have found with Messrs. de M. and de R.

Let us return in detail to this great operation of Nature. Some organized molecules serve in part for the sustenance of the animal, and whatever is superfluous is returned to the organs which prepare the seminal liquid, while unorganized molecules [molécules brutes] are expelled under the ignoble appelation of excrement. The former gather themselves together by a new property which Mr. de Buffon ascribes to Nature: they do not yet form a small body similar to the body of the animal of which they have taken the imprint. For that it is necessary that the two sexes join, in our species as well as in many animals, as opposed to those which reproduce themselves and recast a bud akin to that which serves as a model in the parthenogenic green aphid, in the oyster, or the May-fly [la Perle].

But in man, in order to fix the movement of these organized corpuscles it is necessary that the two seminal fluids mix, and unite. The extract from the eye of the Father must align itself with that from the eye of the Mother. Extracts from the rest of the body must arrive similarly, and the body of the fetus be formed by this coming together. There remains to be formed only the parts of generation which characterize but one of the parents and consequently originate in that parent in which the seminal matter is found more abundantly. The rest of what happens in the formation of animals is nothing more than unfolding which takes place in the uterus of the mother, or in the egg.

I have for Mr. de Buffon all the sentiments which equity and the love of letters inspire for those who work at extending the empire of letters. No vicious reason prevents me from accepting his ideas. I would do that if I could, but can I? To consider things according to principles, has Mr. de Buffon demonstrated that there is female seed? I am aware of the experiments which he has made on the liquid of the corpus luteum,[3] and I shall speak of them soon. But observations on that liquid do not suffice to prove a mixture of two seminal fluids provided by the Father and the Mother. It is extremely doubtful that the woman gives out any sort of fluid at the moment of conception. The effusion of glutinous matter in the region of the urethra has nothing to do with the matter; it is quite unconnected with, and hardly necessary for, conception, for many quite fecund women never exude such matter. But there is no evidence for the effusion of a fluid in the uterus of the woman or the female. Cruel accidents to women have furnished Mr. Ruysch with experimental evidence that there is found in the uterus only the quite distinct seed of the man which filled the cavity of the uterus and that of the Fallopian tube [la trompe], without any mixture of a fluid which could belong to the woman. Add to that the complete uncertainty of women, who never know if they have conceived, or who only know that there is no connection between the effusion of a female fluid and proof of conception. Woman ought to be perfectly similar to man, who is quite aware of his own emissions. Woman, so inviting, often so voluptuous: how could she not know what was contributed on her part? There are, besides, other animals in great number in which the chaste and timid females flee the approach of the male. Harvey is a witness to this fact. The English deer, the fallow-deer, and all this class of animals resort to violence to subdue their females, who fear the male and avoid him. If they had a female fluid to contribute, wouldn't the analogy demand that it be done with pleasure? And wouldn't nature teach them to search out that sweet and easy lesson which, for their own happiness, is to be learned here?

In the system of Mr. de Buffon, the <u>corpus luteum</u> ought to be found in all fecund females between the ages of 12 and 50 years. Since the seminal liquid makes up half of the fetal matter, and since healthy women are disposed by nature to conceive in the age range I have noted, by the laws of the system they ought to have the <u>corpus luteum</u> with as much certainty as men have seminal vesicles and testicles, since their liquid is as necessary for generation as is the seed of the male. But such is entirely contrary to experience. As the result of a quite considerable number of dissections on hundreds of women and girls, I can assure the reader without fear of being contradicted that I have never found the <u>corpus luteum</u> except in pregnant women, or those who have recently given birth. No girl, no woman, dying under other circumstances has ever shown to my eyes the <u>corpus luteum</u>, and I am persuaded that all anatomists must make the same comment.

This observation creates a difficulty to which I cannot find a reply in favor of Mr. de Buffon. It appears that the <u>corpus luteum</u> is not a necessary cause for conception, or an organ requisite for it. The fact of the matter is that this body only appears after conception, and gradually disappears. That is the reason why one never finds but one <u>corpus luteum</u> in a woman, while in the system of Mr. de Buffon it would be impossible to fix their number. Animals are less appropriate for research in such matters, for they conceive quite frequently, while their virginity and the interval between their deliveries is quite short, and thus their <u>corpus luteum</u> scarcely has time to disappear before another replaces it. Thus it is that one almost always finds it, and it is this experience which had led Mr. de Buffon astray when he assures us [see p. 196] that one finds these glandular bodies formed in all females who have attained puberty.

Let us make a brief comment on a critique of Mr. de Buffon's on the supposed little man of Dalempazius, which begins by being a worm, and then casts aside his skin in order to make visible his human figure. This whole critique only amounts to badinage. This Dalempazius is no one but Mr. de la Plantade of the Societe des Sciences of Montpellier, and his experiment is only raillery, pure and simple.[4] Mr. Astruc passed this anecdote on to me.

Let us move on to the principal part of the work, to the experiments which Mr. de Buffon has made in concert with Needham, d'Aubenton, and Dalibard. Whatever kind of hypothesis Mr. de Buffon may make, the experiments maintain their value. They are the gold which remains in the bottom of the crucible when the alloy dissipates. These experiments clarify and bring to completion those of Leeuwenhoek and they lead to a more general and more uniform system.

The role of those spermatic animals which Hamm had shown to Leeuwenhoek, and which Hartsoeker has seen independently and which all of Europe believed it had seen after these naturalists is less significant as a result of the new experiments of Mr. de Buffon. In the seminal fluid of many species of male animals there seem to be filaments which are almost cylindrical, which soon change shape and direction, puff up at intervals, and produce small, sometimes oval globules which hold to the filament by an extremely fine tail [queue]. They are removed from this [filament], after which the tail detaches from the filament and remains attached to the globule. They then swim in the fluid. These are the animals of Leeuwenhoek. They have then an oscillatory movement, and one of rotation. But in several hours they abandon their tails, their movement nevertheless continuing, being even more lively. They are divided some times, and make two globules. Finally, they perish, becoming smaller and forming small groups [chopelets] arranged in a polygon about a center, in which the cylindrical filament generators gather themselves together. In other experiments they change figure and appear to shed an exterior envelope, decrease in size, and become gradually immobile. At times they retain their movement for as much as 20 days.

These tails, which are the essential feature of the spermatic animals of Leeuwenhoek are, according to Mr. de Buffon, foreign to the globule or to the organic part. They are quite small, they hamper such movement as there is in the molecule, their size is inconstant, and they leave the globule without appearing to deprive it of life. This is an essential point, which ought to be decisive concerning the nature of these microscopic tadpoles, and by which one can decide if they are animals or machines. But another, and equally powerful, reason proceeds to the overthrow of the system of Leeuwenhoek.

The liquid contained in the corpus luteum of the ovary of female animals is full of organized particles exactly similar to those of males. They cannot thus be considered as the exclusive possession of the male's liquid, and the two sexes appear to contribute about equal measures to the grand work of generation. Yet there are slightly fewer organized particles in the female's seminal liquid. This is the reason why more males are born than females.

But the following observations diminish greatly the value of what one might wish to conclude concerning these matters. One finds approximately similar globules in the gelatinous infusion of roast veal, in the infusion of carnations, in the milt of fishes, in various infusions of animal flesh, and in several kinds of grains. All these different liquids have provided, after a certain time, globules in motion, which resemble animals, but which change shape, become successively smaller, and which sometimes have filaments and at other times are without them. Mr. de Buffon believes it to be reasonable to refuse to call all these globules animals. According to him, these are the very particles, superfluous to nutrition, molded on all the parts of the human body, and sent back to the testicles and ovaries, of which we have already spoken; and I have repeated most of his reasons for him. But I do not know how to rid myself of the idea that the moving bodies, which stay clear of each other, which move forward and backward, which are born, change shape, and die and lose movement once more, appear to form with the genuine little animals with which vegetable infusions swarm, a continuous chain of animals. That which Mr. Joblot has seen under so many different shapes appears to me to be of the same kind, of the most simple class and the most uniform kind of animal, from which worms differ only in degree, principally by virtue of the fact that they eat. I do not know of any character capable of designating an animal which would not also characterize worms. It even seems to me that polyps in general, because of their tail by which they are attached, but which is capable of being detached, also resemble animals. It is true that my opinion only differs moderately from that of Mr. de Buffon, who regards these particles as the first assemblage of organic molecules, the living outline of animals. But the consequence which each of us draws from this is quite different. Buffon regards these particles as the essence of the semen, as that which will in time form man, animals, and plants. But to me they only appear to be creatures [insectes] which live in animal fluids carrying some materials which serve them for sustenance. It appears inconceivable to me that organic beings destined to form man, and consequently extremely delicate, could be found in gelatinous infusions or in roast flesh, and that their delicate structure would not be destroyed by the action of fire, which is the universal enemy of animals and germs. Adding to this reason those which we have related as counting against the existence of a seminal fluid in the corpus luteum, and against the system of molds, it appears to me that the experiments of Mr. de Buffon have resulted in destroying the idea of a seminal worm which in turn develops into a man. He also appears to have demonstrated that this worm, the dwelling place of the seed, is but a host indifferently attached to all fluids filled with animal or vegetable substance, a host which is perhaps the result of putrefaction, or of its beginnings. I add by way of explaining what I mean, that these creatures [insectes] only live in liquids wherein this beginning of decay is manifest. But these creatures, which are found in the seminal fluid of man, appear to me no more destined to form a man than do the animals of an infusion of pepper appear to be destined to form a pepper plant. They are insects found everywhere, which never become either plant or animal, but live in scum and infusions.

Everything else does not seem to me to be able to offer a refuge against some objections which I am now going to make, or against those which more clever persons could still draw out from several ideas which I have not yet sufficiently matured to risk presenting. I can in fact only speak of them with fear. It is as if I were to dispute with a man who had been in China, on some point of natural history which should appear to me to go against what I have seen in Europe. I only ask to be better instructed.

According to these premises, it is easy to see that I could not further subscribe to the consequences that Mr. de Buffon draws from his premises. There exists, he says, in animals and vegetables an organic and living substance, which is common to them. It produces an animal or a vegetable according to which mold it finds to its inclination. When it is plentiful in an animal body, it forms there other animals, such as the tapeworm, round worms, liverworms, and those parasites of the brain, which it produces by its assemblage. It is this perhaps which produces the poison of the mad dog, or the viper. Since it is the residue of the food which produces the organic particles of the semen, it is obvious that the quadrupeds do not become fecund until after their growth ceases. In the case of fish, their growth lasts for an extremely long time, although they have their milt and eggs formed much earlier. And Mr. de Buffon finds the reason for this in the cartilagenous nature of their skeleton, and another reason for their fecundity in the absence of transpiration, which creates the scaly nature of their teguments. (But the Armadillo and the Scaly Anteater [Pangolin] do not multiply as rapidly as others though they are decked with scales.) The large animals are less fecund than the small, because they require more nourishment. (It is necessary here to except further the large oviparous fishes, such as the Sturgeon and other species of marine fish.) A great number of animals and vegetables [p. 207] are formed by the fortuitous coming together of organic molecules in the mixture of the two sexes. The organic material circulates, and forms by turns animals and vegetables. The life of animals and vegetables is nothing but the result of all the small particular lives of the active molecules, whose life cannot be destroyed. According to their manner of coming together, they are equally capable of forming an animal or a plant. Polyps are many similar organized bodies, brought together under a common envelope. The parts which characterize the sexes appear to be the base of operations about which the rest of the organic particles of the body of the fetus arrange themselves.

Does this correspond to the generation of man?. Isn't the head almost formed with the heart, and the spinal medulla, when the sex is still imperceptible? If there is a part which is first established, wouldn't that be the heart, without which nothing else might be activated, and consequently nothing put in order? There are no hermaphrodites, because the extracts of the genital parts of the woman are not capable of ordering themselves in the face of the quite dissimilar extract of the male genital organs. It is possible that similar bodies and fetuses are formed in the scrotum of the male because the seminal liquid of the corpus luteum could flow into the genital parts of the male, and there mix with the seminal liquid.

The wishes of the Mother are without effect, and the characters which are commonly attributed to such are nothing but the products of imagination, which searches out a cause for every ill that befalls the infant. I am simply repeating all that after Mr. de Buffon.

If I do not address myself to the natural history of man, it is not because that subject is not quite worthy of curiosity. But this summary is rather long, perhaps too long. Nevertheless, it is necessary to stretch it out yet a bit, in order to make a further reflection. Is this system dangerous? Is it in the interest of religion that it should turn out not to be true? Ought a book be suppressed which proposes such a system?

In general, I am not opposed to the efforts made by the clergy to prevent the publication of books which tend to the destruction of morals or religion. I assume, and I must assume, that truth is permeated by Revelation. To tear away from man this torch, the only one that can light his way to port, is to cause him to shipwreck himself and to lead him to an evil eternity. Is it not part of the humanity of a leader of the people, and an obligation of a man placed so as to see clearly, that he prevent the deadly progress of irreligion or immorality? Does this not amount to preventing the progress of a malady which, when more advanced, can produce an Aloisia Sigen[5] or the polished writings of a V[oltaire?] Such maladies only kill men destined to die, and change the date of their death, inevitable in any case, when they do not bring it about. But the apostles of the system of Epicurus lead men who might have been happy without the deadly effects of such doctrine head down, to a miserable eternity.

Has it been necessary, upon these principles, to condemn Mr. de Buffon's book, as it is said the Sorbonne might wish to do? And likewise to condemn the system of spontaneous generation of new animals without primordial germs?--Is all this so opposed to religion? Does this system give to Nature, to chance, to necessary and essential causes in Nature, the power to create the same machines, in which we believe to discern the hand of an infinitiely wise workman? Does this not take away one of the principal arguments of Christians, that by which the necessary existence of a supreme God is demonstrated? He, the

Author of all this superhuman beauty which is exhibited in the construction of plants and animals?

I do not believe that the theory of Mr. de Buffon tends toward such an evil end. Belief in God, and the exaltation of the wisdom of the Creator, flourished before the system of preformation [dévéloppement] was known. And the school of Aristotle, which has been the school of humankind, has not been impious for having maintained that putrefaction produced creatures, even the ant and the toad.

When matter was supposed to possess a power that produced the construction of a plant with the same necessity as that by which gravity drew a stone downward, or that elasticity returned to a straight line a stick which one had bent, I do not believe that any lover of religion was alarmed thereby. Eh! To be alarmed because sea-salt always crystallizes in cubical forms, or that the juice of sugar cane forms kinds of houses composed of squares of different sizes, always arranged equally, and which are always the same solid! I shall elaborate: matter may tend toward certain configurations without having thereby the power to do so of itself. And the system of Aristotle, or of Hippocrates, which is that of Mr. de Buffon, does not differ so essentially from the system of unfolding [of germs], as it might appear at first approach.

In this last theory, a grain of Fennel-seed is planted in a flower-pot. It is true that the plant exists then in miniature, but in fact only quite imperfectly. A small [spermatic] worm of Leeuwenhoek's is quite far from being a man. It becomes, nevertheless, a man thorough a concourse of causes absolutely common to matter. The fluidity of liquids and the contractile power of the heart, or the heat surrounding plants and the attraction of the root tendrils does the rest. These give to the germ or the animalcule, a structure infinitely more complex and more perfect than their own. The heart of a chicken begins by being no more than an aorta. This aorta folds itself under the eyes of the observer, its pleats are joined together, becoming a heart with two cavities, one of which holds to the aorta, and the other to the vena cava. There is much more: a pulmonary artery interrupts the primitive continuity of the vena cava with the aorta, joins itself with a pulmonary vein which arises at the same time, and forms a lung, which has not existed up until this point, since the continuity of the vena cava with the aorta has been visible until now in the punctum saliens.

From the theory of Leeuwenhoek to that of Hippocrates there is thus essentially only a difference of degree. Anatomy forces us to recognize that part of our organs arise from in a viscous material which congeals and configures itself right under our eyes. Leeuwenhoek's followers assert, then, that a large part of the animal structure forms itself without being developed, without having been delineated in the semen. Mr. de Buffon affirms of the entire animal what the rest of anatomists have acknowledged for the largest part of its organs. I have seen the cellular envelope arise innumerable times from the sticky paste [glu] which seeped from the heart or the lungs, and even from the pus which fills up the lower abdomen [bas-ventre]. The human body is only, or at least is principally, nothing more than cellular tissue. A cause which has brought about cellular tissue could produce the human body.

I see nothing but truth in this, which I have explained elsewhere, and the truth does not know how to be impious. Matter does not possess its powers of itself. It could have been without gravity, without elasticity, without irritability --a novel quality, but one essential to the structure of plants and animals. These qualities are not part of the essence of matter. They are alien to it, not being common to all parts of matter. Light and fire do not have weight, water is not elastic, minerals are not irritable. A first cause has thus dispensed to the various classes of matter powers and forces calculated according to a general plan, and it is in this that we may recognize the hand of the Creator.

The gravity and centrifugal force of the heavenly bodies are tempered in an admirable fashion which restrains them mutually without destroying them, and which in fact issues in a uniform and regular movement, which has always been deemed one of the best arguments proving the existence of God. The infinitely superior regularity with which matter, gifted with several extremely simple forces, configures itself and becomes with regularity a carnation a horse, or a man, is an argument of the same sort, but much stronger yet, because the combinations which are set in opposition to the human structure are much more numerous than those which resist the regular orbit traced by a planet. It is not chance which forms man from that viscid substance of the seminal vesicles. Chance is not constant: it can produce with equal facility a horse, a rose, or rock salt. There are forces alloted to this viscid substance which under prescribed, necessary, and invariable circumstances structure it and produce a worm, a fish, a man, a Caesar or a Newton.

This invariable constancy of species which allows for a prodigious variety but which never removes essential characters, appears to me to be one of the most obvious proofs of the hand of the Creator. The matter which forms a carnation, may make it red, or yellow, or bluish, according to the concourse of external circumstances. But it never makes a tulip of it. It is chance which would make of it a tulip. It is God who allows a double carnation to be produced, varying from the original crimson carnation of the Alpine Lakes, but always a carnation, which through a thousand generations never provides other than the seed of carnation. All is not by chance, for if it were carnations might become tulips. All is not by necessity, for if it were carnations would remain always just like the first carnation.

The modern author [La Mettrie] who has claimed to re-establish the opinions of Lucretius has made us see quite clearly just how this system is unsupportable. Man is born only from the seminal fluid of man. The earth has been covered with water throughout all its surface, from which [d'ou] the first man appeared, who peopled the earth after the waters dried up. Man sprang forth from the earth, says the author of this account, the earth gave birth to an egg, from which a man and a woman came forth, who were suckled by a beast, and then they reproduced. And why did the earth not produce more of these eggs? It is an old hen, beyond the age of egg-laying. Therein lies all the wisdom that Lucretius and La M[ettrie] possess.

As to M. de Buffon, he adopts the molds, from which animal and vegetable bodies are constructed, and upon which living, organized matter models and forms itself. These molds have a beginning in the first man who was born without having been molded after another. By what was he formed, if not by the Creator?

This work has been translated into German, and published with a "Preface" by Mr. Haller at Hamburg.[6] This edition has been produced with great precision. In the "Preface" Mr. Haller defends these hypotheses, and his reasons have a great resemblance to those of Mr. de Mairan (in the "Preface" to his Dissertation sur la Glace).

One cannot avoid thinking that he imitated it. Mr. Kaestner, a sharp-witted mathematician of Leipzig, has accompanied the translation with comments, which are not always of the same mind as the author. Mr. Zinck directed the translation of the first volume, and the third is about to appear.

NOTES

*Bibliothèque raisonnée des ouvrages des savans del'Europe (pour les mois de Janvier, Fevrier, et Mars, 1751.) Article 1: [Review of Buffon's Histoire naturelle, générale & particulière, avec la description du cabinet du Roi. Paris. De l'Imprimerie Royale; 1749.]

[1][See Lucretius, On the Nature of the Universe (Baltimore: Penguin Books; 1968), p. 51 (Book I). The original exposition of this theory is by Anaxagoras.]

[2][Reading vide for vuide.]

[3][Les corps jaunes, "yellow bodies": Translated throughout as corpus luteum. On this see above, p. 208.]

[4][François de Plantade (1670-1741), French astronomer and physician of Montpellier, writing under the pseudonym of "Dalempatius," in 1699 perpetrated the hoax in question. See

Jacques Roger, <u>Les Sciences de la vie dans la pensée Française du XVIIIe siècle</u> (Paris: Armand Colin, 1971), pp. 318-19, 794.]

⁵[<u>Aloisiae Sigeae Toletanae satira soladica de arcanis amoris</u>, obscene work of Nicolas Chorier (1612-1692), produced sometime before 1680. See <u>La Grande encyclopédie</u> (Paris: H. Lamirault et Cie, n.d.), XI, 233-34.]

⁶[There are some obvious similarities in this review to the second preface by Albrecht von Haller which appears below.]

13. The Sorbonne's Condemnation of the *Histoire naturelle* (1751)
John Lyon

On the seventeenth of January, 1751, the syndics and deputies of the Faculty of Theology at Paris (the Sorbonne) notified Buffon that they had observed that some of the pronouncements in the first three volumes of the Natural History did not accord with appropriate sections of sacred scripture or an orthodox dogmatic stance. The Sorbonne's attention to Buffon's dereliction or intentional heterodoxy may have been aroused in part by the critical reviews in the Nouvelles ecclésiastiques on Feb. 6 and 13, 1750, which have been translated above.

Buffon's prompt and submissive response to the notification of doctrinal variation in various parts of the Natural History was hailed as heartfelt compliance on his part by the Sorbonne in its reply. But the very same response seemed to be the most crass sort of Nicodemism to others (see the extended comments on Buffon's religious stance in Herault de Sechelles' "Visit to Buffon," in Part V, below). Among the latter was the author of subsequent reviews in the Nouvelles ecclésiastiques (June 26 and July 3, 1754). "what a disgrace it is to the Faculty [of Theology]," the reviewer noted, "to commend such a declaration as [Buffon's]. It claims to give us assurance that he has had only upright intentions while so grossly deviating from the Mosaic account; but to believe that everyone would have to be as dull-witted as the Sorbonnic carcass has been."[1] Nevertheless, Buffon's work would be issued in subsequent editions with this set of exchanges attached.

NOTE

[1] As cited by Jean Piveteau, "La Pensée religeuse de Buffon," in Leon Bertin, et al., Buffon (Paris: Museum d'histoire naturelle; 1952), p. 127.

LETTER OF THE DEPUTIES AND SYNDIC OF THE FACULTY OF

THEOLOGY OF PARIS TO M. de BUFFON

(January 17, 1751)*

Translated by John Lyon

Sir:

We have been informed by one of our members speaking on your behalf of your commendable resolution. When you learned that the <u>Natural History</u> of which you are the author was one of the works which had been chosen by order of the Faculty of Theology to be examined and censured as containing principles and maxims which are not in conformity with those of religion, you declared to him that you never had the intention of deviating in such matters, and were disposed to satisfy The Faculty on each article which it might find reprehensible in the work under question. Monsieur, we are not able to give adequate praise to such a Christian resolution; and in order to afford you the opportunity to execute your resolve, we forward to you the propositions extracted from your work which have appeared to us to be contrary to the belief of the Church.

With complete esteem, we have the honor of being, Sir,

Your most humble and obedient servants,

The Deputies & Syndic of the Faculty of
Theology of Paris

In the House of the Faculty
the 15th of January, 1751

PROPOSITIONS EXTRACTED FROM A WORK ENTITLED <u>NATURAL HISTORY</u> WHICH HAVE APPEARED REPREHENSIBLE TO THE DEPUTIES OF THE FACULTY OF THEOLOGY OF PARIS

I. It is the waters of the Sea which have produced the mountains, the valleys of the earth.... It is the waters above the earth which bring everything back to sea level, which one day will return this earth to the sea, which then will take possession of it while opening to discovery new continents similar to those we inhabit (edit. in-4°, Vol. I, p. 124; in-12, Vol. I, p. 181).

II. Isn't is possible to imagine... that a comet falling upon the surface of the sun could have displaced this star and separated several small parts from it, to which it would communicate a movement of impulsion...in such a manner that the planets which formerly would have belonged to the body of the sun, and which would have been detached from it, etc. (edit. in-4°, p. 133; in-12, p. 193).

III. We see in what state they (the planets and above all the earth) were, after having been separated from the mass of the Sun, (edit. in-4°, p. 143; in-12, p. 208).

IV. The sun will probably be extinguished... due to lack of combustible material... the earth, upon its exit from the sun was therefore on fire and in a state of liquefaction (edit. in-4°, p. 149; in-12, p. 217).

V. The word "truth" only gives birth to a vague idea... and the definition itself, taken in a general and absolute sense, is no more than an abstraction, which only exists in virtue of some supposition (Edit in-4°, Vol. I, p. 53; in-12, Vol. 1, p. 76).

VI. There are many kinds of truths, and we are accustomed to putting in the first order those of mathematics. Nevertheless, these are only truths of definition. These definitions rest upon simple but abstract suppositions and all truths of this sort are but the universally abstract consequences compounded of these definitions (Idem).

VII. The significance of the term "truth" is vague and complicated. Therefore it has not been possible to define it in general. It has been necessary to distinguish its genres, as we shall come to show, in order to form a clear idea of it (edit. in-4°, Vol. I, p. 55; in-12, Vol. I, p. 79).

VIII. I shall not speak at all of other orders of truths, those of morality, for example, which are in part real and in part arbitrary... they only have as their object matters of decorum and probabilities (edit. in -4°, Vol. I, p. 55; in -12, Vol. I, p. 79).

IX. Mathematical evidence and physical certitude are, then, the only two aspects under which we ought to consider truth. Insofar as it is distanced from one or the other, it is no more than verisimilitude and probability, (edit. in-4°, p. 55; in-12, p. 80).

X. The existence of our soul has been demonstrated to us, or rather, we ourselves are identical with it (edit in-4°, Vol. II, p. 432; in-12, Vol. IV, p. 154).

XI. The existence of our body and other exterior objects is doubtful for whoever reasons without prejudice. For, this extension in length, breadth, and depth, which we call <u>our body</u>, and which appears to belong to us so imtimately what else is it except an account rendered by our senses? (edit. in-4°, Vol. II, p. 432; in-12, Vol. IV, p. 155).

XII. We are capable of believing that there are some things external to us, but we are not sure [of their existence], whereas we are certain of the actual existence of all that which is within us. The existence of our soul is thus certain, while that of our body appears doubtful, as soon as one comes to think that matter might well be no more than a mode of our soul, one of its manners of perceiving (edit. in-4°, Vol. II, p. 434; in-12, Vol. IV, p. 157).

XIII. Our soul will understand in a still quite different fashion after our death and after all that which causes its sensations presently, matter in general, can, indeed exist no longer for it, for our own body will no longer be anything to us (edit. in-4°, _idem_; in-12, p. 158).

XIV. The soul ... is essentially impassible (edit. in-4°, Vol. II, p. 430; in-12, Vol. IV, p. 152).

RESPONSE OF M. De BUFFON, TO THE DEPUTIES

AND SYNDIC OF THE FACULTY OF THEOLOGY

Gentlemen:

I have received the letter which you did me the honor of addressing to me, with the propositions which have been extracted from my book, and I thank you for having put me in a position to explain them in a manner which may leave no doubt or uncertainty concerning the rectitude of my intentions. Gentlemen, if you wish it, I quite voluntarily shall publish in the first volume of my work next to appear the explications which I have the honor to send you. I am with respect, Gentlemen
Your quite humble and obedient servant,

BUFFON

March 12, 1751

I declare:
First: That I have never had any intention of contradicting the text of Scripture; that I believe quite firmly all that is related there concerning creation, be it concerning the order of times or the circumstances of events; and I abandon that which, in my book, concerns the formation of the earth, and, in general, all that which may be contrary to the narration of Moses, having only presented my hypothesis on the formation of planets as a pure philosophical conjecture.
Second: That with regard to that expression, <u>the word truth only gives birth to a vague idea</u>, I have not meant anything more than what is meant in the schools by "general idea," which has no existence in itself but solely in those species in which it has an actual existence. And, consequently, there are really some truths unquestionable in themselves, as I explain in the next article.
Third: That beyond truths of inference and assumption, there are first principles which are absolutely true and certain in every instance independently of all suppositions, and that the consequences evidentially deduced from those principles are not arbitrary truths, but eternal and evident truths. Thus I have only meant by truths of definition mathematical truths alone.

Fourth: That there are some evident principles and consequences in many sciences, and above all in metaphysics and morals. In metaphysics, some such would be: the existence of God: his principle attributes; and the existence, spiritual nature, and immortality of our soul. In morals, some such principles and consequences would be: the obligation to worship God; to give to each that which is due to him, and consequently, to avoid larceny, homicide, and other actions which reason condemns.

Fifth: That the objects of our Faith are most certain, without being evident, and that God who has revealed them (Who reason itself tells me is incapable of deceiving me) guarantees to me the truth and certitude of them. These objects are for me truths of the first order, be they concerned with dogma or matters of practical moral application. The latter is of course an order of things which I expressly said I would speak of not at all, for my subject did not require it.

Sixth: That when I said that moral truths have as their object and end only matters of decorum and probability, I never wished to speak of actual truths, such as the precepts of the Divine Law, and those of the natural. I only mean by arbitrary moral truths those laws which depend on the will of man and which are different in different lands, and, consequently vary with the constitution of various states.

Seventh: That it is not true that the existence of our soul and ourselves is identical in the sense that man is a purely spiritual being, and not one composed of body and soul. The existence of our body and other external objects is a certain truth, since not only does Faith make this known to us, but further the wisdom and goodness of God do not allow us to think that he wished to keep man under a perpetual and general illusion in such matters. For this reason, this extension in length, breadth, and depth (that is, our body) is not simply an account given by our senses.

Eighth: That, consequently, we are quite sure that there is something outside of us, and that the belief which we have in revealed truths presupposes and comprises the existence of many external objects. Thus, we cannot believe that it [body, matter] is but a modification of our soul, even in the sense that our sensations truly exist, but that the objects which appear to excite them do not really exist.

Ninth: That whatever the manner may be in which the soul will perceive [verra] itself in the state where it shall find itself from death to the last judgment, it will be certain of the existence of bodies, and in particular of that of its own body, the future state of which will always interest it, as Scripture points out to us.

Tenth: That when I said that the soul was impassible [impassable] by its own essence, I did not claim to say anything else, except that the soul is, by its nature, not susceptible to impressions from outside, which could destroy it. I did not believe that by the power of God it might not be made susceptible to sentiments of sorrow, which Faith makes known to us shall be felt in the other life as the punishment for sin and the torment of the wicked.

The twelfth of March, 1751

                                            Signed: Buffon

SECOND LETTER OF THE DEPUTIES & SYNDIC OF THE FACULTY OF

THEOLOGY TO M. De BUFFON*

Sir:

We have received the explications which you sent us of the propositions which we had found reprehensible in your work, entitled <u>Natural History</u>. After having them read in our special committee [assemblée particulière], we presented them to the Faculty in its general session [assemblée générale] of April 1, 1751. After having heard a lecture on the matter the faculty accepted your explications by its deliberation and conclusion on the said day.

We have also taken into consideration at the same time, Sir, with the faculty, the promise which you made to have these explications printed in the first work which you shall next give to the public, should the Faculty so desire. The Faculty received this proposition with extreme delight, and hopes that you would be so good as to carry out your promise. We have the honor of being, with sentiments of the most complete respect,

Sir,

Your most humble and
obedient servants

The Deputies and Syndic of the
Faculty of Theology of Paris

NOTE

*Georges Louis LeClerc, Comte de Buffon <u>Histoire naturelle générale et particulière, avec la description du Cabinet du Roi</u>. Paris: L'Imprimerie Royale; 1753, Tome IV, pp. V-XVI. (In Sir Harold Hartley and Duane H.D. Roller, eds., <u>Landmarks of Science</u>. New York: Readex Microprint, 1969).

Figure 17 - Albrecht von Haller, by Freudenberger

14. Buffon on Hypotheses: The Haller Preface to the German Translation of the *Histoire naturelle* (1750)
Phillip R. Sloan

Buffon's work received its most extensive German language discussion in two lengthy prefaces to the first German edition of the Natural History, written by the Swiss physician and naturalist Albrecht von Haller (1708-1777), who was at the time the editor of the prestigious journal of the Royal Academy of Sciences at Göttingen, the Göttingische Anzeigen von gelehrten Sachen. As the probable author of the initial short reviews of Buffon's work in that journal,[1] Haller saved his extended discussion for the prefaces of the German translation, initiated shortly after its first publication by scholars centered around the University of Leipzig, and overseen by the Leipzig mathematician and translator of several foreign works into German, Abraham Kaestner.

Rather than focusing on the specific scientific, theological, or epistemological issues posed by Buffon's work, a task carried out by Kaestner's running commentary in the footnotes to the first three volumes, Haller chose to situate the work within the framework of a much larger methodological issue--the role of speculation and hypothesis formation in science. This discussion, often cited by Haller's contemporaries and later reprinted with the title "On the Utility of Hypothesis,"[2] gave a clear articulation to a growing methodological option in eighteenth century philosophy of science. Threading his way between the phenomenalistic positivism of the Continental Newtonians, who had taken Newton at his most literal word when he proclaimed that he did not "frame hypotheses,"[3] and the alternative presented by the Cartesian approach to science in which rational principles and "clear and distinct" ideas were deemed sufficient to guarantee natural knowledge, Haller sees hypotheses, such as are manifest in Buffon's work, as dynamic guiding principles, necessary to direct and organize empirical inquiry into nature. Published at this date in a widely-read source by a leading publisher of scientific works, Haller's reflections on "hypotheticalism" focused interest on this methodological issue in German science and philosophy. Many of the arguments made here were only to be deepened and expanded by Immanuel Kant in the Critique of Pure Reason of 1781,[4] and it is probable that Kant was aware of Haller's discussion from his early years, in that he used Buffon's work extensively in his early course on Physical Geography.[5]

Haller's discussions in this selection also provide a valuable context within which to read Haller's own theories and speculations, as they were stimulated by Buffon's work. In light of this general defense of hypothesis-making, Haller was not so much concerned with the simple empirical truth or falsity of Buffon's specific theories, but with a defense of the legitimacy of proposing such speculations, even when they were incorrect. Within this context, Haller's discussions both constituted a strong defense of Buffon's work, while providing justification for arguing against him on specific points with contrary hypotheses of his own. This becomes of particular importance in the second selection on theories of generation that prefaced the second volume of the German edition.

NOTES

[1] See unsigned reviews for January 1750, pp. 51-55; 59-62; April, 1750, pp. 283-87; January 1751, pp. 3-4; September 1752, p. 939. Several passages in these reviews reappear almost verbatim in the Haller prefaces.

[2] Haller, "Vom Nützen der Hypothesen," Tagebuch seiner Beobachtungen über Schriftsteller und über sich selbst (Bern: 1787), I, 95-118. This contains minor clarifying revisions, and some substantial omissions.

[3] See discussion of the roots of this positivism in the "General Introduction," above, pp. 11-14.

[4] I. Kant, "The Regulative Employment of the Ideas of Pure Reason," Critique of Pure Reason: Transcendental Dialectic trans. N.K. Smith (London: Macmillan, 1963), pp. 532 ff.

[5] Buffon's Natural History is one of three works listed as forming the basis for Kant's course in Physical Geography in its original prospectus of 1757. Although it is unclear from citations which edition he is consulting, it seems improbable that he would not have been familiar with the authoritative Haller edition, the only edition in German until the 1760's. See the prospectus to Kant's course in Kant, Frühschriften, ed. G. Klaus (Berlin: Akademie-Verlag, 1961) 2 vols., I, 284.

ALBRECHT VON HALLER'S, "FORWARD"*

TO THE NATURAL HISTORY

Translated by Phillip R. Sloan

In all conjectures[1] of men, there is a ruling fashion, a custom which is unreflected upon and changeable, which entire peoples follow without being able to give a reason for their obedience to it. These [conjectures] are changeable exactly because they have not been built on true foundations. It is the direct claim of the Truth that it eternally endures.

Almost a hundred years have elapsed in Europe since the explanation of natural events, and the construction of arbitrary systems,[2] became the esteemed priorities of great scholars. Accordingly, once Rene Descartes put forth the organization and construction of the world by means of mechanism, and had taken the liberty to give whatever figure and motions to the smallest parts of matter he needed for his explanations, all of Europe regarded this constructive power as an inalienable privilege of natural philosophy.[3] Worlds were constructed, elements, vortices and screw-like particles were created, and it was believed that this generally was the best thing to do, when the actual events in Nature allowed themselves to be explained only in part through the alleged constructs which had appeared.

But this easy practice did not last as long as the lazy theoreticians of nature no doubt would have wished. The discoveries of the imagination are like a synthetic metal, which can possess the color, but never the density and indestructable solidity which Nature gives to gold itself. A counterfeit coin can be passed, because its newness gives it some luster. But time reveals its rustability, and its counterfeit origin. The controversies, which the natural arrogance and ambition of men inevitably provoke, were the first means by which the weak points of hypotheses[4] were uncovered. A young natural philosopher found an easy way to fame in the refutation of a famous man, and it was much easier for him to find this weakness, than to put up something better in the place of the demolished systems. An ordinary test[5] detects the copper present in noble metals, but making gold is another matter. From this followed a general war between the learned, and since nothing in their conjectures were founded upon Nature, nothing was left remaining of that which had been built up with such great wonder. The Cartesian pushed aside the disciple of the Peripatetic sect; the Gassendist found the weakness of the Cartesian; and a general neglect has since that time buried the disputing theoreticians. Both the conquering and conquered conjectures have returned to the undifferentiated chaos[6] from which the imagination, without the sanction of Nature, had spun them.

A great superiority of modern times has been the continually increasing skill of the craftsman who manufactured devices for the unveiling of Nature. Better telescopes, rounder glass lenses,[7] more accurate units of measurement, syringes and scalpels, have done more for the augmentation of the kingdom of the sciences, than have the creative spirit of Descartes, the father of order Aristotle, or the erudite Gassendi. At every step which drew one closer to Nature, one found a picture unlike that the natural philosophers had made for us. The disdain of hypotheses grew with the conviction that they were as unlikely to be correct as a head of Aeneas, Romulus or Paramond, painted from the imagination, could be similar to the true archetype. Neither the artist nor the natural philosopher would have ever known the archetype.

As mathematical discipline spread over Europe, it taught us to creep rather slowly toward the truth, where we previously wanted to fly, rather than rapidly depart from it. The difficult rule was proposed to man to believe nothing except what could be proven, and little by little it was accepted by cultivated people. England began this, and Boerhaave and Holland followed afterward, and Germany submitted to it. France, so unwilling to forget its countryman, and [finding it] so difficult to deny the authority of the imagination, in which it took precedence over its neighbors, finally was embarrassed, and made in its Academy, in the persons of Reaumur, Maupertuis and Clairaut, a long-overdue apology to the Truth.

The middle path is the most difficult way for man. He will much more readily move from unbelief to superstition, or from a hedonistic life to that of a Trappist monk, than continue living moderately in a reasonable Christianity. The middle path is a line, a way without breadth. [But] who would wish to hold himself on this? As little as man's heart can be held on this middle way, so slightly can his understanding. On one side man flies to the heights with the wings of an eagle, and becomes a Pelagian. On the other side he descends, and becomes, at the hands of the Jansenists, a machine. The same thing occurs in the theory of Nature. One found error in arbitrary explanations, and became a sceptic. The [Platonic] Academy, which wanted to preserve itself from error, sank always deeper in it, and finally believed no longer in anything, so as not to err.

I believe I can with justification attribute the practice of rejecting all hypotheses and all systems to the excesses and extravagances (the fortunate language of our moderate Fatherland has no exact words for [the French] excés and caprice) of human understanding, a practice which always increases more and more, and which can become more detrimental to mankind than the dreams of the philosophers of the schools could ever have been.

Man is by nature lazy. His sluggish life-force[8] inclines him, with a continuous power, to inertness. All wild peoples, who are abandoned defenseless to the instincts[9] of Nature, are extremely discontent with all work. They squat on the ground, smoke tobacco, sleep in their horses bed, and would never arise, if not driven from their comfort by hunger and necessity.

Europeans have more elastic life-forces[10] which their sleep restores. Ambition, example, modesty, admonition, and desire for novelty do not lie so fallow in their understanding, as they do in peoples who do not even have a name for honor and wisdom. But all of these instincts are still scarcely strong enough to stimulate us toward the difficult work required by the investigation of the truth. In sharp-sensing Italy, and in deep-thinking Spain, thousands and thousands of capable minds rest under the shade of superstitions and customs, and sleep and dream away their life-forces.

Yet the soul of the European seethes with the desire for honor, and with the love of novelty, which Linnaeus recognized as the essential privilege of man, by means of which he raises himself over the animals, and by means of which he became the conqueror of both the natural world, and that other world which he has constructed, namely theory.

According to the wish of many new philosophers, a time is now envisioned when in all Europe, all arbitrary opinions and all hypotheses will be entirely banished. The propositions of these geometers, incorrectly informed on human feelings, are accepted; namely, that the inner nature of things cannot be known by man; that we have nothing to hope for but the observation of appearances and that the truth lies in an abyss to which we have no ladder.

What will be the consequence of this language of despair, if it should win the upper hand? Just that which acknowledging the infertility of a new land has. As soon as the daring explorer no longer hoped for gold or other commodities from the uphospitable shore, he would abandon the original discovery, [and] no one else would follow his first explorations. The land remains uncultivated, and its mere name remains in the indifferent remembrance of posterity.

I fear very much that the realm of truth will experience this fate, once we no longer hope to make productive discoveries in it, [and] as soon as our curiosity and ambition no longer expect any reward from it.

When the way to the truth is made so distant for us, so uncertain and so difficult, and we are told that we can only proceed with a sounding lead in hand, and at the same time are informed that we will, in spite of all caution, fall at any moment; or if all our endeavors lead us nowhere except from a vulgar to a learned ignorance, will we really motivate ourselves, will we embark on a difficult journey, from which we have no more hopes? Will not Comfort, like a new Ennius,[11] say in the ear of every learned Pyrrhus: "why do you wish to renounce the certain delights of pleasure and tranquility, and with a chimaerical knight-errantry proceed to defend fruitlessly the rights of the Truth, without the least likelihood of accomplishing something? When you have done all this, you will again and again be only further into uncertainty."

If someone wanted to refute me with the example of mathematics, from which all arbitrariness, all half-truth and indemonstrability is believed to have been banished, my replies to this objection would still remain sufficient to remove any differences between me and my opponent.[12]

Mathematics is concerned with extremely simple things--with lines, triangles, quadrangles, numbers--whose properties are few in number, and are completely determined. It is concerned with these simple dimensions, and seeks relations and connections of the same. No other human science has this advantage, and no other is permitted to employ the same rigor.

[However], since I am writing principally of natural philosophy[13] in its entire extent, it is acknowledged that most of the bodies which constitute Nature, and the motions which constitute its inherent forces, are unknown to us. A theoretician of mathematics begins from such simple things as the point and line, whose complete definition he has at hand. But where does the natural philosopher begin? The elements of bodies are completely hidden. The first seeds of matter, the primeval forces of gravity, elasticity, electrical and magnetic essences, light and fire, which arise from the elements, are only known to us on occasion, and in an incomplete and piecemeal way. The larger structure of animals and plants, the structure revealed to us by the magnifying glass only as a heap of elements, is still only rarely revealed in individual bodies. Even the [knowledge of] the still grosser structure, which can be dissected by the scalpel, measured by the straight edge, and separated by the refining furnace, is so incomplete and uncertain, that the greatest mathematicians, if they should have to write on the faculties[14] of animals [that] have arranged a feather, have demanded that dimensions and angles, and a first principle, must be provided upon which they could build. Can one then demand of us mathematical rigor? Can a totality of concepts become certain, if the individual ones are still indeterminate?

In truth, this austere beauty, mathematics, is not as hostile to hypotheses as it pretends to be. It regards them as a weakness of which it is ashamed, and still cannot entirely do without them, and it acknowledges their temporary value. The great excellence of today's higher mathematics, this dazzling art of measuring the unbounded, is established on a pure hypothesis. Newton, the destroyer of arbitrary conjectures, has not entirely been able to do without them. Just as on the body of Achilles, there must still be a vulnerable spot on his heel, so an Euler, and even a Mahler[15] or a Gautier[16] could refute him. Was his ether[17] the medium of light, sound, sensation and elasticity, not a hypothesis? And when this

Prometheus turned to practical applications,[18] and wanted to measure the times, and fix boundary conditions on events, was he not forced to set as a foundation arbitrary and very doubtful conjectures?

After a Newton, no one will be completely ashamed to propose something which is not completely demonstrable. If such a respectable intellectual has used the probable as currency, then it cannot be entirely without value. It is [a currency], a paper money, which serves merely to sustain trade between the learned. Certainty is a prohibited currency, whose value can never be depreciated. It would be pleasing to us, if we had so much of it that we could dispense with arbitrary currency.

However, this is not feasible. Without such arbitrary currency we must remain silent on almost all of natural philosophy. All the parts of human science would become nothing but <u>fragments</u> and independent pieces without connection and unification, if we did not fill in the missing parts with probabilities, and construct a building instead of a ruin. I intentionally make use of this comparison. I have seen books by mathematical theorists who have written on the structure of the human body, and who believed themselves to be obliged to banish everything inexplicable from their work. How inadequately, how disjointedly, how generally, and how vaguely do they have to speak in order to avoid the probable.

Thus I come to the true utility of hypotheses, which are, to be sure, still not the truth, but which lead to it. Furthermore, I would assert that men have still found no route which would have led them there more successfully. They are the clues which lead to novelty and truth, and no discoverer comes to my mind who has not made use of them. When Kepler wished to ascertain the laws of the course of the planets, he formed for himself a conjecture, an improbable conjecture, which has been proved to be unfounded. And yet, this conjecture led him to the wonderful law, which posterity has confirmed, concerning the connection of the [law relating] the periodic revolutions of the planets and their distances from the sun, to a first principle[19] which was firm enough for Newton to build upon.

The Alchemists created fictions for themselves--mountains of gold, and metamorphoses greater than Ovid's. They worked to reach these phantoms, and found on the way many useful truths, perhaps more useful ones than a means to make lead into gold, which in a short time would make us, with all this conceivable gold, impoverished, and out of necessity [gold] would be replaced with diamonds, or with some other sufficiently scarce and permanent measure of value for conducting our business.

The great Lawgivers of Botany have themselves made arbitrary, ruling principles, according to which they have formed the Classes, Genera[20] and Species[21] and by means of which they united or divided the plants. All of these laws are arbitrary, [and] they have in addition all been found unreliable. But they have, nevertheless, done us an unbelievable service. The innumerable multitude of plants has now been placed in such order, that we distinguish ten thousand plants more easily and with infinitely greater certitude, than the Ancients did their six hundred. The hypotheses which have been admitted, have truly discovered demonstrable similarities. They have still not led us entirely to the truth, but nevertheless much closer to it. Each new theoretical system leads us somewhat nearer to it, and without these we would not have taken a single step.

I find this comparison so clear for my purpose that I do not yet wish to leave [the matter]. If one used no hypotheses in botany, and only, like Clusius[22] or Bauhin,[23] sought entirely to describe individual plants, then accordingly this mathematical method, which our opponents praise, would have done nothing at all of benefit for this science.

Clusius and [Caspar] Bauhin were great herbalists and learned men. Their understanding and industry are without reproach. But they had no goal and no order in their work. Although they had no theoretical system, they needed to have one. Thus their plants remained without order. The dissimilar ones were mixed together, and the similar ones separated, contradicting Nature, and so spoiling the nomenclature, that this great aid for the memory more deceived than served it. If Veronica, Onagra, Scutellaria, Epilobium, Peplis and Lysimachia are all called Lysimachia,[24] and all bear the same name, then according to the rules of the [botanical] characteristic, they are similar beings. However, since this name has been distributed unsystematically, it designates both similar and dissimilar entities, and thus leads the learned astray by its formal similarity.

Since these great men had no system, and assigned no part of the plant a special priority that was to be the distinguishing character of the species,[25] they frequently did not note at all the shape, number, and position of the blossoms, nor their sepals, stamens and pistils, nectar-containers, or the partitions of the fruit. Their descriptions were thus so incomplete, that they very often could not be used at all, if it had not been not for the industriousness of modern [workers], who had cut away the covering of similar plants, and determined the size and number [of these structures].

Already in their lifetime, Cesalpino[26] had appeared, a man who had embraced the love of order and the desire for total explanation of the Peripatetic school. He was a mediocre herbalist, and Clusius had discovered more plants than Cesalpino was acquainted with, since his herbal, according to the testimony of Micheli,[27] described not even 900 species.

With such a small collection Cesalpino set out to create systematic Botany. It was almost impossible that he could, with his slight knowledge of the individual parts, have surveyed and arranged the whole flawlessly. He took the fruit, and indeed mainly only that part bearing the seeds, as the clue, and yet approached nearer to the truth, and determined more true similarities, and more natural Classes, than all the herbalists from Theophrastus[28] until Tournefort.[29]

The latter regarded plants from another perspective—in terms of the flower—a perspective from which the French are accustomed to regard all things. He selected a few shapes of the petals, which were neither sufficiently defined nor differentiated, and by means of these incomplete hypotheses erected a structure which incurred the wonder of all Europe. Even his opponent, Ray,[30] made use of his illumination to clear up the arrangement he prescribed for the herbs in his old age. Tournefort's renown lasted almost thirty years after his death, and he was a law-giver in his science. But, as we have already said, hypotheses are a scaffolding by which truth is approached, and they may not be allowed to remain. Linnaeus appeared in the Northlands. He selected for himself new axioms, and grounded his hypothesis on another hypothesis, [namely], on the artificial arrangement of the plants according to their stamens and pistils, which according to a purely probable conjecture, had some similarity to the male and female parts of animals. This new theoretical system performed the greatest service, and set everything astir. All parts of the flower and fruit of all the Herbs were described with the greatest exactitude, because they had, by this time, been noted. Botany at present advances its lead over all other sciences. Not only is it nearest to completion, and has little by little determined the nature of almost all its Classes and resemblances; it has [also] imparted its laws to the entire Kingdom of Nature. Zoologists and minerologists have received their laws from it and accepted them as the Romans did from the [council of the] Areopagus (sic).

Up to now I have established my thesis with examples. It will be just as easy for me to support myself with abstract concepts. When men act, they become more efficient according to the relationships between the [different] strengths of their instincts. Their natural inactivity is overcome by ambition, and other similar agitations of their elastic force.[31] Life, money, and repose are readily abandoned as soon as their ruling instinct demands it. And [the concept of] the ruling instinct is essentially a hypothesis, and accordingly has the pleasure of destroying another [hypothesis].

A theoretical system that should bear our good name, a conjecture which has arisen from our life-forces, does for intellectuals what ambition did for Alexander. Effort, expense, time, experience, art and mechanisms, and all the powers of the will and understanding, are employed freely[32] and without conflict, if we have at the same time a goal, and if our theoretical system thereby becomes more probable, more certain and more acceptable. Who would have counted and measured the stamens in almost innumerable plants and flowers, if they had not been for Linnaeus the essential [structure] in his theoretical system, and thus the chief means to complete and establish a general rule[33] in botany? Once Newton had thought

of decomposing the rays of light, he spared no costs, [and] an
artisan had to do his utmost to supply him with devices which
were delicate enough for his purpose. He undertook the most
laborious and difficult measurements and separations [of light
rays], because it concerned the truth of his theory.

Yet hypotheses have a still more serious utility, which
they would maintain in the most unreceptive natural philosophy,[34] if indeed one should arise which loved the truth
purely on account of its beauty, without regard to fame.
Namely, they pose questions, which without an hypothesis would
not [otherwise] have occurred to us, whose answers are exacted
from experience. [This is] a result which has untold benefit
in the sciences. The least of men would have sufficient keensightedness in and of himself to direct questions, and be able
to perceive the most useful [things] from one view on a subject. But a system, or the destruction of one, poses an innumerable quantity of questions which we can pose to Nature,
and which it very often answers. Thus the Ptolemaic, Tychonic
and Copernican systems of the heavens have directed the
astronomers to what they would have to observe, and discriminated for them the observations by which the truth could be
discerned.

Each and every probability possesses a portion of the
individual truths, which in conjunction with still others that
we still lack, constitute a general principle. Thus we perceive with precision from those which we do possess, those
things which we lack, and we uncover a set[35] of experiments and
observations which, if we possessed them, would turn our probability into certainty. A theoretician of Nature acts like a
land surveyor, who begins a map on which he has determined some
locations, but lacks the positions of other places in between.
[He] nevertheless makes an outline, and according to half-certain reports, indicates the remaining towns, of which he
still has no mathematical knowledge. If he had made absolutely
no sketch in which he combined the certain and uncertain [components] in one composition, then his work of determining more
exactly the locations and boundaries which still remained would
be much more difficult and almost impossible. Indeed, it would
not be possible, because [the work] would have no coherence,
and would constitute no whole.

Finally, the merely probable theoretical systems are also
of the greatest utility because they generate jealousy and competition among the learned. A runner who, for the sake of a
prize, runs against a competitor, exerts himself in a wholly
different way, and runs much faster, than if he ran by himself
alone. The prize is the glory of those who possess the Truth,
and the common good enjoys the fruits of the efforts of the
conflicting [parties].

If I might myself give an example, which can be done without conceit. Assuming Boerhaave's and my own opinion on
respiration would never have been doubted, then I would have
satisfied myself with one or two basic principles, and my

conviction would not increase. The extent of the sciences is immeasurable, [and] one does not know where he is supposed to begin to work in a field whose breadth and fertility are equally great. However, controversy teaches us to select a portion of the field on which we work more diligently, and if controversy is raised about it, to delimit it rigorously. [Thus], I was compelled to make new investigations, and to repeat these often, and I found not only the truth of the things which I maintained, but I also found new foundations, and persuaded myself that no ground could any longer remain by which one could doubt a theory whose correctness I had seen [verified] with my own eyes in so many ways.

This is a minor example. However, that concerning the figure of the earth is a major one. Newton's opinion of [its size] was somewhat greater, and Cassini's somewhat less, than [was] probable. A controversy ensued, and this controversy brought about two splendid expeditions, one to the [North] Pole, and the other to the Equator, which henceforth determined the case to be not only in accord with Newton's principles, but also determined it more exactly and certainly than a purely general calculation would have been able to do.

Pure geometrical propositions, or generally and completely true explanations of natural appearances, would generate neither controversy, nor rivalry. Hypotheses do this. Indeed, they put many things in a position in which they can be defended, but much still remains that can be assailed. Disputing sects are like flint and steel, which indeed generate fire, but thereby also illuminating light.

No one will, to be sure, believe that my discourse in defense of hypotheses is so vulgar, that I wish to place probability on the side of truth. No. The moon will never shine like the sun. But still its weak and cold glimmer is useful to us. The Ptolemaic system of the world was false, [and] no one at present doubts its groundlessness. Many valid observations lay mixed together beneath still more untrue opinions. And yet, the world has used this hypothesis with great utility for so many years, and it has thereby in ordinary life almost an advantage which we seldom have from the truth. Finally, the light dawned, and the defects of this theoretical system--the crystalline heavens, the presumptuous location of the earth[36] in the middle of cosmos,[37] and the unnecessary velocity of the sun and fixed stars--have [been] excluded from the true [account]. If one had prematurely demanded general certainty of the astronomers, which they were not in a position to supply, then for so many centuries there would have been absolutely no concept of the proximity or distance of the stars from us, of their order and movement, or of the relations of the parts of the cosmos among themselves. One would have been unable to form any concept of the commonest appearances, the eclipses, and would have remained in barbarous ignorance, from which Ptolemy nevertheless has indeed delivered us.

Therefore, concerning assumed and unproven theoretical propositions, no one will be able to complain if we grant Truth its eternal priority, and give probability only the value that it actually has. No one will be deceived if we supplement the deficiencies of truth[38] with the probable, or if we build bridges with it over the abyss of ignorance, if at the same time we caution that they are only trustworthy to a limited degree. We can assume what we wish, if we openly remind the reader concerning the assumed things, that our probable conjecture is still greatly, moderately, or only slightly removed from the truth, when we admit that in order to be convinced, we are still lacking some unperformed experiment, or measurement, or [knowledge of] the structure of still undetermined parts. Can anyone complain if small change is declared to be small change, and its value is established no higher than the value of the silver in it? He alone is deceived who accepts it for pure silver.

I believe that from these considerations, those who exert themselves to explain the actions and works of Nature will be given the freedom to construct something unproven, something probable, along side the proven, and thus to erect a structure whose main supports are indeed strong, but whose parts are not all of the same unshakeable strength. Experience has taught us that such hypothetical components have, for the most part, been the cause of their subsequent replacement by the most reliable material, either by their proponent, their opponents, or by posterity. And I believe that it is henceforth probable that hypotheses, not only with regard to the understanding, but also to the will, are beneficial for the growth of the sciences.

This consideration is all the more natural, since Mr. Buffon, as the chief author of the present work, has taken the liberty to expound hypotheses, and to some extent wholly new, wholly foreign, and to an uninitiated reader, improbable theoretical principles.[39] For this freedom also has its great utility. If one travels along the same path as his predecessors, a new way is not sought, and nothing is discovered. It is true that the discoverers of new worlds are not always happy, some suffering shipwreck, and others returning without success. Still others discover infertile regions which we are uninterested in knowing about. However, if no Columbus or no Magellan would have set sail from Spain, many shipwrecks would have been avoided, but the New World would also not have been discovered.

Mr. Buffon appears to be on such voyages, which very well might discover new seas and new worlds, and thereby is neither spared any [of the] pains of sea voyage, nor the peril of shipwreck, because an erroneous theory is a shipwreck for an investigator [of Nature]. In the three parts of the work which I introduce here there is none in which he has not proposed an unusual[40] hypothesis, and indeed more than one. His companion, Mr. Daubenton, appears to restrict himself more to Nature itself, and yet he has, in the comparison of the generative parts of the two sexes, undertaken to demonstrate as something novel, an accepted, and indeed ancient, conjecture.

The title of this work promised something less than it delivered. One expected merely a catalogue, along with a description of the curiosities of the Royal Cabinet. I use the word "merely," although such a description [would] indeed constitute a very large and useful, and almost inimitable work. Consider the size of France, the number of its flourishing plantations [41] in America, Asia and Africa, and the general instructions, given long ago and often repeated, that from all districts everything remarkable displayed by Nature in these regions [was] to be sent to the Royal Academy of Sciences and the Royal Collection. Or consider the zeal and the ambition of the French, with which they burn as soon as the name of their king is heard, and their desire to enter his circle of acquaintance. Or reflect upon how many great naturalist's collections have been combined in the Royal collection. Then one would expect from the present work nothing else than something extraordinary, and we do not doubt that the fulfillment of this hope, as great as it is, will be completed and assisted.

The [three] parts of this great and magnificent work we have on hand, belonging to the description of the Cabinet, will be followed afterwards by a fourth and a fifth, in which all the Quadrupeds, and all the animals living [both] on land in in water, will be described. In the sixth will be described the Fishes; in the seventh the Molluscs; in the eighth the Insects; in the ninth the Birds; and in the tenth, eleventh and twelfth, the Plants. In the thirteenth, fourteenth and fifteenth, the Minerals will be described. In this order will be disclosed[42] the whole extent of Nature, and the entire Kingdom of Creation that is subject to man and created for his contemplation.

However, I have said that this work contains much more in it. Indeed, several demonstrations[43] of this will follow in the succeeding parts. In the fifth volume will be found a physiology of animals.[44] In the eleventh, a treatise on agriculture and the cultivation of plants; and in the thirteenth, another on the formation of stones and metals. Yet already the first, second, and part of the third [volumes] contain much more than the presumable content of a catalog. And precisely here lies the foundation of our defense of conjectures which are not completely proven. Thus, these parts contain the thoughts of Mr. Buffon on the origin and formation of the world, the generation of animals, and other hypotheses which pertain to the general structure of the terrestrial globe and the lesser cosmos.

[Such conjectures] are precisely of the kind which need a justification, but they are also capable of one. Everywhere one finds much information about things, many experiments, and much insight. Yet the author always goes somewhat further than his information, experiments, and insight. I would be glad to converse on this with the reader, and speak my own thoughts to you about those of Mr. Buffon with freedom and moderation, if it were possible to express them briefly, or [if] my time were sufficient. [This must be] elsewhere than in a preface to a book. Since these two conditions cannot be met, I can do nothing more than admonish the reader to read this work with philosophical attention. He will find much [in it] that is new and unusual, and there will be few readers who will not have learned something thereby. They will, however, also encounter such propositions as must be regarded with the restrictions of which I have spoken above.

Delivered at Göttingen, 23 Sept., 1750.

NOTES

*Albrecht von Haller, "Vorrede," in Buffon, Allgemeine Historie der Natur nach allen ihren besondern Theilen abgehandelt; nebst einer Beschreibung der Naturalienkammer Gr. Majestat des Königes von Frankreich (Hamburg und Leipzig: Grund und Holle, 1750). Thl. I, pp. ix-xxii.

[1] [Meynungen.]

[2] [Lehrgebäude]

[3] [Weltweis]

[4] [Hypothesen]

[5] [Probestein]

[6] [unparteyisches Nichts]

[7] [Glasstropfen]

[8] [Kraft]

[9] [Trieben]

[10] [Elastische Kräfte]

[11] [Quintus Ennius, Roman poet.]

[12] [meinen Gegner mit mir zu vereinigen.]

[13] [Naturlehre]

[14] [Kräften]

Buffon on Hypotheses   309

15[Possibly Johann Andreas Mahler, German physician and chemist.]

16[Most likely Abbé Joseph Gautier (1714-1776), professor of mathematics at Metz.]

17[Allgemeine Materie]

18[sich näher zur Erde]

19[Gründe. The reference is apparently to Kepler's hypothesis of the sun's attractive virtue.]

20[Geschlecter]

21[Gattungen]

22[Charles de l'Ecluse (1526-1609), Renaissance herbalist.]

23[Caspar Bauhin (1550-1624), Swiss Renaissance botanist and herbalist and author of the influential Prodromus theatri botanici (Frankfurt, 1620).]

24[These are species of European willows.]

25[Art]

26[Andreas Cesalpinus (1519-1603), Professor of Medicine at the University of Pisa, and author of the influential De plantis libri xvi (Florence, 1583).]

27[Pier' Antonio Micheli (1679-1737), director of the botanic garden at Florence.]

28[Pupil of Aristotle and author of the main text on botany to survive from Antiquity.]

29[Joseph Pitton de Tournefort (1656-1708), leading French Botanist and main eighteenth century rival of Linnaeus in botanical systematics. Author of the influential Élémens de botanique (Paris, 1694).]

30[John Ray, (1627-1705), English Botanist and Zoologist author, among other important works, of the Methodus plantarum nova, (London, 1704), being referred to here by Haller.]

31[Schnellkraft. Haller is implying in this section that each of the main human drives is constituted by a distinct force or Kraft.]

32[mit Luft]

[33][Reich]

[34][Weltweis]

[35][Verzeichnis]

[36][Erde]

[37][Welt]

[38][Wahren]

[39][Lehrsätze]

[40][Reading eigene for eigne.]

[41][Pflanzstätte]

[42][und folglich findet man.]

[43][Beweisthumer]

[44]["Im fünften Bande eine Physiologie der Thiere." This a curious claim. In the original prospectus of the Histoire naturelle, published in the Journal des sçavans for October, 1748, pp. 639-40, mention is made of all other topics outlined in this paragraph except for an intended "physiology" of animals. The fifth volume is stated to deal only with the amphibious quadrupeds and cetaceans.]

15. Haller on Buffon's Theory of Generation (1751) (selected)
Phillip R. Sloan

Haller's preface to the second volume of the <u>Historie der Natur</u>, dated March of 1752, had in fact been published separately in a French translation a year previously as an explicit discussion of Buffon's theory of generation and the issues raised by the controversial experiments of Buffon and Needham, as these had been reported in 1749. The translator of this French edition, which apparently appeared both at Paris and Geneva in 1751, is not indicated, and a contemporary review speaks only of a "young bilingual scholar" as responsible for it.[1]

This document is significant in that it represents an extended discussion of Buffon's theories by Europe's leading physiologist and medical theorist, Haller. Haller's discussion has drawn the attention of several scholars because of the role this discussion appeared to play in Haller's own restatement of the preformation theory in embryology after a period when he had advocated the epigenetic option.[2]

Briefly summarized, Haller had originally been an adherent of the "spermist" version of the preformation (or more correctly, "preexistence")[3] theory of generation advocated by his teacher Hermann Boerhaave (1668-1738). In this theory, the new generation is assumed to arise from a preexisting minature encapsulated in the head of the spermatozoa. By the mid 1740's, however, Haller had abandoned this theory due to its inability to explain regeneration phenomena, especially the regenerative ability of the fresh-water hydra described by Abraham Trembley (1700-1784) in 1741.[4] In its place Haller adopted a theory of the successive formation of parts (epigenesis). By the latter 1750's, however, Haller had abandoned epigenetic theories and returned to a subtler version of the preformation theory of the "ovist" variety.[5] Through the exposition of this theory in his highly influential medical writings, and particularly his multi-volumed <u>Elementa physiologiae corporis humani</u> (1757-66), the standard text on medical theory for late eighteenth-century Europe, the preformation

311

theory, supported by further endorsements by Haller's contemporaries Charles Bonnet (1720-1793) and Lazzaro Spallanzani (1729-1799), continued as the main account of generation into the early nineteenth century.[6]

Scholarly controversy has concerned itself with the relative importance of empirical observations and larger speculative theory in bringing about Haller's important change of position, and the role which Buffon's own theories played in this change.[7] As this text reveals, Haller seems particularly concerned with the need for teleological forces to account for the embryonic formation, particularly when the whole problem of variation is considered. The issue of preformation and epigenetic development would thus seem to be subordinate first to the issue of whether Buffon's non-teleological internal molds could adequately deal with the problem of organization in light of variation, and secondly, whether the immanent forces involved could be anything other than teleologically directed forces. As a consequence of Haller's conclusions here, the options open to him would then seem to be either a looser form of preformationism, in which the range of variation Haller describes could be reconciled with some kind of prior organization of parts before fertilization (the position he eventually adopts), or else the assumption that development is governed by special teleological forces, an option that later becomes widespread in the German tradition, but is not in the end endorsed by Haller.

Of more general philosophical significance is the question Haller raises concerning the theological issues that seem to be at stake in allowing that microforces of some kind could account not simply for the organization of inorganic bodies, but also of living things. This issue, which had constituted the larger philosophical core of the preformationist-epigeneticist conflict since Descartes,[8] persistently raised a theological issue concerning God's action in the generation and even origin of living beings. The preformationist theories of generation, in their many varieties, seemed to many most compatible with two larger metaphysical framework-principles: the doctrine of creation, and the assumption of nature as a universal mechanism.

Once it was admitted, however, that matter contained inherent self-organizing powers, a more complicated picture emerged. On the one hand, questions could be raised, as they had been previously by Leibniz, over the adequacy of the mechanical philosophy itself. On the other hand, if it was allowed that matter was no longer inert, but had within itself active principles, the possibility of a secondary causal order, acting independently of God even in the coming into being of organisms, became a significant issue. Consequently, the issue of the natural generation of organisms became involved with the possibility of a Deistic or even atheistic natural philosophy.

It also provided an arsenal of arguments in favor of a non-mechanical conception of nature. Haller's discussion reveals much of the issue at stake in this conflict.

Because the French text appeared first, we have selected it for translation, rather than the German. Material significantly differing between the two texts is inserted between asterisks.

NOTES

[1] Unsigned review, Göttingische Anzeigen von gelehrten Sachen (24 February, 1752), p. 181.

[2] Recent discussion of this issue can be found in Shirley A. Roe, "The Development of Albrecht von Haller's Views on Embryology," Journal of the History of Biology 8(1975), 167-90; and François Duchesneau, "Haller et les theories de Buffon et C.F. Wolff sur l'epigénése," The History and Philosophy of the Life Sciences 1 (1979): 65-100.

[3] The important distinction between a preformation of parts prior to fertilization, and the preexistence of the complete embryo in miniature, dating from the creation of the world, has been made by Jacques Roger, Les sciences de la vie dans la pensée française du XVIIIe siècle, 2nd. ed. (Paris: Colin, 1971). The reigning theories of generation of the latter seventeenth and early eighteenth century are correctly designated as predominantly preexistence theories.

[4] Abraham Trembley, Memoire pour servir a l'histoire d'un genre de polypes d'eau douce (Paris: Durand, 1744) 2 Vols. For discussion on this see especially Roger, op. cit., pp. 390-96. See also the valuable discussion in C.W. Bodemer, "Regeneration and the Decline of Preformationism in Eighteenth Century Embryology," Bull. Hist. Med. 38 (1964), 20-31. The critical text of Haller on this is the notes he has added to Hermann Boerhaave's Praelectiones academicae, ed. Haller (1744), vol. V, translated in H.B. Adelmann, Marcello Malpighi and the Evolution of Embryology (Ithaca: Cornell University Press, 1966), II, 893-900.

[5] Roe, op. cit.

[6] See Elizabeth Gasking, Investigations into Generation, 1651-1828 (Baltimore: Johns Hopkins, 1967).

[7] See references in Note 2 above.

[8] See Roger, op. cit.

ALBRECHT VON HALLER, REFLECTIONS ON THE THEORY

OF GENERATION OF MR. BUFFON*

Translated by Phillip R. Sloan

It is with hesitation that I address the public at this time. An author suffers who is obliged to ask the patience of his readers.

My physical illness weakens my soul, and deprives my mind of a portion of its powers at a time when my undertaking demands them all. I must make a critique of a man whose nation, so slow in recognizing superiority, has judged him meriting its praises. This undertaking demands superior knowledge and judgement, but time is pressing, and by necessity this preface is to be printed at a fixed date.

Although it is contrary to my ordinary [practice], I cannot avoid reducing to two principal points my reflections concerning Mr. Buffon's system of generation. First I will state some reasons which prevent me from entirely adopting the opinions of this ingenious Frenchman. Secondly, I will examine whether his reestablishment of generation from putrefaction is not prejudical to revelation. As it is impossible for me to keep my identity a secret, I very much sense that I could incur the resentment of the author insofar as my opinion is found different from his own. However, I rely on the equity I have endeavored to pursue in my critique, on the esteem I have for the talents and works of Mr. Buffon, and on the undoubted presumption that the qualities of his heart are equal to those of his intellect. I have only too often myself suffered on behalf of the truth, and my true opinions have not been recognized, when my zeal for the divine rights of the truth has been attributed to the desire to tarnish the fame of others. [This is] an unjust suspicion which ought to be dispelled, if the dispassionate justice that I universally render to all those whose works have populated or extended the kingdom of the sciences would be considered.

Mr. Buffon has himself reduced to a summary his conclusions and experiments on page 420 of the second volume of his work. I will reprint them here so that my remarks will be sufficiently based on his own words:[1]

. . . .

Such is the theory of Mr. Buffon. I would offend the intelligence of my reader, if I tried to tell him that Mr. Buffon's conjecture concerning the collection of the seminal fluid from all parts of the body, has a great similarity to the ancient doctrine of Hippocrates, but that it has some pecularities, and is quite different from the generally received theory of development with regard to that organized matter,[2] which is equally suited to become a man, animal or plant. Indisputably, this opinion derives its greatest probability from the universal conformity of Nature as a whole. The laws of gravity, attraction and elasticity, whose dominion extends to infinite distance, seem to demonstrate a great inclination in nature to govern several bodies by the same forces, and accomplish several effects by the same laws. One easily discovers the traces of a Creative Mind[3] in this artistry, producing such different conflicting and complicated effects by the same causes. And one finds in this intelligent economy proofs of the all-governing Divinity which always chooses the shortest means, and is never so extravagant as to employ two laws where a single one could suffice.

The simplest formation that we would be acquainted with is that of salts, whose structure resembles that of crystals. In a solution of salt exposed to the cold, a multitude of angular particles separate out from the water, in spite of its uniform appearance, which, according to the differences of the salts, form triangular, quadriangular, and multi-angular crystals. These crystals form, by their mutual attachment and connections, different kinds of regular bodies. Everyone is acquainted with the cubical particles of common salt and sugar, and the triangular spires of saltpeter and rock crystal. The great masses of rock crystal, of which I have myself seen specimens *in the Alps*[4] weighing almost seven hundredweight, and the almost invisible crystals of salts, are [both] composed of particles which are entirely similar in themselves, and in the large masses they compose. The <u>Homeomeria</u> of Anaxagoras evidently govern this portion of nature, wherein one sees particles form into a whole with constancy and regularity, without the least suspicion of a semen or seed being involved.

From salts to snowflakes, to the trees of Diana,[5] to the feathery plumes of ice, there extends an uninterrupted chain of organized beings which without any other artistry, are produced by the force of attraction alone.

Is it a great distance from there to the <u>Conferva</u>[6] which can be either short or long, knobbed or unknobbed, depending on the greater or lesser movement of the water, and forms itself before our eyes from a green scum? And is there not a great affinity between this most simple of all plants, and the genus of the Mushrooms, and thence with the entire plant kingdom? Is it so distant from these organized beings of which we have spoken, and which are devoid of all awareness,[7] to the most simple animals, in which all the parts consist only of a homogeneous and uniform gelatine *and which arise in all kinds of rounded shapes*[8] in the scum of stagnant water, or regenerate themselves, after being cut up by the naturalist, from a moist and sticky jelly into which they return shortly afterward? Where does the dominion of general laws end? Where is the point at which their formative power terminates and beyond which they are impotent?

These reflections we have made must singularly prepare us to find less paradoxical Mr. Buffon's doctrine. But we see that his experiments prove it for him. Buffon, Daubenton and Needham have observed many times, with expert vision, that fermenting grain contains a liquour which swells into the shape of horns and spires, and divides itself at the extremities, and emits from these fissures small mobile bodies, oval in figure, and entirely similar to other microscopic animals. These bodies are not the production of some invisible mites, because boiling water, which is a mortal poison for all animals, and their eggs and germs, does not inhibit this productive force.

At this point Mr. Needham provided a link in the great chain of the plant kingdom, and soon afterwards, Mr. Buffon joined to it another drawn from the animal kingdom. With the help of the microscope, knotted filaments are observed in the semen of all kinds of animals, from whose knots moving globules are seen to leave, and swim in the semen, and which have a very distinct similarity to the moving globules which arise from cornmeal. At this point the plant and animal kingdoms, the vegetative and the generative force, are united. Life is a higher stage than vegetation, and that a level higher than crystallization. An unbroken chain of organized beings extends from the organization of an Alexander to [the formation] of a snowflake.[9]

I do not believe that I can be accused of partiality in the exposition of the teaching and principles of Mr. Buffon, and I will continue to expound his opinions.

The spermatic animals of Hamm or Hartsoeker,[10] known to everyone, and ordinarily attributed to Leeuwenhoek, because he had observed them in the greatest number of animals and described them with the greatest care, are not, properly speaking, animals, according to Mr. Buffon. They are organized particles of productive matter, and are seen to arise from the knots on the filaments in the semen. They change their shape, and instead of increasing, they diminish in volume. Little by little they shed their tails, which do not essentially belong to them. They cannot be animals, since they are found in the infusion of roast meat, where heat would not have failed to destroy all the germs of life it would have contained. Finally, they are not specific to male animals; they are also observed, although in smaller number, in the humour of the corpora luteii[11] which are found in the female ovaries. Both sexes thus have a semen, with moving organized particles, which generate the fetus by their union. Mr. Buffon here approaches the hypothesis of the Ancients, which had prevailed until the time of Steno.[12]

These particles are entirely similar to all the particles of the father and mother, and they have acquired their shape from having been lodged in their interstices. Nature, that expert artist, has separated them from the crude *and unshaped*[13] parts, and organized them from the human humours, and has imprinted upon them the image of all the parts of the body of the father *and mother*.[14] From this arises the resemblance of the child to its parents, the blending of the traits of the father and the mother in their offspring, and the mottling of animals, whose parents are of different colors, *and the Mulattoes intermediate between whites and blacks.*[15] Finally, a number of questions which are almost insoluble on the theory of the unfolding [of preexistent germs] are here answered. If it is asked by what means these particles are able to assume the interior structure of the father's body, since to speak properly, they could only be the impressions of hollow vessels, Mr. Buffon answers that we do not know all the forces of Nature, [and] it has reserved to itself, to the exclusion of man, the art of continually fashioning machines, which exactly manifest the internal form of the mold.[16]

I have now said enough for my purposes, and have let Mr. Buffon speak. It is now time that I express my own thoughts to my reader.

I do not doubt that Mr. Buffon merits the prize that is due all those who have planted Truth on the debris of a generally received error. By his experiments, as well as those of Mr. Needham, it would appear to be incontestably proven that the spermatic animals are not a peculiar characteristic of man, but a common genus of some mechanisms[17], which are found in the substance of all kinds of animals and vegetables placed under certain circumstances. Indeed, an individual well-versed in the use of microscopes, who has repeatedly observed all the signs of life in the medium of the seminal fluid, confirms my idea that these machines even might be true animals. Mr. Needham himself is here at odds with his friend, and grants the attributes of life and spontaneous movement to the spermatic animals. *Furthermore, two skilled experts on lower creatures[18] have, after repeating Buffon's experiments, returned even these little worms to the Class of animals.*

However, is it not possible that these worms could be nothing more than creatures which arise in all rotting humours, and are only found in great quantity in the seminal fluid, precisely because the seminal vesicles, and the region of the large intestine are the most proper location for putrefaction? Furthermore, isn't that volatile, alkaline odor which is emitted by all putrefying things found in the semen of most animals? Would it even be probable that these worms would have ever existed in the body of the mother and father in the character of organic particles? Here I find it impossible to defer to the opinion of Mr. Buffon, and the eternal rights of the Truth cause me to abandon his hypothesis. A multitude of objections simultaneously present themselves to my mind and contest for first consideration *and other naturalists have likewise brought forth others*.[19]

I will begin with the internal molds. What is there that could represent itself by something similar? Is it possible that nature, from a viscous matter, could produce an infinitely miniature creature, perfectly similar to the parent, and in which the blood, for example, infinitely surpasses in refinement that which flows in the veins of the parent? Is this matter susceptible of taking a form other than that which it takes from the interstices of nutritive particles between which it occurs, and from which, according to Mr. Buffon, it has expelled its appropriate surplus? Is it these elementary interstices which constitute the specific form of man? Is it from these that one takes his large nose, and another his large mouth? But perhaps these objections, and some others that have been made to Mr. Buffon do not have sufficient force. Also, I will not stop to elaborate them. I prefer simply to deny to Mr. Buffon that off-spring resemble their parents. If I prove this point, the offspring are no longer images of their parents, and the remainder of the edifice will collapse upon itself.

We pass over [the fact] that for the examples that can be adduced of offspring which have resemblance to their parents, there are always a greater number of them which have acquired neither their traits nor likeness. My thoughts go still further. There is no man who is similar to another in the internal structure of his body, and as a consequence no child which is [exactly] similar to its parent.

It is anatomy which has taught me such a troublesome truth, and has only too often increased my work. If men were [exactly] similar to one another, one would need only a single description and a single diagram of the arteries of the hand, for example. If these drawings were once [exactly] similar to the original, they would remain so forever. But nature is far distant from such a convenient uniformity. There have never been two men in whom all the nerves, arteries, veins, and even all the muscles and bones, were not infinitely different. After having made fifty descriptions of the arteries of the arms, head or heart, I have found all fifty entirely different. The most boring work in the world is certainly that of summarizing the arterial [system] in one general and uniform account. This variety rules in all of nature. No plant has ever been [exactly] similar to that from whose seed it has come. However, according to Mr. Buffon, this must happen, since there is here no mixture of the seminal fluids of the male and female, in which one would have been able to disturb the structure of the other.

This variety is much greater than one has been accustomed to believe in the ordinary manner of teaching anatomy. It is so especially great and infinite in the nerves and veins that it is almost impossible to make a description of them, and one is almost tempted to believe that nature not only had no model in the formation of the animals, but even that it worked without plan. This would, indeed, be to push doubt too far. Not only is there a continual difference in the size of the branches, in their angles, locations, divisions, the placement of the valvules, (and) in the extremities of the small branches, but even the number of parts is different in each individual. The great branches often vary, the medium ones always, and in the smallest this variation constantly extends to the two sides of the same body. The child is thus not the [exact] image of its parent. If it were, would it have parts missing in its parent? It is a commonplace for anatomists that thousands, and thousands of millions, of vessels are still found in the fetus which no longer occur in mature adults. The fetus has two umbilical arteries and an umbilical vein, an [umbilical] ureter,[20] a mammary gland, a <u>foramen ovale</u>,[21] and many other parts lacking in its parent. There is, for example, a double row of teeth, while its parent has only a single one.

However, anatomy isn't a light which enlightens everyone. Therefore, let us light the flame of Nature, which casts its rays even on the eyes of the least learned. Let us consider a Hottentot who has only one testicle, or a Switzerlander in whom one of the testicles has been cut off in youth because of the modesty so common in this industrious people. This occurs a long time before the time that according to Mr. Buffon himself the abundant particles would be transmitted to form a seminal fluid. But that Hottentot or Switzerlander produce offspring who are not deprived of any parts, and have both testicles. A man who has lost a hand, a leg, or an eye, does not fail to produce whole offspring. If Mr. Buffon was tempted to attribute the hand and eye of the child, missing in the father, to the mother, at least the testicle would be beyond the capabilities of the mother, and Mr. Buffon would have no remaining recourse than a universal adultery in the case of all these nations, an accusation which is much too harsh and improbable. Doesn't one see everyday that female dogs, confined with a single male, and with both deprived of ears,[22] produce offspring with complete ears? Does one not see that young colts bear incisor teeth that the mare as well as the stallion have lost a long time before mating?

After this example I do not need to mention that lame, deformed and disfigured parents generate healthy offspring, in which the spine has not the least similarity to that of its parents. The foregoing example has sufficient force, and frees us from citing others.

The child is thus not the [exact] image of its parents, just as the plant is not that of the one which supplied its seed. It entirely differs in all its [fine] internal structure, and very often in all the larger parts, and the number of its organs is always more abundant than that of its parents.

The second difficulty is no less significant than the first, and I am no less curious to know how the ingenious author will resolve it. Even if we would presume for a moment that the exact copies of the interstices of the eyes, ears, *and bones*²³ were able to assemble themselves in the seminal fluid; or even if we could presume that they there preserve the resemblance to the body from which they arise, we would see, however, that these organized particles swim without order in the seminal fluid; and Mr. Buffon has not yet made known the cause which puts them in order, and which joins the eye particles of the father with the eye particles of the mother, [and] those of the right side with right, and those of the left side with the left. [Nor the cause] which puts the ear particles in their place and at the proper distance, and which measures the situation and proportion of all the parts with exactitude, (and) adjusts thousands and thousands of separate parts of arteries in order to make a complete canal continuous through the length of the body. In brief, [what is the cause] which arranges the human body in such a way that an eye is is never attached to the knee, an ear is never connected to the hand, a toe never wanders to the neck, or a finger is never placed at the extremity of the foot, as happens in the crystallisation of salts, where one finds at all times some of the spires sometimes similar, and sometimes different, [and] often without form, and in reverse order? *I cannot imagine that there could be between the organized particles of the seminal fluid a differentiation, a form, which distinguishes one from the other, and which separates the elements of the foot from the elements of the eye; and even if I would presume that some veins and microscopic nerves arise in the seminal fluid*,²⁴ I could not, however, find a natural force which could join, according to a plan traced from all eternity, the separated parts of the body, those thousands and millions of veins, nerves, fibers and bones. It seems to me that Mr. Buffon has entirely passed over this great difficulty, similar to Timanthes, who in place of painting the sorrow of Agammemnon, believed he had excused himself by covering the portrait with a veil. Mr. Buffon needs a force which has foresight, which can make a choice, which has a goal, which, against all the laws of blind combination, always and unfailingly brings about the same end. Since most animals conceive on the first mating, and always form organized animals, in comparison with which the number of monsters is so rare that it vanishes when examined by the rules of calculation, I would wish that Mr. Buffon would do me the honor of reading and resolving this objection, which assuredly has weighed upon me. There are some minds who, similar to the heroes of Vergil, lift weights which several men of ordinary strength would not know how to move.

There still remains one doubt for me which appears no less important, and which I would entrust to the examination of the reader. Mr. Buffon does not, for a moment, hesitate to postulate the [existence of] seminal fluid in females. Half of his edifice is built upon this foundation, and in his theory he absolutely cannot dispense with it, since without a female humour, the organized particles of the seminal fluid of the father would only ever be able to produce male offspring. But I do not find the least proof of the existence of this seminal fluid. I find nothing which would be able to convince me that the fair sex possess one, nor that it emits it and intermixes it with that from the male. We admit in fact that the humour of the <u>corpus luteum</u> is filled by moving particles. *We will accept the experiments of Mr. Buffon with esteem*.[25] But it contains nothing which is not found as well in the other human humours. Even meat broth has similar ones. However, it is from the <u>corpus luteum</u> itself that I will draw an argument against Mr. Buffon.

The testicles of the male are peculiar to him from earliest infancy. They have attained their degree of maturity when he mates, and the prolific humour that he emits for the great work of generation draws its origin from the testicles, which have been prepared for a long time to furnish it.

But young females, and especially virgins, do not have the <u>corpus luteum</u>. All women who have died without conceiving have never had one. At the time that a young, healthy, and marriageable beauty has conceived, she is found entirely without the apparatus for the claimed seminal fluid. Thereby, from where will this seminal fluid be drawn? Here Mr. Buffon commits an anatomical error that we will willingly pardon him for. We ought to be indebted to him for having such great knowledge, in spite of the time that he has devoted to military service, rather than reproach his insights in some arts which were so far removed from his occupations. However, the rights of the truth are invariant, although the fault of those who violate them would be greater or lesser, according to the greater or lesser occasion and facility they had to be instructed. It is the animals which reproduce very rapidly and at short intervals that have made Mr. Buffon believe that all females capable of generation have the <u>corpus luteum</u>, and as a consequence, seminal fluid and organized particles. But it is incontestable that these glandular bodies are not the cause of fertility, but its consequence. They arise in the woman only after conception, and they are retained only for a certain time after delivery, disappearing little by little, and never to be replaced by other similar glandular bodies, at least unless the woman conceives anew.[26]

. . . .

Certain supporters of Providence consider the hypothesis of Mr. Buffon and Mr. Needham as dangerous. Matter, according to these learned men, has the power to form itself. By certain universally distributed expansive and attractive forces, is produced the divine structure of a Theresa[27] or a Newton. The force which can create men is likewise suitable to construct the planets, and these necessary and eternal forces of Nature free us from a Creator. They are sufficient without him to develop for us the order and beauty of the world. To banish this proof of Divinity is to deprive mankind of a conviction which, by its clarity, has been apparent to all peoples.

This has excited the fear of a portion of the *Sorbonne*[28] clergy, who have for some time fixed their attention on the work of Mr. Buffon. But is this fear founded, and would faith be lost if from experience we were to grant productive[29] forces to Nature?

I am without fear on this matter. The existence of God is grounded equally on the material world, and on revelation. The former demands that the atheist supply a creator, and the latter furnishes an unending series of proofs, which claim mutual support for themselves through the relation of the prophecies to their fulfillment, by the miracles, and by the connection of contemporary Christianity with its beginnings.

It is true that one could charge us with too much liberality if the time came that we granted to freethinkers that matter can be formed and constructed by certain forces inherent in it, which Mr. Needham has limited to two--the forces of repulsion and attraction. But the still greatly distant proofs of the reality of these forces do not trouble my mental repose. Truth is upheld from all sides; all concurs to support its edifice. It is only error which collapses when one removes the single support that it seems to have.

Clearly we observe that salts, crystals, [and] metals are formed by certain general forces, without the least suspicion of a semen or germ being involved. Two forces, very similar to the forces of Mr. Needham, govern the movements of the heavenly bodies, and who would draw from this proofs against the existence of a Creator? Do we take from the domain of Providence the modern (or the ancient and revived) belief in final causes? Is it possible that a theory could deprive us of that clear conviction that the eye is made for seeing, whether its origin is from a [preexistent] germ or is formed without one? From the moment that an eye, with all its membranes, humours, dimensions and proportions, and in all the diversity of its structure accomodated to the diversity of animals, is given to them for seeing in their different situations, must we not then recognize the *providential*[30] will of a Creator? [It is he] who distributes everything, who has furnished man with hands, denying to him the natural weapons that he has given to all the animals, depriving him of the long jaws so suitable to the brutes, and finally, who has denied to him all the advantages which he has liberally bestowed on all the animals, necessary to the brutes for their preservation, but which the hands of man would render superfluous.

Does matter have intentions? Is it by an act of its own intelligence that it has given the fish, which must live in a more dense medium, a much rounder crystalline [lens], than man, who must breathe a more subtle air? Has it forseen that man will walk on his feet at the time that it already doubled, by a hard additional layer of skin, the soles of the feet in the fetus, just as it prepared in the dog, while in the womb of its mother, the callosities on which it must walk after birth? Must one have recourse to the discretion of an artful gravity, and to a penetrating elasticity, when one wishes to unravel the reason why man, endowed with speech and capable of knowledge, has so little refinement in smell and taste, while the animals, which must be instructed on the beneficial or harmful properties of their food by their own experiences, have the same senses and organs much more finished and perfected? Is it to the choice of matter, which has been initiated into the mysteries of sublime geometry, that must be attributed the proportion observed in the length of man's fingers? [Does this] determine that the fingers found at the extremities are the smallest, just as the extreme [latitudinal] segments terminating the Orient and Occident of a world globe are the smallest, while those [longitudinal ones] which pass over the poles, and must embrace them, are greater, as are the fingers from the middle? Would it be inevitably necessary that animals would produce milk at the very time they give birth, and that they would have the number of mammaries proportional to that of their young *with the dog and pig having several mammaries, while the cow, goat and horse, with single birth, have only two*?[31] From the chance play of blind matter, would absolutely no other structure result except that which has such an admirable relationship to the nourishment of the new-born animal?

It is thus not properly the development or the means of generation that furnishes us with proofs of the existence of the Divinity. We find the traces of the wise power of a Creator in the workmanship which is most evident in the marvelous relation of structure to its purposes.

If matter has forces which make it suitable to form *something*,[32] we do not believe that it possesses them through blind destiny. They are confined between eternal limits, [and] they always form with perfection not mechanically identical beings, but similar beings which have been prescribed according to an inviolable plan, but with a variability which excludes all constraint of a blindly operating matter. I have proven that there never are two men, [or] two animals [exactly] similar to one another in their structure, although they have a perfect harmony among their principal parts. Who has given to the matter of the seminal fluid the possibility to produce greater or smaller vessels, to form greater or smaller nerves, or to double or omit their branchings? And who has, at the same time, prescribed the law for always and infallibly producing a great artery, a heart, the great sympathetic nerves, muscles, and everything which serves not only the life of the animal, but also can contribute to its happiness? If nature were not the instrument of a superior Wisdom, one would observe no less difference in the general plan than in the smallest parts of the human body, in place of the variation that always rules in the latter category, but never in the former.

*Finally, what has given these forces to matter? If it possesses them by inherent right, then why does fire have no gravity, water have no elasticity, and metals have no irritability? And why do the different classes of matter possess different forces, which do not, by their essence, exclude one another, but are found in one place united, in another separated, and in other portions of matter absolutely lacking?*33

Who makes these forces so wise, so constant in the generation of animals? If only the repulsive and attractive forces in the seminal fluid form a man or deer, [and] if it is only chance which combines them, then why does that matter, which according to the conjecture even of Mr. Buffon is equally indifferent to all kinds of shapes, never produce an ape in place of a man, which he so closely resembles? How is it possible that from a sticky humour there always (since I have already said that animals almost never mate without conceiving) arises an animal, and an animal of the same species as its parents? This constancy is sufficient to convince me, in opposition to the experiments of Mr. Needham, that there is something anteriorly *preformed and structured*34 in the prolific semen of man and animals, although one cannot say that this would be a complete miniature of the entire body. *Nor is the caterpillar a prefigured butterfly*.35 The invariant reproduction of always similar and divinely constructed animals, seems to be beyond these simple forces, which can produce only a <u>Conferva</u>, or a salt crystal, or an egg-shaped microscopic animal, lacking a heart and member, whose life consists only in irritability, and whose entire form is indistinct, *and changeable at every moment*.36 Even the crystallization of salts appears originally to be founded on the existing form of salt particles, and not on a simple attractive force, since liquid saltpeter is still saltpeter in taste and in all the other properties, although the crystals are dissolved in water. But let us leave this matter, which I cannot expand here. I digress too much from my subject.

It is sufficient to state that Mr. Buffon and even Mr. Needham, bear no more prejudice to religion than Newton, who by means of two forces, has unfolded the wonderful system of the universe, and the hidden laws of the revolution of the celestial bodies. The doctrine of Mr. Buffon is still less dangerous *than that of Mr. Needham.*37 His organized matter is molded in man himself in order to become a man. But after the time when the earth was encompassed by fire, and inundated by a universal sea, the [first] men needed to be produced without a mold, since water and fire would not have been able to produce an *ancient primordial* father,38 from which they could have arisen. Their structure, that general model of the human genus, was therefore, according to Mr. Buffon himself, immediately produced by the hand of God at the time that the earth had become dry. His theory offers no other plan for the beginning of mankind. Nature does not itself supply this form, according to Mr. Buffon, it only makes copies of molds already created.

We can thus calmly anticipate whether the experiments of the learned will *confirm or*[39] oppose the vegetative and animal forces of Mr. Needham. By new lights we will always approach the truth, and God by means of the truth.[40]

NOTES

*Albrecht von Haller, Réflexions sur le systême de la génération de M. de Buffon (Geneva: Barrillot et Fils, 1751). With only slight changes, this also appears as Haller's "Vorrede" to Buffon, Allgemeine Historie der Natur (Hamburg und Leipsig: Grund und Holle, 1752) Th. II, Bd. 1. Passages of significant difference in the two texts are inserted between asterisks.

[1][See text omitted here as given above, pp. 205-207.]

[2][German reads allgemeinen gebildeten und bildbaren Materie.]

[3][German: unendlichen Verstandes.]

[4][In German only.]

[5][A branching crystalline structure forming in some saturated solutions.]

[6][Aquatic filamentous algae, generally assumed in the period to arise by spontaneous generation.]

[7][Connaissance]

[8][In German text only.]

[9][German: Von der Bildung eines Alexanders zu der Entstehung eines Schneeflockens. . . .]

[10][Reference is to the original, and apparently independent, discovery of spermatozoa by Jan Hamm and Nicholas Hartsoeker in the 1670's.]

[11][French: corps glanduleux, German: gelben Drusen. See discussion above, p. 208. Haller's discussion displays some of the persistent confusions between the Graffian follicles and the corpus luteum evident in the controversy. Haller is generally speaking of the corpus luteum in its strict anatomical sense, whereas Buffon was not.]

[12][Nicholas Steno (Niels Steensen) (1638-1686), Danish physician and geologist, and one of the main formulators of the preformation theory in the seventeenth century.]

[13][German only.]

[14][German only.]

[15][German only.]

[16][German: innere Modelle und innere Abdrücke.]

[17][French: machines; German: lebenden Dingen.]

[18][German: Insekten. Passage in German only.]

[19][German only.]

[20][The urachus, a structure involved in fetal waste disposal.]

[21][The open passageway between the right and left auricles in the fetus of mammals which closes upon birth.]

[22][German: Die so wenig als der Vater ein Ohr haben.]

[23][German only.]

[24][French only.]

[25][German only.]

[26][Deleted here is a technical discussion on the nature of the female "semen."]

[27][Reference unclear. Probably Maria-Theresa of Austria (1717-80).]

[28][German only. See text of the Sorbonne's reaction above, pp. 283ff.]

[29][German: bauenden.]

[30][German only.]

[31][German only.]

[32][German only.]

[33][French text only.]

34[German only. The passage reads: <u>etwas</u> <u>vorgebildetes</u> <u>und</u> <u>gebautes</u> <u>im</u> <u>befruchtenden</u> <u>Säfte</u> <u>der</u> <u>Menschen</u> <u>und</u> <u>Thiere</u>. . . .]

35[German only. Haller is apparently making reference here to Jan Swammerdam's observation in 1667 of the preformed adult butterfly within the pupa. His report of this had been reprinted in 1737.]

36[German only.]

37[French only.]

38[German: <u>ältern</u> <u>Stämmvater</u>.]

39[German only.]

40[German text is dated at this point March 30, 1752.]

Figure 18 - Houdon's bust of Malesherbes

16. Malesherbes' *Observations* on Buffon's Natural History (1749, 1798), (selected)
John Lyon

Chrétien-Guillaume de Lamoignon-Malesherbes was born in Paris Dec. 6, 1721, the son of Guillaume de Lamoignon, Premier President of the Cour des Aides, and Chancellor of France, 1750-1768. Educated by the Jesuits at the College Louis-le-Grand, Chrétien-Guillaume was destined for the public life of a magistrate. Though thoroughly enamoured of literature and natural history, especially botany, he entered public life in 1741, and soon was appointed as Director of the Library by his father. In this position he had the supervision of the censors, who were named by the Chancellor.[1]

Upon his father's appointment as Chancellor, Chrétien-Guillaume became First President of the Cour des Aides, an appellate court whose sovereignty extended over cases concerned with all "subsidiary" income of the Crown, including income generated by indirect imposts such as the Gabelles and tailles.[2] He was exiled to his estate by a lettre de cachet, April 8, 1771, for his opposition to the Crown and his attempt to preserve the integrity of the Cour des Aides, and remained there for three years. The Cour des Aides was then abolished also, but was re-established under the new King, Louis XVI, in 1775, and Lamoignon-Malesherbes was reappointed First President.[3]

An opponent of the tyranny of the Crown, Chrétien-Guillaume was nevertheless loyal to the principle of Monarchy, and came to Paris in defense of the King during the Convention. "Compromised" by his royalist sentiments and the activities of his relatives, Lamoignon-Malesherbes was arrested in December, 1793, and guillotined with his daughter and her husband on April 22, 1794.[4]

In 1750 Chrétien-Guillaume became an honorary member of the Académie des Sciences. In 1759 he was admitted to the Académie des Inscriptions, and in 1775 to the Académie Française.[5] Passionately devoted to "natural history," both as theorist and critic and as a practicing "improving" landowner, Lamoignon-Malesherbes was a friend and devoted student of the naturalist Bernard de Jussieu (1699-1777).[6]

Lamoignon-Malesherbes apparently composed his criticism of the first volumes of Buffon's work as soon as it appeared (1749), but never published the two tomes of his Observations sur l'histoire naturelle. Instead, the manuscript came by a circuitous route to the hands of Paul Abeille, who published it in 1798.[7] Though most of the work is concerned with demolishing Buffon's argument about the formation of the earth, it is Buffon's penchant for metaphysics, as exemplified in the "Initial Discourse" to the Natural History, which Lamoignon-Malesherbes reserves for some of his most telling criticism.[8] It is the "long aside" about metaphysics, truth, and method in the first volume of Lamoignon-Malesherbes work which is translated here, with all its hybrid scholasticism and touches of Descartes and Port-Royal.

NOTES

[1]See Pierre Grosclaude, Malesherbes, temoin et interprete de son temps (Paris: Librairie Fischbacher; n.d. [ca 1962], Table I, p. 21; pp. 41-44, 60, 67, 109. See also: Chambers Biographical Dictionary, rev. ed., ed. J. O. Thorne, (Newark: St. Martins Press, 1969), pp. 837-38; and Nouvelle biographie générale (Paris: Firmin Didot; 1863), Vol. 33, Cols. 22-30 (hereafter, NBG).

[2]Grosclaude, op. cit., p. 210.

[3]Ibid., pp. 309-310.

[4]See Chambers, NBG, and Grosclaude, op. cit., pp. 737-752.

[5]NBG, Vo. 33, col. 29.

[6]Grosclaude, op. cit., pp. 495-497.

[7]See Paul Abeille's "Introduction," to C.-G. de Lamoignon-Malesherbes, Observations sur l'histoire naturelle générale et particulière de Buffon et Daubenton (Paris: Pougens; An VI [1798]), I, iv-viii.

[8]Grosclaude, op. cit., 487-488.

OBSERVATIONS OF C.G. LAMOIGNON-MALESHERBES ON THE

NATURAL HISTORY OF BUFFON AND DAUBENTON*

Translated by John Lyon

This discourse on the manner of studying and expounding natural history ends with a metaphysical dissertation on truth, and on the various orders of truth.

Although I have confidently attacked M. de Buffon on several matters regarding natural history, it is only with hesitation that I enter the lists with him over a question of pure metaphysics. Forewarned as I am of his talents in this area, I am not able to see in this piece of work anything but grand words which only suggest a vague sense, and rambling phrases which I have not been able to discover either the connection between or the application of. That has led me to transcribe a fairly long textual passage of Buffon's. As a matter of fact, I have taken pains to extract here a line of reasoning which I have never been able to comprehend.

> Truth, that metaphysical entity of which everyone believes himself to have a clear idea, seems to me to be confounded with such a great number of strange objects to which its name is applied that I am not at all surprised that it is hard to recognize. Prejudices and false applications are multiplied in proportion as our hypotheses have become more learned, more abstract, and more perfected. It is thus more difficult than ever to recognize what we can know, and to distinguish clearly what we ought to ignore. The following reflections will serve at least as advice on this important subject.
>
> The word truth gives rise to only a vague idea; it never has had a precise definition. And the definition itself, taken in a general and absolute sense, is but an abstraction which exists only by virtue of some supposition. Instead of trying to form a definition of truth, let us rather try to make an enumeration of truths. Let us look closely at what are commonly called truths and try to form clear ideas of them.

There are many kinds of truths, and customarily placed in the first order are those of mathematics, which are, however, only truths of definition. These definitions are concerned with simple but abstract suppositions, and all the truths of this sort are nothing more than the worked-out and always abstract consequences of these definitions. We have made the suppositions, and we have combined them in all sorts of ways. The body of combinations that results is the science of mathematics. There is, then, no more in that science than what we have put into it, and the truths which are drawn from it can only be different expressions under which the suppositions which we have used are presented. Thus, mathematical truths are only the exact repetitions of definitions or suppositions. The last consequence is true only because it is identical with that which preceded it, and this latter in its turn with its antecedent. Thus one may proceed backward right to the first presupposition. And since definitions are the sole principles upon which everything is established, and since they are arbitrary and relative, all the consequences which can be deduced from them are equally arbitrary and relative. Hence, that which we call mathematical truth is thus reduced to the identity of ideas, and has nothing of the real about it. We make suppositions, we reason on the basis of our suppositions, we draw the consequences of them, we come to conclusions. The conclusion, or the last consequence, is a proposition which is true in proportion as our supposition was true. But the truth of this proposition cannot exceed that of the supposition itself. This is hardly the place to discourse on the methods of the science of mathematics, or on the abuse of such methods. It is sufficient for our purposes to have proved that mathematical truths are only truths of definition or, if you prefer, different expressions of the same thing, and that they are only truths in relation to the very definitions with which we started. It is for this reason that they have the advantage of always being precise and conclusive, but abstract, intellectual, and arbitrary.[1]

The idea that we have of truth is one of the most simple and most distinct of our ideas. And the difficulty in defining it only results from the fact that it is clearer than those ideas that could be substituted for it, or by which one would explain it.

Truth is the reality of relations enunciated in a proposition, and in this sense truth is one and ought not be divided. But that which is signified by a truth is different from what is signified by the truth. A truth is nothing but a true proposition. In this sense it is possible to have various orders of propositions. But it is not as truths, but solely as propositions, that they vary. Their difference amounts to the difference that exists between the objects of our ideas. And the same difference will be found to prevail between mathematical errors and physical errors as prevails between mathematical truths and physical truths. The word truth thus does not give birth to a vague idea, everyone has a clear idea of it. And the enumeration of various orders of truths is useless for knowing what truth is. This enumeration begins with mathematical verities, and it is said that these verities are only different expressions under which suppositions are presented; that they are only exact repetitions of definitions; that the final consequence is only true because it is identical with that which precedes it, this one with its predecessor, and thus serially back to the initial proposition.

The disjoined phrases and the chopped style which M. de Buffon makes use of are actually the fashionable style. I indeed agree that this style is the most appropriate for certain readers who cannot adapt themselves to the formal method of an author who defines his terms, and who sets principles in order to deduce consequences from them. I believe, however, that this latter method, though hardly brilliant, is the most appropriate for a discussion of metaphysics wherein the aim is to ascend to basic ideas, and wherein most errors arise from the abuse of words. So, in order to refute M. de Buffon, I shall be obliged to use the form of reasoning and even the terms of scholastic philosophy. This sort of dialetic is banished today from most science. Not that it isn't the most proper of all for giving clear ideas; but it commonly seems only to prove propositions which are self-evident. There are however certain captious forms of reasoning from which it is hard to disentangle the falsity without the assistance of this method. In a word, I shall be adequately justified in my use of this method if I succeed in reducing M. de Buffon's processes of reasoning to their just value.

I shall begin then by recalling to the reader, and to M. de Buffon himself, the distinction which logicians have recognized between three sorts of propositions: universal propositions, particular propositions, and singular propositions.[2]

Singular propositions are those in which the subject is one or many individuals. For example, I say that a certain house is built of bricks, or, that such and such houses are built of bricks. These are singular propositions.

But the subjects of most propositions do not have a determinative signification. For example, if I should say that some houses are built of bricks, my proposition would not have a fixed and certain subject. And this is what is called a particular proposition. One easily sees that a similar proposition presumes the notion of what is called a house; but, this notion once established, if I should say that all houses are built of bricks,[3] my proposition would have a subject more extended than the two others, and this would be a universal or general proposition.

However different these three kinds of propositions may be, I do not believe that anyone has claimed that universal and particular propositions have an object less real than singular propositions have, nor that general verities are exact repetitions of definitions.

That being understood, let us reflect a moment on the principle of abstractions and mathematical suppositions. I shall begin by remarking that all ideas contain abstractions, except singular ideas, that is to say, those which represent individuals.

All universal or particular ideas have as objects species, or collections in which the individuals are put together on the basis of certain characteristics, abstracted from others.

The idea that I have of man, iron, or a house is abstracted from other properties which each man, each house, and each piece of iron may have. This is so true that it is impossible for me to imagine a man, or a house, without adding to the image characteristics particular to such a man, such a house, characteristics foreign to the general idea of man, or house.

There are then no propositions which do not contain abstractions, except those which assert the identity of persons, because they are the only propositions in which the subject and the predicate may be singular ideas.

Singular propositions themselves contain an abstraction when they affirm or when they deny that such individuals have such a quality, since the quality affirmed or denied is an abstraction.

As to general propositions, not only do they contain an abstraction, but it could even be noted that there is no difference with regard to the real signification between a general proposition and an abstract proposition. An abstract proposition is one in which the subject is an abstract being, that is to say, a quality, a modification, or a relation. A general proposition is one in which the subject is a collection of individuals which resemble each other in some quality, modification, or relation. Now, to assert that such a quality is never found without another such quality, or to assert that all beings endowed with one such quality have another, is to say the same thing. For example, to say that the form of roundness is the most appropriate to movement is to say that round bodies are those which, all things being equal, move with the most facility. The difference in expression makes of the proposition either an abstract or a general proposition.

I say further, and I maintain, that to generalize a proposition is to apply to a species what has only been applied to one or to many individuals, or to increase the collection of individuals which are the subject of the proposition. And this collection can be increased only by eliminating from it characteristics which are necessary in order to distinguish it. Thus, if I wish to apply to the whole human race something which may have been said of Negroes in particular, I have no other choice but to detach from the characteristics which I have attached to my idea, those which distinguish Negroes from other men. And if I wish to extend my proposition to all species of animals, among the characteristics of man I have [to focus on] those which distinguish animals from other beings. All that remains is to separate those from the characteristics which belong particularly to man. It can be seen that in diminishing the number of qualities which are attached to an idea, the number of individuals to which these remaining qualities apply is increased, and that the more general propositions are, the more abstract they become.[4]

When one tries, then, to prove an abstract or general proposition, that is to say, when one reasons about a particular quality, one of the most appropriate methods is to imagine a being endowed with the quality or the qualities which one wishes to examine, abstracting from it all other properties which it might have. That having been posited, one is certain that all the properties which will be discovered to belong to the imagined being will be of the given or imagined property, and, consequently, belong to all the individuals which are parts of the object of the general proposition.

It is not, then, properly speaking, a supposition which is the object of mathematical truths; it is rather a collection of beings to which the mathematical proposition can equally apply, and which are designated by one of the existing or possibly existing beings which compose the collection.

Thus, when one says, "a triangle is half of a parallelogram of the same base and same height," that is to say that all triangular bodies equal in surface area one-half of all squared bodies, or bodies squared along the same base and the same height. And this fashion of reasoning can be employed equally outside of the mathematical sciences, although one expresses oneself in other terms. When I say, for example, that quadrupeds have hearts divided into two ventricles, this is the same thing as if I should say that "I imagine a quadruped," or "let there be a quadruped." I say that it has a heart divided into two ventricles. Indeed, I may not consciously reflect upon this proposition, that all quadrupeds have hearts divided into two ventricles, without attaching to some quadruped which I imagine the idea which belongs to all the individuals of this family. It is true that I can apply this idea successively to different quadrupeds. But likewise one can apply the idea of an isoceles triangle which one imagines in geometry to a triangle placed on paper or on the ground, to a wooden triangle, or to one of metal. And when one says that the suppositions of mathematics are found to be contradictory to the nature of the postulated entity, that amounts to saying that one does not understand the principle of such suppositions. One supposes, for example, a line without breadth and without depth. This is not to say that one is reasoning about a line which has neither breadth nor depth; it is to say that one is reasoning about an extended entity the breadth and the depth of which one does not advert to. When, ordinarily speaking, one says that one road is twice as long as another, this is the same as to consider a line without breadth or depth. I do not believe that M. de Buffon would say that this proposition has no real object. However, I maintain that this proposition, or this truth, differs from mathematical truths only as a singular proposition differs from a universal proposition.

It is true that all mathematical truths are [parts of closed systems with] all inferences mutually interdependent. But the same is the case with all true propositions of natural philosophy, of history, in a word, of all imaginable propositions. For it is evident that everything is connected in Nature; that everything has a cause; that causes hang together and are the causes of each other; and that there is no proposition which cannot be led back to a first cause from which one can derive such other propositions as one wishes. The difference between <u>mathematical truths</u> and other truths consists, then, only in the fact that mathematical verities are those of which one understands the derivation; and this difference is relative to the understanding of each mathematician. But truth is an object independent of the understanding which men have of it.

What has perhaps led M. de Buffon into error is that one commonly says in mathematical demonstrations that two propositions come to the same thing when one is the necessary consequence of the other. But this error manifestly arises from an abuse of language.

I know that the three angles of a triangle taken together are equal to 180 degrees. Thus, if I have named the three angles of a triangle A, B, and C, and if I have come to prove that A and B taken together equal 100 degrees, I can state (what is the same thing) that C equals 80 degrees. That is to say that it is the same thing for me, or that it is a matter of indifference to me to know that A and B are equal to 100 degrees, or that C itself equals 80 degrees, because knowing the one, I easily know the other. But it is not to say that the two propositions are the same thing, since they actually present two completely different ideas. That is so true that if I do not know that the three angles of a triangle are equal to two right angles, or even that the three angles A, B, and C, are of the same triangle, I will not sense the consequence and the imagined identity of the two propositions.

Let us again take up Buffon's text where we left it.

M. de Buffon says that <u>since definitions are arbitrary and relative, the consequences [of them] are equally arbitrary and relative.</u>

In geometry the definitions are arbitrary, but one could not say that they are relative, unless one is willing to consider all general propositions as relative to the various individuals who consider them as objects.

Consequences, on the contrary, are relative to definitions, but are not at all arbitrary.

The author next moves to an examination of physical truths, and the conclusion of the parallel consideration is that in the abstract sciences one arrives at evidence; in the real sciences at certitude; that the word truth takes in the one and the other, and consequently answers to two different ideas; that its signification is vague and complicated; that it thus has not been possible to define it generally.[5]

I maintain to the contrary that the word "truth" takes in neither the one nor the other concept. And in order to maintain this I need not enter into metaphysical reasoning. It suffices to recall the common idea which everyone has of "truth", and one will easily sense that the certitude and evidence of a proposition are entirely relative to us. There are different degrees of knowledge that we can have of truths; but these are not the truths themselves. Truth is entirely independent of us. It exists prior to all our knowledge. It was true that the volume [<u>solidite</u>] of the half-sphere equalled two-thirds of that of a cylinder with the same base and the same height, before Archimedes had demonstrated it, that is, before he had made this proposition <u>evident</u>. Likewise, it was true that the elevation of water in a vacuum depended on an equilibrium of the water with air before Galileo and Torricelli had proved it, that is to say, had rendered this proposition <u>certain</u>.

Finally, the difference between the mathematical and the natural [<u>physiques</u>] sciences amounts to two principal points, which are these:

In the first place, the mathematician considers fewer properties, and, consequently, the subjects of mathematical propositions are more simple. But even in mathematical propositions there are objects which are more complicated than others. A cycloid surely presents a much less simple idea than does a straight line. The more one pushes back the limits of science, the more complicated objects become. Thus on this point, mathematical verities have no different distinctive characteristics from physical truths.

The other difference amounts to the fact that mathematical properties are such that one has clear and distinct ideas of them, while those that the physicist considers are themselves the result of many unknown properties. And the idea one has of them, being founded on sense data, is always an imperfect idea. This makes a quite real difference between the two sciences, but it does not make a difference between the truths themselves. The difference is only relevant to the knowledge we have of them, and to the fashion of conducting our mind in the search for these truths.

If the efforts of M. de Buffon only tend to establish this difference between the sciences susceptible of evidence, and those which are only susceptible of certitude, he has set himself a most worthless task. For nothing is more patent than this difference. It has been pointed out and defined in all treatises on logic.⁶ It is no more than what one reads in the prefaces of numerous treatises on mathematics and natural philosophy. It is what one tries to get students in geometry to understand, those who, beginning this study, have trouble consenting to suppositions and abstractions of which they do not yet understand the applications. And in order to render sensible something so platitudinous, it is not necessary to begin by announcing that truth, that metaphysical being of which one has only a vague idea, has never been well-defined.

Indeed, all this metaphysical digression concludes by distinguishing two kinds of truths and giving the name of truth to two things which are really quite different, and recognized as such. But neither the one nor the other of these two is the truth.

It was, further, quite useless to speak ten or twelve times in succession, in various terms, without trying to prove the matter even once, that the truths of geometry are identical among themselves, and are only different expressions of suppositions. This is a proposition which has as a foundation only the equivocation which is brought about by the use of the expression, that which is the same thing.

This section is followed by many reflections which are quite wise and quite judicious⁷ concerned with the manner of conducting the mind in inquiries in natural philosophy. But at page [127], it appears that the author finds himself in the realm of true and intelligible being. The important thing is, he says, to know how to distinguish what there is of the real actually existing in a subject, from what we place there arbitrarily... If one does not lose sight of this principle, paradoxes will be seen to disappear... [as will] insoluble questions in the abstract sciences... the metaphysics of the sciences will come to be understood, etc.⁸

There is no doubt that many errors do come about only when abstractions are taken for realities. M. de Buffon rightly reproaches Plato for this fault. Pythagoras and some other Ancients are not exempt from this reproach either. But it appears here that M. de Buffon wants to insinuate that it is from this cause alone that our errors proceed--as if passions, prejudices, precipitate action, the good opinion one has of oneself, ignorance of facts, sensory illusions, and, finally, the limitations of intelligence in each thinking being were not also fruitful sources of error, even for those who are distinctly aware of the difference between the real and the abstract.[9] Besides, I do not see either how the decrease of this fault among men would cause paradoxes and insoluble questions in the abstract sciences to disappear, or the advantage which might be found in causing them to disappear.

What is commonly understood by paradox is a truth which appears to be an error. And what might be regarded as an error is seen to be a paradox in the moment that one discovers that it is a truth. In as much as there will be unknown truths, there will be paradoxes, and it is important that such be the case.

The same is true of insoluble questions in the abstract sciences. This designation is given to those questions which cannot be resolved by precise evaluations, but only by approximations. If M. de Buffon gives proof that by the method of reasoning on which we insist, precise evaluations will be given, we would assuredly be indebted to him for a great discovery. But until that proof is given us, I do not see why we should choose to do away with the methods we have. Moreover, when the abstract shall have been distinguished clearly from the concrete, it will be concluded that insoluble questions are abstract like other questions of the abstract sciences, but not that it is necessary to banish them from consideration.

There is another kind of question that can also be called insoluble, and perhaps M. de Buffon had this kind in view. This sort is that which terminates in imaginary numbers [des imaginaires]. But we know what the usage of such imaginary numbers is. We know that they are the proofs of the impossibility of a problem, and this impossibility is a mathematical truth. Real values lead to assent; imaginary values lead to dissent. Both are equally interesting to know.

In indeterminable problems, it is the imaginary numbers which mark the limits of the values which can be given to an unknown variable, and these limits are often the most important element in the solution of these problems.

It is astonishing that it should be the translator of Newton who speaks thus of the insoluble questions of the abstract sciences!

The liberty which M. de Buffon allows himself of attacking in this fashion the greatest men in the various sciences appears to rest on the blame which he imputes to them frequently of being deficient in metaphysics. I believe that, on the contrary, M. de Buffon has fallen into most of the faults to which I have called attention, that he has abused this metaphysics of which he makes such a fuss, and that he has come to believe that with the aid of this metaphysics he can dispense with learning from facts, which are the foundation of most of the sciences. I even believe that he fancies himself as a metaphysician capable of divining by the light of reason alone that which authors had believed they thought, and that upon this foundation he imputed to M. Linnaeus and other naturalists views which they have never had and absurdities of which they are incapable.

This same prejudice in favor of metaphysical reasoning has caused him to seize with avidity an apparent difference between physical truths and truths of mathematics. And, after this deceptive insight, he has not taken the trouble to make more profound reflections which would have infallibly apprised him of the fact that there are no distinct orders of truth, but that the difference which has so struck him is only the difference between <u>evidence</u> and <u>certitude</u>, a distinction continually knocked about in the schools, a distinction which is today nowhere more used than in many preliminary discourses.

All this leads me to try to evaluate this great noun, "Metaphysics," of which so much is heard.

Metaphysics, following its etymology, is the science of supernatural things. It is in that that it differs from natural philosophy. According to this signification, metaphysics, among us, ought to designate nothing other than theology.

But in common usage, and as a part of philosophy, the knowledge of matter and its properties is given over to the physicists, and metaphysics has become a science which has for its object spiritual substances and mental beings. It has been divided in two parts: the first, which is called <u>pneumatology</u> or <u>psychology</u> in the schools, is concerned with the nature of the soul.

There are so few fixed points from which one might set out to reason on this matter, outside those given by revelation, and the questions of psychology are so intimately linked with great theological issues, that it is difficult to separate them from each other. And philosophy has, as yet, little hold in this science.

The second part of metaphysics is <u>ontology</u>. This name has been given to the science of ideas, and the knowledge of the operations of mind.

It is possible to believe that all the sciences end there, since each science is only the collection of ideas which we have on such and such matters. However, ontology has no influence whatever on the other sciences, because it only considers ideas such as they are in us, and independently of the external objects to which they relate. It does not even consider the nature of the ideas and does not enter into the detail of different ideas of the same nature.

The other sciences do not have any effect on ontology, either, for in order to reason about the nature of our ideas, it is not necessary to know the various ideas pertinent to the other sciences. The working of the mind is the same in all men. The thoughts of a peasant, although less complicated than those of a philosopher, are of the same nature. Each man who would apply himself has only to reflect maturely on the operations of his soul, compare them, dissect them—if this phrase is permissible. This is the most certain manner of knowing the nature and the origin of them. Thus ontology remains a science distinct from all the others.

What is it then that has given rise to this noun "metaphysics" which is so fashionable in this century, and which is considered as a science applicable to everything? Here is, I believe, the origin.

There are three kinds of sciences. The first of these are the sciences of pure reason, which are also called abstract sciences. The sciences of facts are second. These are few in number. Although there are many sciences which have facts as their objects, there are few that, in the means which they employ to reach their goal, do not become sciences of [pure] reason if it only be in determining the degree of confidence that one ought to have in those sciences concerning which one would establish facts. The facts, in turn, often depend upon a throng of very recondite reflections on the character of men.

Finally, there is a third sort of science which I might readily call <u>mixed sciences</u>. These sciences hold more or less to one or the other of the preceding kinds of science, as they contain more facts or more reasoning.

The greatest part of the reasoning in these mixed sciences is abstract reasoning, or at least involves an abstraction. Indeed, I have already remarked that all propositions, except singular propositions, contain an abstraction. I believe that I have also shown that the more general propositions are, the more do they become abstract.

When one has generalized to a certain point, or when one has come to a certain point of abstraction, one has arrived at what is called the metaphysics of science, an expression which is used only in mixed sciences. For, in the sciences of facts there is no metaphysics of science, and in the science of pure reason, or the abstract sciences, all is metaphysics of science, or rather, these sciences themselves are the metaphysics of the concrete sciences to which they apply. Thus one could say that rational mechanics is the metaphysics of a good part of physics, and, in pushing the reasoning further, one will see that geometry is the metaphysics of rational mechanics; and algebra, the metaphysics of geometry and, perhaps, of all the other sciences. For it might not be impossible to prove that, by adding a few things to algebraic signs, algebra and logic would be the same science presented under two different aspects.

There, then, is the idea that I have worked out of what is understood today by <u>metaphysics</u>, when the word is conceived in a sense different from what is called metaphysics in the schools, which is a science distinct from and separated from all others.

There are also people who give a different signification to the same word. They confound metaphysics with what is obscure. The same people also make of the word "abstract" a term synonymous with the two others. The difficulty which the study of metaphysics and the abstract sciences[10] entails is apparently what has given rise to this denomination.

These kinds of people have never had any clear idea of either abstraction or metaphysics, or only have an idea of it relative to their intelligence. Happily, we are not concerned with such persons. We are concerned with M. de Buffon who always has clear and distinct ideas of the matters he treats, when they are of such a nature as can be grasped by the sole light of the mind and through a process of reasoning.

To return to the idea which I have conceived of what is called the metaphysics of a science, and what M. de Buffon especially means by this word, I believe that it is the abstract part of science, in which is contained general or first principles.

Let us take grammar as an example, a science, so it appears, of facts, so much so that its goal consists in apprehending words by heart. From the first steps one sees that the declensions give abstract and general ideas. Syntax further augments the number of abstractions. But as soon as one wishes to reflect upon the principles of syntax, and on the analogy of words, one finds that these principles can be prodigiously simplified and reduced to a very small number. And this gives rise to so many recondite reflections and abstract processes of reasoning, which cause it to be said that grammar is one of the most metaphysical sciences.

If one would proceed further and try to compare the informing principles of various languages, one comes to the general and reasoned grammar which Messieurs de Port-Royal have given us, an essay which is one of the best applications of the metaphysical spirit. There one discovers the systematic principles of a general convention by which men easily communicate to each other all their ideas, such as this convention must be and such as it has in fact been by the apparent common consent of men. Without such a convention no one of those whose common interests are with language could know the principles from which he sets out and the end to which he tended.

Metaphysics thus understood is not properly a science. It is rather a particular configuration of the mind which those who possess apply to all the subjects which they consider. It appears that the metaphysician puts himself in an elevated spot from which he can see at a general glance a great extent of country, while other savants, whose genius is more limited, spend all their lives in scrupulously traversing but a portion of this same landscape.

But, however brilliant this metaphysical spirit be, however useful to the rapid progress of science it be, it can be abused.

The most common abuse arises when, sometimes, these universal geniuses believe themselves to possess a detailed knowledge of everything whose general principles they have comprehended. It is as if he who has overlooked a large province from a high bell-tower should pretend to a better knowledge of the paths which lead round a hamlet than those who have lived in the hamlet since their birth.

The second abuse is peculiar to our century, which is a century that bears the stamp of self-satisfaction and frivolity.

This abuse arises from what metaphysicians, or those who fancy themselves such, have scornfully noted sometimes of those concerned with details who have not taken care to grasp the general relations of things. This scorn has been quite handy to presumptuous persons [pour les gens avantageux] who, moreover, don't care to take the trouble of detailed research. It is easy for them to pass off a few big words as great ideas, words that they not only substitute for observation but which embolden them vaguely to critize all those who have taken the trouble to acquire genuine knowledge.

The final abuse of the metaphysical spirit is in the application which is made of it.

In the sciences which are totally abstract, and in those which presuppose only a small number of observations, the higher genius on a few occasions surmounts many obstacles. But when the number of facts involved increases, it is necessary that the metaphysician proceed more warily.

Finally, there are some sciences in which all the processes of reasoning are subordinate to the facts, and the essential facts are so numerous as to make a certain time necessary in which to come to observe them. Newton was completely Newton from the very end of infancy. The first works of Stahl, on the contrary, are, by his admission, only imperfect essays which he would freely disavow in his old age. And most of our naturalists have only brought their work to fruition in the last years of a life filled with labor.

To return to natural history: of what use can the spirit of metaphysics be in this science? I believe that, considering our definition, the metaphysics of natural history is nothing else but what is commonly called "profound views," "analogy," or "general theories." And in this sense one will find that, among the natural philosophers, a Malpighi, or a Boerhaave; [or] among the systematic naturalists, a Gesner, or a Tournefort, have really been very great metaphysicians, without ever having paraded this pompous title.

However long this digression may have been, I have believed it necessary in order to respond to the reproach of neglecting the spirit of metaphysics which M. de Buffon continually makes against the greatest naturalists. This reproach is given its insulting tone by virtue of the idea attached today to that great word. It is a reproach, however, whose only foundation is the fact that the word "metaphysics" was only given in their time to a particular science which had no influence whatsoever on other sciences, and, above all, had no influence on sciences of facts, such as natural history.

NOTES

*Observations de C. G. Lamoignon-Malesherbes sur l'histoire naturelle générale et particulière de Buffon et Daubenton [Paris: Pougens; An VI [1798], Tome Premier, pp. 183-218.

[1][See Buffon's "Premier discours" to the Histoire naturelle (trans. above pp. 122-23).]

[2][For this same division, see Antoine Arnauld, The Art of Thinking (Indianapolis: Bobbs-Merrill, 1964), pp. 109-110. This work is the translation of Arnauld's La Logique, ou l'art de penser (the so-called "Port Royal Logic"), originally published in 1662.]

[3]Whether the proposition is true or false makes no difference.

[4]It is only possible to generalize a proposition by abstraction. But it is possible to abstract without generalizing in one single case, that is, when two qualities are united in such a manner that the one could not exist without the other. Then although one of these qualities be considered separately by abstracting it from the other, what one says [of it] will only fit those beings which have the two qualities equally.
For example: All extended being has length, breadth, and depth; and every being that has one of these dimensions is an extended being. Consequently, it also has the two other [dimensions]. Thus, if I consider a line, that is to say the length of a body, or a portion of space, without considering its breadth or its depth, what I shall say of it will not be applicable to a greater number of beings than what I might say should I consider the whole solid, that is to say, if I should consider space, or matter according to its three dimensions.

[5][For this passage, which is in the paragraph following the ones quoted on the first page of this article, see pp. 123-24, above.]

[6]I only ask that one compare what M. de Buffon says here about mathematical evidence and physical certitude to the definitions of metaphysical, physical, and moral certitude, and to the definition of evidence, upon which the young maintain and defend theses.

⁷There is however an item [article] on which I am not completely of a mind with M. de Buffon. I refer to that contention in which he appears to disapprove of the use made in courses of experimental physics of proving by experiment that which is demonstrated by geometry. Apparently he has not paid attention to the fact that many people of necessity teach themselves experimental physics without having preliminary knowledge of geometry. Furthermore, those very ones who are capable of moving back to principles understand with more rapidity, and without the same struggle, that which takes place before their eyes. Finally, one of the aims of geometry is its application to the phenomena of physics which one comes across. And for that purpose it is advantageous to be accustomed to considering the relation that exists between the most simple reasoning in geometry and the most common physical events. For there is habitual facility to be acquired in the very sciences of reasoning.

⁸[See the "Initial Discourse," above, p. 127.]

⁹I make use here of M. de Buffon's terms, but I have shown elsewhere that the abstract is as real as the concrete.

¹⁰All abstract ideas are not as difficult to comprehend as they are ordinarily imagined to be. Often the terms in which they are set forth are frightening and make up a part of the difficulty. I believe that there are no such ideas which could be more obscure than to consider a line without breadth or depth, if one were not accustomed to do the latter. However, there is no peasant who could not have a quite clear idea of the length of a road quite independently of its breadth.

In like fashion students of geometry need a demonstration in order to be genuinely convinced that similar triangles have homologous sides proportional. However, everyone feels the proportion which exists between the different parts of the facade of a building on its premises, and the same parts represented in an exact plan of the building. No one fails to understand easily the conventions of a scale attached to a drawing, or to a geographical map. No one, in a word, fails to understand perfectly the sense of this proposition: this is reduced in scale here which out there stands proportionately larger. Thus that which is axiomatic for all men who have the use of reason is the general, and that which the masters of mathematics look for ways to prove by demonstration, even rather indirectly is but the particular.

PART IV:

Buffon in Retrospect: His Style and Glory

Figure 19 - Hérault de Séchelles

17. Hérault de Séchelles' Visit to Buffon (1785)
John Lyon

Marie-Jean Hérault de Séchelles was born in Paris on October 20, 1759,[1] and baptized in the Church of Saint-Sulpice. Hérault's parentage was suspect, the nominal father having died two months earlier than the child's birth as the result of wounds incurred at the battle of Minden, August 1, 1759.[2] Educated by the Oratorians at Juilly, he had among his classmates Jean-Baptiste (Anacharsis) Cloots and Louis de Bonald.[3] In December, 1777, he was made King's Counsel (Avocat du roi au Châtelet), as his grandfather had been, and in July of 1785 he became Counsel-General to the Parlement of Paris. It was in the Fall of this year that he made his controversial visit to Buffon at Montbard.

Born to a considerable fortune, Hérault's family connections with the Polignacs and Contades assisted his rapid rise in society. Through the Duchess of Polignac he was presented to the Queen, who took him under her protection.[4] Hérault consciously cultivated his rhetorical skills, and developed an apparently justly deserved reputation as an orator. Handsome, ironic, a member of a new breed now gone stale - the permanent adolescent - Hérault became "un homme à la mode," while his reputation as an epicurean developed.[5] Almost everything in his manner, particularly his elaborately stylistic dress, made him an unlikely candidate for prolonged favor in the impending reign of Sansculottisme.[6]

Hérault was apparently at the Bastille on July 14, 1789. Many of his near relations fled France that month, and his connections with them were to be held against him at later junctures of events. He was elected a judge of the first arrondissement of Paris in December, 1790, and in 1791 became a commissioner to Alsace, a member of the Jacobin Club, King's commissioner to the Court of Cassation, and then an elected member of the Legislative Assembly (whereupon he resigned his royal commission). Michelet calls him a weak man, a pompous actor, servitor to cowardice [l'homme faible, le pompeux actor, qui servait aux lâchetes], and a handsome but empty-headed man.[7]

His political preference fluctuated, moving successively from the Feuillants to the Girondins to the Montagnards, but his fortune proceeded.[8] He was elected President of the Jacobin Club in June, 1792, and then Vice-President (August, 1792) and President (Sept. 5, 1792) of the National Assembly.[9] An elected deputy to the Convention, in November, 1792, he became President. Sent next on a mission to Savoy and thus absent from Paris during the trial of the King, he let his condemnation of "Louis Capet" be known by letter.

Upon his return to Paris, Hérault was several times President of the Convention including the 24th of June, 1793, when the Constitution of the Year 1793 was proclaimed, of which Condorcet was the principal author, and Hérault the chief redactor. To this document was also appended the equally famous Declaration of the Rights of Man and the Citizen.[10]

Elected to the reconstituted Committee of Public Safety of the Convention, July 10, 1793, Hérault's work was largely in the diplomatic section. He had been associated with the Diplomatic Committee since March, 1792. During his mission to the Upper Rhine (Fall, 1793) he organized the "Terror" there as the only means of consolidating the Republic, as he had earlier contributed to the first organic decree of the Terror in September 1792, and later would contribute to its career in Nantes.[11] The revolution seems to have become for him "a matter of faith as well as a sensuous stimulus."[12]

Accused by Fabre d'Eglantine on the 14th of October 1793 as implicated in the suppositious "Conspiration de l'Étranger," Hérault later was cited similarly by Amar and Bourdon de l'Oise and recalled from the Upper Rhine.[13] On March 17, 1794, Saint Just called for Hérault's life.[14] Suspected for his past, his relatives, his Dantonist leanings, and his "affected clothing," Hérault was accused of treason, imprisoned in the Luxembourg, and executed, along with Camille Desmoulins and Danton, April 5, 1794.[15]

A notorious anti-cleric, a collector of erotica, an aristocrat by birth, an epicurean by taste, a tribune of the people by force of circumstances, Hérault was one of those audacious parlementarians who, Juin tells us, "stood at the crossroads of two universes," with "feudalism engrained in their hides and the future ensconced in their heads," speaking boldly but defending – at least for themselves – the priveleges of birth and blood.[16] Between the Revolution and him there was, Juin tells us, a misunderstanding.[17] Or again: "He was a terrorist in lace."[18]

A man of letters, with a somewhat theatrical cast, a formalist, Hérault was, it seems, satisfied with signs and symbols, effects and applause, with little concern for the things beyond signs and audiences.[19] In 1779 he had published his impertinent Éloge de Suger, which had been unsuccessful in a competition sponsored by the Académie Française. It brought him a certain amount of notoreity, but introduced his name into the circle of Parisian salons.[20] He had given the address at the re-opening of Parliment in 1786, and in 1788 had printed his Codicille politique et pratique d'un jeune habitant d'Épone.[21] He sought glory,[22] and Buffon had a near monopoly of that quality. This is what brought Hérault to Montbard.

The tone of the Voyage à Montbard is self-serving and somewhat frivolous. In fact, Sainte-Beuve speaks of Hérault at Montbard as an "espion léger, infidèle et moqueur." But Sainte-Beuve then unfortunately contrasts Hérault's tone with the supposed self-abasement of Rousseau during his visit to Buffon, for the details of which he has to rely apparently on Hérault's account.[23]

Just how much time Hérault ever spent with Buffon is doubtful. A letter from Buffon's son to Mme. Necker dated October 30, 1785, speaks of Hérault's request to be allowed to visit, the permission granted, the Count's illness, and Hérault's periodic sessions with the Count "de temps en temps lorsque son état le lui permet,"[24] just as Hérault seems to record it. Yet Henri Nadault de Buffon, the naturalist's grand-nephew, says that Hérault's request to visit in 1785 was denied due to Buffon's illness and that, stung by a refusal the cause of which he failed to recognize, Hérault came to Montbard anyway and published subsequently an account of his visit in which truth and falsity were cleverly intermixed.[25] Humbert-Bazile, Buffon's secretary, says that Hérault was never received at Montbard.[26] Quite a bit of Hérault's analysis of Buffon's character and regime finds independent verification in Humbert-Bazile's account (except for those supposed facets of Buffon's character which might be considered rakish). In fact, Humbert often sounds as if he were writing with a copy of Hérault's Voyage before him. Hubert Juin says that the younger Buffon's testimony is contestable, that the controversy is unsettleable, and that its issue is, literarily at least, beside the point. The Visite à Buffon became a model, a masterpiece.[27]

The soundness of Hérault's account and the fidelity of it to his supposed observations have been denied, notably by Henri Nadault de Buffon, who thought the account not worth discussing.[28] Humbert-Bazile's account of Buffon's religious stance contradicts Hérault's rather directly.[29] He also suggests that Hérault's description of the relations between Buffon and Marie-Madeleine Blesseau was calumnious.[30] It would be disproportinate to enter into this controversy in an introduction as brief as this one necessarily is. Suffice it to say that it is not at all obvious that Hérault's account is unduly exaggerated. However, one might wish to make the sort of distinctions that Jean Piveteau makes concerning Buffon's religion, for instance, and while refusing to bracket Buffon among the materialists, accord him a position as a sort of anti-metaphysical and a-dogmatic nominal "believer" (or "spiritualist") who, while for various practical reasons not wishing to offend the authorities by writing against Scripture and its account of creation, nevertheless had no real intention of conforming himself to any credal formulary.[31]

Such as it is, a testimony to the glory and, perhaps, vainglory, of a monumental naturalizer and a natural monumentalizer, the Voyage à Montbard follows.[32]

NOTES

[1]Hubert Juin, "Portrait du Conventionnel Hérault de Séchelles," in Hérault de Séchelles: Oeuvres litteraires et politiques, ed. by Hubert Juin (editions Rencontre, 1970), p. 11. The Biographie universelle (Paris: Michaud; 1817) T. Vingtieme, 222, gives the year 1760, without further specification. The Ency. Brit. (1961), Vol. 11, p. 474, gives Sept. 20, 1759, as does the 11th edition (1910), Vol. 23, p. 332.

[2]"Portrait," 11.

[3]"Indications Biographiques," in Juin, ed., op. cit., p. 25.

[4]Jules Michelet, Histoire de la Revolution Française, ed. Gerard Walter (Paris: Gallimard; Angers: Editions de l'Oeust; 1952), II, 354, and "Table Analytique: Personnages," 1419.

[5]"Portrait," 8-9.

[6]See the article on Hérault in the Dictionnaire biographique et historique des hommes marquans de la fin du dix-huitieme siècle (Londres: 1800), T. 2, p. 189. For a good brief description of Hérault, see Herman Wendel, Danton (New Haven, Yale University Press, 1935), pp. 21, 176-77, 259-60.

[7]Michelet, II, 393, 421

[8]Ibid., II, 1420

[9]"Indications," 31-33

[10]Biog. univ., T. 5, 224; "Portrait," 21; E.B., 11th Ed., Vol. 13, 333.

[11]Biog. univ., T. 5, 225, 224; "Indications," 37, 38; Michelet, II, 718. Alison Patrick, The Men of the First French

Hérault de Séchelles' Visit to Buffon   353

Republic (Baltimore: Johns Hopkins; 1972), 374, lists Hérault as a member of "The Executive Committee of the Terror" from May 30, 1793, to January, 1974.

¹²Wendel, p. 35

¹³"Indications," 39-40

¹⁴Michelet, II, 771

¹⁵ M.A. Thiers, Histoire de la revolution française (Paris: Furne, Jouvet et Cie. 1867), II, pp. 475, 481. See also Biog. univ., 226, E.B., 11th Ed., Vol. 13, 333; Dict. biog., T. 2, 190.

¹⁶"Portrait," 7, 14.

¹⁷Ibid., 15.

¹⁸Ibid., 10.

¹⁹Ibid., 9, 10.

²⁰Ibid., 14.

²¹Ibid., 17.

²²Ibid., 15.

²³Sainte-Beuve, Les Grands Ecrivains français: XVIIIᵉ siecle: philosophes et savants (Paris: Garnier Frères; n.d.) Vol. I, p. 78. See also Maurice Allem's Annotations to this Volume, p. 286. A significant part of Sainte-Beuve's account seems to rely on Hérault's Visite.

²⁴Cited in "Indications," 26.

²⁵Henri Nadault de Buffon (ed.), Buffon: sa famille, ses collaborateurs, et ses familiers (Paris: Jules Renouard; 1863), "Preface," pp. vii-viii.

²⁶In a letter to M. Faujas de Saint-Fond dated April 25, 1839, cited in a footnote, ibid., xii. This was, of course, more than half a century after the event, and given from memory.

²⁷"Indications," 15-16.

²⁸Nadault de Buffon ed., op. cit., "Preface," viii.

²⁹Humbert-Bazile, in ibid, pp. 49-58.

[30] Humbert-Bazile, in *ibid.*, p. 420.

[31] See Jean Piveteau, "La Pensée religeuse de Buffon," in Roger Heim, ed., *Buffon* (Paris: Le Museum National d'Histoire Naturelle Publications Françaises, 1952), pp. 125-132. See also the Sorbonne's condemnation of sections of the *Histoire naturelle*, and Buffon's reply to the Sorbonne, above, pp. 283-91. See also the review of the first volumes of the *Histoire naturelle* in the Jansenist *Nouvelles ecclésiastiques*, above, pp. 235-52.

[32] I wish to thank my friends and colleagues Teresa and Michel Marcy, of St. Mary's College, and Dr. Robert Nuner, of the University of Notre Dame, for their patient and careful scrutiny of my translation, and for their numerous suggestions and corrections.

Figure 20 - The Tower and Study at Montbard

HÉRAULT DE SÉCHELLES' <u>JOURNEY TO MONTBARD</u>*

Translated by John Lyon

> The old man Aristonous, living alone,
> on the island of Delos. Playing a
> golden lyre, he sings of the revolutions
> of the celestial spheres, the marvels of
> nature, the graces, friendship, virtue.
> - Fenelon, <u>Aventures d'Aristonous</u> -

For a long time I had a great desire to know M. de Buffon. Apprised of my desire, he kindly consented to send me a very courteous letter, wherein he anticipated my impatience, and invited me to spend as much time as possible at his chateau.

It is proper, as will appear in a moment, that I make mention here of the letter by which I responded. It ended with these words: "But whatever my desire may be, Monsieur le Comte, to see and hear you, I shall keep in mind your work, which occupies a great part of your day. I know that, clad in glory, you work unceasingly, devoting your genius to its tasks from the rising of the sun above the tower of Montbard, and often retiring from them only when evening comes. It is only after such work is done that I would dare request the honor of conversing with you and consulting you. I shall regard that <u>epoch</u> as the most glorious of my life should you consent to honor me with a bit of friendship, if <u>the interpreter of nature would deign</u> sometime to share his thoughts with him who would be <u>the interpreter of society</u>."

I proceeded to Montbard, indeed, but on my way, at Semur, which is only three leagues distant, I learned that M. de Buffon was suffering excessively from the stone, that he gnashed his teeth and stamped his feet, he who had always prided himself on being superior to pain. I also heard that he had withdrawn to his chamber, and would see absolutely no one, not even his servants. He suffered none of his relatives to come to him, neither sister, nor brother-in-law, while just barely allowing his son to enter for a few minutes. I thus decided to remain a few days in Semur, not even daring to inquire of news of the sick man, for fear that the announcement of my arrival would disturb him.

Despite my precautions, I remained only three days at Semur. M. de Buffon was apprised of my departure for his vicinity by a letter from Paris. He had immediately, in the very midst of his sufferings, the thoughtfulness to send me a messenger, who was instructed to say to me that, although he [Buffon] was seeing no one, he wished to see me, and that he would wait for me at his house and receive me in the intervals when his illness was not upon him. I left immediately. What palpitations of joy seized me when I saw in the distance the tower of Montbard, and the terraces and gardens which surround it! I observed the layout of the grounds - the hill on which the tower stands, the mountains and hillsides which dominate it, and the skies which cover it. I eagerly looked for the Chateau; but eagerness did not allow me to espy the residence of the celebrated man with whom I was going to speak. One can only see the chateau when one is upon it: except that, instead of a Chateau, you imagine that you are entering some Parisian house. M. de Buffon's house is not striking. It is situated in a street in Montbard, which is a small town. For the rest, it has a very pleasant appearance.

Upon arriving, I found M. le comte de Buffon's son, a young officer in the guards, who came to meet me and conduct me to his father.[1] With what lively emotion was I seized while mounting the stairs and crossing the room ornamented by all the richly painted birds, such as one sees them in the great edition of the <u>Histoire</u> <u>naturelle</u>! Then I was in the chamber of Buffon. He came out from another room, and I ought not fail to mention an incident which struck me, because it illustrates his character. He opened the door, and although he realized that there was a stranger in his room, he quite tranquilly turned around and took his time about closing it. Then he came to me. Would this be due to a spirit of order which treats everything with the same exactitude? Such is the manner of M. de Buffon. Would it be due to the unhurriedness of a man who, satiated with praise, waits for it rather than seeks it. One might suppose so. Finally, would it be due to a bit of adroitness on the part of a celebrated man, who, pleased with the great desire people have to meet him, cleverly increases further this avidity by delaying, even though it be only for a moment, precisely when he satisfies this desire, thus making himself all the less available precisely when you seek him? This artifice would not be totally unthinkable in M. de Buffon. He majestically approached me, with both arms open. I stammered out a few words, careful to say, "M. Le Comte," for any other form of address he would find deficient. I had been made aware that he did not dislike this manner of address. He replied to me by embracing me: "I ought to think of you as an old acquaintance, for you have signified a desire of seeing me, and I too have desired to know you. We have been searching for this meeting for quite some time."

I saw a handsome figure, noble and calm. Despite his age of 78 years, he could easily pass for 60; and what is even more remarkable is that he has just passed sixteen sleepless nights, undergoing the most extraordinary pain, which had not yet left him. Yet he was as fresh as an infant and as relaxed as one in good health. I was assured that such was his character. All his life he drove himself to rise above his own ailments, never out of sorts, never impatient. His bust, by Houdon, is that which seems to me to most resemble him. But the sculptor was unable to render in stone those black eyebrows which shade the quick, black eyes beneath his handsome white hair. It was curled when I saw it, though he was ill. This was one of his idiosyncracies, and he admits it. He has it put in curl papers daily, and has the curling iron applied twice, rather than once; that is, in the past, after having it curled in the morning, he as often as not had it curled again before supper. It was done up in five small floating ringlets. His hair, tied at the back, hung down to the middle of his back. He wore a yellow lounging robe, sprinkled with white stripes and blue flowers. He had me sit down, spoke to me of his health, and complimented me on the bit of recognition with which he claimed the public favored me, on eloquence, and on oratorical discourse. As for me, I spoke to him of his glory, and did not allow myself to grow tired of noticing his features. The conversation having fallen upon the desirability of knowing at an early age the state to which one is destined, he read me at once two pages which he had written on this subject in one of his works. His manner of reciting is infinitely simple and common. He spoke in the tone of everyday conversation, without affectation, raising first one hand and then another, speaking as things came to him, only adding several inflections. His voice is quite strong for his age, and is one of great intimacy. In general, when he speaks, his eyes are not fixed on anything, but wander randomly, either because he is shortsighted or, more probably, because such is his manner. His favorite words are: "all that" and "By God!", which recur continually. His conversation appears to contain nothing outstanding, but when one pays attention to it, it is noticeable that he speaks well, that there are even things quite well set forth, and that, from time to time, it is sprinkled with interesting views. One of the prime traits of his character is his vanity. It is total, but frank, and in good faith. A traveller (M. Target)[2] said of him: "There is a man who has a great deal of vanity, all of it in the service of his pride."

The reader might be curious as to some examples of this vanity. I said to him that, in preparing to see him, I had read quite a number of his works. "What did you read?," he said to me. "The 'Views of Nature'," I said. He immediately replied: "There one finds samples of the highest eloquence!" Then, contrary to his usage, he spoke of recent developments

Figure 21 - Buffon by Houdon, 1781

Figure 22 - Buffon by Houdon, 1783. (It is uncertain which bust Sechelles is referring to in the text)

and of politics, which gave him occasion to have me read a letter from M. le Comte de Maillebois,[3] concerned with events in Holland. He then came, after a moment, to the death of the unfortunate M. Thomas,[4] and had me read a letter which his son had received from Mme. Necker,[5] a strange letter, in which Mme. Necker appeared to be already consoled at the loss of her intimate friend, despite the bombast and enthusiasm which went into her description while quoting M. de Buffon, whom she extolled in even more fulsome tones. There was a sentence which he showed me with satisfaction. Mme. Necker, drawing a parallel between her two friends, spoke of M. Thomas as "the man of the century," and of M. de Buffon as "the man of all the centuries."

M. de Buffon's son had just raised a monument to his father in the gardens of Montbard. Near the tower, which is very high, he had had a column placed with this inscription on it:

As this humble column is to an
    exalted tower,
So the son to his father, Buffon, 1785.[6]

I was told that the Father had been reduced to tears by this homage. He said to his son: "My son, that will be a credit to thee."

He ended our first interview, for his suffering from the stone recurred. He added that his son would take me about, and show me the gardens and the column. The young Comte de Buffon first showed me about the house, which is very well kept and quite well furnished. It has twelve complete suites. But it was built without regularity, and though this defect ought to render it more convenient than handsome, it has some beauty nevertheless.

After the house we visited the gardens, which are above the house. They are built on thirteen terraces, as irregular in their own way as is the house in its. But the view from them is immense and magnificent: prairies cut by rivers; vineyards; hillsides sparkling from the labor of man; and the entire town of Montbard. The gardens are mixed with trees in groups of five--with pines, plane trees, sycamores--with hedgerows, and, everywhere, there are flowers among the trees. I saw the great aviaries where Buffon raised foreign birds which he wished to study and describe. I also saw the pit, now filled-in, where he had formerly kept lions and bears.[7] Finally, I saw what I really wanted to see - the study where this great man works. It is in a detached building, which is called the Tower of St. Louis. One goes up a stairway and enters through a green double-door. But one is astonished indeed to see the simplicity of the laboratory. Beneath a fairly high vault, which somewhat resembles the vaults of those churches and ancient chapels whose walls are painted in green, he has had a miserable wooden secretary set in the middle of the tiled room; and in front of the secretary is an arm-chair.

That's all. No books, no papers. But isn't this starkness a bit striking? One calls to mind here Buffon's ornate pages, the magnificence of his style, and the admiration which it inspires! However, this is not the study where he did most of his work. He goes here only in the heat of summer, for the spot is extremely cool.[8] There is another sanctuary where he composed almost all his works. "The bower of the <u>Histoire naturelle</u>," as it was called by Prince Henri, who wished to visit it. It was here that J.J. Rousseau knelt down and kissed the door step. I spoke of this to M. de Buffon. "Yes," he said to me, "Rousseau did homage there."

This study, like the first one, has a green double door. On the inside there are screens on each side of the door. The study is parqueted,[9] panelled, and papered with images of birds and a few quadrupeds from the <u>Histoire naturelle</u>. There is a couch there, several old chairs covered with black leather, a table with manuscripts on it, and another small black table. These are all the furnishings. The secretary where he works is at the end of the room, near the fireplace. It is a rough piece of walnut furniture. It was open, and the only thing visible was the manuscript on which Buffon was currently working, a "Treatise on the Magnet." Next to it was his pen. Above the secretary was a cap of gray silk which he wears. Facing the secretary is the armchair where he sits, an old miserable armchair upon which is thrown a dressing gown, gray with white stripes. In front of him, on the wall, is a print of Newton. There Buffon has spent the greatest and the best portion of his life. There almost all of his works were born. Indeed, he dwelt much at Montbard, staying there eight months out of the year. In this fashion he has been living for more than forty years. He used to spend four months out of the year in Paris in order to take care of his affairs and those of the Jardin du Roi, and returned to throw himself into his studies. He told me himself that it was his greatest and dominant pleasure, along with an extreme passion for glory.

His example and his conversation confirmed me in my belief that whoever passionately loves glory ends by obtaining it, or at least approximating it. But one must desire it - not just once, but constantly. I have heard tell of a man who was Marshall of France and a great general, who used to stride about his room for a quarter of an hour each morning, while saying to himself: "I want to be Marshall of France and a great general."[10] M. de Buffon spoke a quite striking word to me on this subject, one of those words capable of bringing out the whole man: "Genius is nothing but a greater aptitude for patience." Indeed, it suffices to have received this quality from nature. With it one is able to patiently observe objects, and succeed in penetrating them. This calls to mind the words of Newton. Someone asked Newton, "How have you been able to make so many discoveries?" "By always searching," he replied, "by searching patiently." Note that patience is necessary for everything: patience in searching out one's object, patience in resisting all that distracts one from this, patience in putting up with everything which would crush an ordinary man.

I shall take my examples of patience from M. de Buffon himself. When in Paris as a young man, he often returned from dinners at 2:00 A.M. At five in the morning a servant[11] came to pull him out of bed by his feet and put him on the floor, with orders to do him violence should he get angry at the treatment. He also told me that he worked until six o'clock in the evening. "I had then," he said to me, "my little mistress, whom I adored. Well! I had to do my utmost to wait until six o'clock before going to see her, often thus taking the risk of missing her." At Montbard, after his work, he called in a little girl, for he was always enamored of them. But he got up exactly at 5:00. He only saw little girls, not wishing to have women who might cause him to fritter away his time.[12]

After this fashion was his day arranged; and it is even possible to specify the allocation of his time further. At five A.M. he rose, dressed, had his hair done, dictated letters, and put his affairs in order. At six o'clock he went up to his study, which is at the end of the gardens, at a distance of almost a quarter of a league. This distance is all the more taxing because one always has to open gates and climb from terrace to terrace. There he either wrote in his study or walked about the lanes in its vicinity. He forbade anyone to come near him, dismissing any of his servants who would come to disturb him.

His habit is to re-read often what he has composed, leaving it to lie dormant for several days, or even longer. He said to me: "It is essential not to hurry, for this gives one the chance to see things with clearer eyes, and you always have to add or change things." First, he wrote. When his manuscript was too full of erasures, he gave it to his secretary for copying until he was satisfied with it. For example, he confessed to the canon teaching theology at Semur, a man of wit and his friend, that it was in this fashion that he re-wrote eighteen times his Époques de la nature, a work he had been turning over in his mind for fifty years.

I must not forget that M. de Buffon, who was very orderly, placed his study at such a distance from his dwelling not solely so that he would not be distracted,[13] but because he wished to separate his work from his business. "I burn everything," he said to me. "Not a single paper will be found when I die. I have taken this course of action realizing that otherwise I would never make it. One would be buried in one's papers." He only saved poetry praising him, of which I shall have occasion to speak in a moment. Likewise, in his bedroom, one finds only his bed, which is, like the wall paper, of white satin, with a flowery design in it. Near the fireplace is a secretary; on the shelf above it only one book, which is apparently his book of thoughts. Near the secretary, which is always open, is the armchair in which he is always seated; and in a corner of the room is a small black table for his copyist.

He only takes up his pen after long having meditated on his subject, and, furthermore, scarcely has any other paper about than that on which he is writing. This orderliness in his papers is more important than one might think. M. Necker seriously recommends it in his book; the Abbe Terray[14] worked in this same fashion. The order that we see in these things about us is indeed reflected in our productions. If a writer as celebrated [as Buffon], and above all if two controllers-general as assiduous as these have given similar examples, it would be very difficult to discover pretexts for not imitating them.

I return to M. de Buffon's daily activities. At 9:00, breakfast is brought to him in his study, or sometimes he eats it while dressing. Breakfast consists of two glasses of wine and a bit of bread. He then works until one or two o'clock, at which time he returns home. He then dines, loving to do so leisurely. It is at dinner that he puts his intelligence and his genius aside. There he abandons himself to levities of all sorts, as they pass through his mind. He greatly enjoys telling dirty jokes, which are peculiarly enjoyable since they always contrast with his imperturbable character. His laughter, his age, present a striking contrast with the seriousness and gravity which are natural to him. His jokes are often so raw that the women present find it necessary to leave.

In general, Buffon's conversation is quite offhand.[15] When this was pointed out to him he replied that this was his hour of repose, and that it mattered little whether his words were polished or not. It isn't that he doesn't say excellent things, when one mentions style in his presence, or natural history. He is, further, of much interest when he speaks of himself, and often does so with much self-praise.

For myself, who have witnessed his discourses, I assure you that, far from being scandalized by them, I find them pleasing. It is not pride, it is not vanity on his part. It is his conscience one hears speaking. He is self conscious, and does service to himself. Let us agree, then, to have great men, now and then, on such terms. Any man who does not know his own strength is not strong. Let us not demand of superior beings a modesty which could not be other than false. There is perhaps more spirit and skill in concealing, in veiling one's merit; there is more good nature and benefit in displaying it.[16]

For the rest, he did not praise, but rather passed judgment, on himself. He judged himself as he will be judged by posterity, with this difference, that an author is more cognizant than anyone else of the secret of his works. He said to me: "Everyday I learn how to write. In my later works there is infinitely more perfection than in my first ones. I often have my works re-read to me, and I then find ideas that I shall change, or amend. There are other pieces that I simply cannot do better."

This frank good faith is precious, original, a relic of former ages, and is seductive. It is possible, moreover, to rely on M. de Buffon in this respect, for no one is more severe than he when it comes to style, precision of ideas (which he considers the prime characteristic of a great writer), and the appropriateness and exactness of contrasts which one must make between ideas if they are to be clearly seen, or developments which are necessary to show these ideas to advantage. I have heard him argue about entire pages most rationally and sensibly, but at the same time analyze them inexorably. "I have been obliged," he said to me, "to speak in all sorts of fashions in my work. One must know what rung of the ladder it is necessary to reach."

By a natural progression, he demands that an author be of frank good faith, of propriety in the texture of his opinions, and above all that he be consistent. Buffon never forgave Rousseau his contradictions. Thus, it could be said that he calculated his sentences and his thought, as he calculated everything. This was a remarkable quality, which perhaps was born in his mind through mathematics and the habit of applying it. He told me that he had studied mathematics at an early age with intensity, first in the works of Euclid, and then in those of the Marquis de l'Hôpital.[17] When he was twenty years old, he discovered the binomial theorem of Newton, without knowing that it had been discovered by Newton; and this vain man never published his discovery. I told him I would be quite pleased to know why. "Because no one is obliged to believe me about it," he replied. Thus, there is this difference between Buffon's vanity and that of others, namely, that his is grounded on achievement, if one could express oneself thus. This difference comes from the temper of his soul, an upright soul, which strives for frank good faith, and proscribes inconsistency.

Speaking of Rousseau, he said to me: "I loved him quite a bit; but when I saw his Confessions, I ceased respecting him. His soul disgusted me; and, in the case of Jean-Jacques, the opposite of what usually happens took place: after his death, I began to hold him in low esteem." A severe judgment, I should say even unjust; for I have to say that the Confessions of Jean-Jacques did not produce that effect on me. But it could be that M. de Buffon lacked in his heart that element by which one ought to judge Rousseau. I would be tempted to believe that nature had not given him the sort of sensibility necessary to appreciate the charm, or rather the piquancy, of that aberrant life, of that existence given over to chance and to the passions. This severity, or rather this fault, which could perhaps be found in the soul of M. de Buffon, was itself the sign, in another way, of beauty, and even simplicity of soul. Consequently, by a natural progression, he is easy to dupe, whatever order he may put in his affairs. Recently, we received proof of that.

A year ago, the director of his ironworks caused him to lose 120,000 livres. For three years M. de Buffon agreed to forego the payment due him, and came to believe all the pretexts and subterfuge with which this fraud was tinged. Fortunately, this event did not alter his serenity at all; nor did it have any effect on his expenditures and his way of life. He said to his son: "I only grieve for you. I wanted to purchase a piece of land for you. Now it will be necessary to postpone this for awhile." He always had a year's income ahead. It is believed that he has 50,000 ecus in rent. His ironworks must have enriched him greatly. He produced 800,000 pounds of iron annually. But, on the other hand, he has enormous expenses. This considerable establishment cost him 100,000 ecus to build. It stands idle today, because of legal proceedings against the director; but when it is in production, 400 workers are employed in it.

It is not astonishing that M. de Buffon, with so simple a soul, believes everything that is told him. Furthermore, he loves to hear reports and gossip. This great man is at times a bit of a busybody; at least one hour each day is, you have to admit it, given over to gossip. During the time devoted to his toilet, he has his hairdresser and his servants tell him of all that transpires in the town of Montbard, and all the events of his house. However he might appear to devote himself to his deep thoughts, no one knows better then he the petty events which surround his life. Perhaps this is also due to the taste he has always had for women, or rather for little girls. He loves the scandal sheet; and to be kept informed about such news in a small region is to take in almost all its history.

This preference for little girls, or rather the fear of being dominated, led him to put all his confidence in a peasant woman of Montbard whom he has set up as governess, and who wound up governing him. Her name is Mme. Blesseau.[18] She is a woman of forty, well-made, and must once have been rather pretty. For almost twenty years she has been with M. de Buffon. She looks after him with great zeal. She shares in the administration of the house: and, as often happens in such cases, she is detested by the servants. Mme. de Buffon, dead these many years, did not like this woman at all. She adored her husband, and it is claimed that she was extremely jealous over this relation. Mme. Blesseau is not the only one who controls this great man.

The empire of his house she shares with another "original," a Capuchin, who is called Père Ignace.[19] Let me pause here a moment for the history of Ignace Bougot, born in Dijon. This monk possesses to an eminent degree the valuable art of his order, of developing in others the habit of giving, so that whoever gives appears to owe much gratitude to him. "Not everyone who is willing can give," he often says. With this talent he was able to succeed in rebuilding the Capuchin nest of Semur. This talent is rather common in Churchmen. I have seen a cure who was a rival of Ignace's in this sort of petty begging. He captivated old women to such a point that they believed themselves only too happy to give him what they had, and often more than they had. Men of such character are also men of intelligence. They love to have a finger in everything; they have a good sense of business and carry out assignments faithfully - such activity is not at all foreign to them. They are as careful not to appear displeasing to the lackeys, whose forgiveness they need for the profits which they steal from them, as they are careful to please the masters whose favor they are trying to obtain. Such is Ignace.

If you would have an idea of his person, imagine a huge man with a round head, a bit like one of the masks of Harlequin in Italian comedy. This comparison seems to me to be all the more appropriate, as he speaks precisely as does Carlin - same accent, same wheedling ways. It is to this Reverend Father, cure of the village of Buffon, which is two leagues from Montbard, that M. de Buffon entrusts a great part of his confidence, and even his conscience, if others' words may be taken for it. Indeed, Ignace is Buffon's confessor. He is everything at his house, and gives himself the title of Buffon's Capuchin. He will tell you, should you wish, that one day M. de Buffon took him to the Academie Française, where he was the object of everyone's attention; that he placed him in a chair among the forty [members of the French Academy]; and that M. de Buffon, after giving his discourse, brought him to his carriage in full view of the public, which had eyes only for him. M. de Buffon has cited him as his friend in the article on the canary.

He is also Buffon's lackey. I have seen him following Buffon on a walk, hobbling along behind him, for he is lame-- the whole scene suitable for a painting; while the author of the Natural History marches proudly along, head held high, hat in the air, always alone, scarcely deigning to notice the earth, absorbed in thought, resembling the man whom he depicted in his history of man - a portrait no doubt patterned after himself - holding a cane in his right hand, his other hand resting with majesty on his left hip. I have seen Ignace, when Buffon's valets were absent, take away Buffon's napkin and the small table upon which he had been dining. "I thank you, my dear child," Buffon said to him. And Ignace, taking a humble attitude, had a more servile air than the servants themselves.

The same Ignace, Capuchin-lackey, is also the confessor-lackey of M. de Buffon. He told me that thirty years earlier, knowing that he was going to preach during Lent at Montbard, the author of the Époques de la nature had Ignace come to him at Easter time, and had him hear his confession in his laboratory. Here, in the very place where he developed materialism! Here, in the same place where Jean-Jacques would come, some years later, to devotedly kiss the doorstep! Ignace told me that M. de Buffon had hesitated a bit before submitting himself to this ceremony. "A result of human weakness," he added. Buffon then wanted his valet confessed before him. All this may astonish you a bit. Yes! Buffon, when he is at Montbard, takes Communion at Easter, each year, in the Seigneurial Chapel. Every Sunday he attends high Mass there; during Mass he leaves sometimes to walk through the gardens which are nearby, returning for interesting passages of the liturgy. Every Sunday he gave the value of a <u>Louis</u> to various alms-collectors. It is in this chapel that his wife is buried. She was a charming woman, to whom he was married at forty-five. He was genuinely attached to her; and she always adored him, despite the numerous infidelities in which he was involved.[20] She had been relegated to a Convent at Montbard, for though of good birth, she was without fortune. He courted her for two years, and at the end of this period espoused her, despite his father, who was still living and who, being financially ruined, was opposed to the marriage of his son out of concern for financial advantage. Her name was Mlle. de Saint-Belin.[21]

I heard M. de Buffon say that, as a matter of policy, he respects religion;[22] that it is necessary for the lower class to have some; that, in small towns one is observed by everyone, and he did not wish to shock anyone. He said to me: "I am persuaded that in your addresses you are careful not to put forth anything which might be offensive in this respect. I have always had this concern in my works. I only published them serially, in order that ordinary men would not be able to grasp the chain of my ideas. I have always mentioned the Creator; but one need only remove this word, and mentally substitute in its place the power of Nature, which results from the two great laws, attraction and impulsion.[23] When the Sorbonne imposed their pettifoggery on me, I had no difficulty in giving them every satisfaction they desired. It is only quibbling. But men are so stupid as to be contented with it. For the same reason, when I fall dangerously ill and shall feel my end approaching, I shall scarcely hesitate to send for the sacraments.[24] One has duties to the public cult. Those who act otherwise are fools. It is never necessary to run square against public opinion, as Voltaire, Diderot, and Helvétius have done. The last of these was my friend. He spent more than four years at Montbard, at various times. I recommended such moderation to him, and if he had believed me, he would have been better off."

It is indeed possible to judge of the efficacy of this policy of Buffon's. It is obvious that his works set forth materialism; however, they are published by the Imprimerie Royale.

"My first volumes appeared," he added, "at the same time as the <u>Spirit of the Laws</u>. M. de Montesquieu and I were tormented by the Sorbonne. Furthermore, we underwent a barrage of criticism. The President was furious. 'What are you going to reply?' he asked me. – "Nothing at all, President,' I replied. And he was simply unable to comprehend my composure."

One night I read to M. de Buffon some lines of poetry by M. Thomas on the immortality of the Soul. He laughed:

"Pardieu, religion would offer us a fine present, if all that were true!" He criticized the poems severely, but justly, for he is unrelenting on matters of style, and above all concerning poetry, which he does not like. He maintains that it is impossible in our language to write four consecutive lines of poetry without committing an error, without violating either the appropriateness of terms or the fittingness of ideas.[25] He advised me never to write poetry. "I could have written it, like everyone else," he said to me. "But I quite quickly abandoned the game where reason is enchained. She labors under enough handicap already, without having new ones imposed on her."

These verses call to mind a pleasant fit of vanity which followed them. On the morning of the day of which I speak, M. de Buffon, under the pretext that his health would not allow him to become fatigued by doing paper work, asked me to read aloud to him a multitude of verses which had been addressed to him. He saved almost all of these, though they were almost always mediocre. When, in one, he was addressed as "creative genius, sublime spirit" - "Eh! Eh!," he responded complaisantly. "There is some substance, there's something there." That evening, while listening to the verse of M. Thomas, he said to me with charming naivete: "All that is nothing compared with this morning's verses."

I take the liberty of inserting here another trait of his of the same kind. He said to me: "One day, when I had worked long and had discovered a very ingenious system to explain generation, I opened Aristotle and what do you think? I found all my ideas in that rascal Aristotle! Indeed, by God, that was Aristotle at his best."

The first Sunday during my stay at Montbard, the author of the Histoire naturelle asked his son to stay up with him the night before. He had a long conference with him, and I learned that the purpose of it was to get me to consent to go to Mass the next day. When his son spoke to me of this, I replied that I would do so quite voluntarily and that plotting was not necessary to get me to fulfil an action of civil life. This response charmed M. de Buffon. When I came back from high Mass, which Buffon's ailment with the stone kept him from attending, he thanked me a million times for having been able to put up with three-quarters of an hour of boredom. He said to me once more that in a small town such as Montbard Mass was a matter of obligation.

When Buffon left the services, he loved to stroll about the square, escorted by his son and surrounded by his peasants. It pleased him above all to appear in the midst of them in elaborate attire. He made a great thing of ornamentation, hair dressing, and good clothes. He himself always dressed like a great lord, and scolded his son when he went about only in a contemporary dress-coat. I knew this preference of his, and when I went to meet him I wore an outfit decorated with braid and a short coat charged with gold. I was told that my precaution was marvelously successful. He cited me to his son as an example. "There is a man!" he exclaimed. And in vain did his son reply that the fashion was out of style. He paid no attention.

Indeed, it is Buffon who wrote, at the beginning of his "Natural History of Man" that our clothes constitute part of ourselves. Our machine [body] is so constructed that we are at once predisposed in favor of the person who appears elegant to us. One does not initially separate the clothes from the man, and the mind grasps the whole, clothes and the man, and judges the merit of the second by the appearance of the first. This is so true, that M. de Buffon himself was finally caught by this trick; and I, with my dress, worked upon him the illusion which he wished to communicate to others. What would be the case, in particular, if we already know the person whom we approach, and are apprised of his glory and his talents? Then genius and gold conspire together to dazzle us, and the glitter appears to come from the genius of the man himself!

Buffon was so accustomed to this magnificence that he said one day that he could work only when he felt himself to be well attired and well-set. A great writer sits himself down at his work-table in the same way we all do when we put on our most magnificent adornments to perform our most solemn actions. He is alone; but he has in his minds' eye the universe and posterity. Thus, Gorgias and the Sophists of Greece, who astonished shallow people by the eloquence of their discourse, never appeared in public except when adorned with a purple robe.

I must finish with M. de Buffon's daily schedule. After his dinner, he scarcely troubles himself about those who live in his Chateau, or with those who have come to see him. He takes a nap for half an hour in his room, then goes for a walk, always alone, and at 5:00 he returns to his study, working there until 7:00. Then he comes back again to the salon, has his works read, explains them, admires them, takes pleasure in correcting the works which others have sent to him, and on which he has been consulted. Such has been his life for fifty years.[26] To someone who was astonished at his renown he said: "I have spent fifty years at my desk."

At nine in the evening he goes to bed, and never takes supper. This indefatigable writer still led this laborious life up to the moment when I arrived at Montbard, that is, when he was seventy-eight years old. But, as the sharp pains of the stone were upon him, he had to suspend his work. Then, for several days, he was shut up in his room, alone, walking from time to time, receiving no one whatsoever of his family, not even his sister, and only allowing his son a minute's visit per day. I was the only one he would much allow about him. I always found him beautiful and serene in his suffering, hair curled, even very well dressed. He complained gently of his health, claiming to prove, by the most cogent arguments, that the illness enfeebled his ideas. As his pains were continuous, as well as the irritation of his needs, he often asked me to leave him after about a quarter of an hour, sending for me again a few moments after. Little by little the quarter-hours became entire hours.

This good old man fondly opened his heart to me. At one time he had me read the most recent work he was composing, namely, a <u>Treatise on the Magnet</u>, and, listening to me, he mentally reworked all his ideas, to which he gave some new directions, or else changed their order, or withdrew some superfluous details. At other times he would send for a volume of his works and have me read from them some of the most beautiful passages, such as that discourse on the first man wherein he describes the history of his sensations, or the description of the Arabian desert, in the article on the Camel, or another picture even more well-done, according to him, in the article on the bird "Kamichi." At yet other times he explained to me his system concerning the formation of the world, or the generation of creatures, or the "internal molds," etc. Yet again he would recite to me entire pieces of his works, for he knew by heart all that he had written. And this is a proof of the power of his memory, or rather of the extreme care with which he worked his compositions. He would hear all the objections one could make to him, would appraise them, and admit it when they were justified. He has, furthermore, a fairly good fashion of determining whether writings are going to be successful. This consists of having them read from time to time from the manuscript itself. Thus, if despite the erasures the reader is not held up, he concludes that the work is easy to follow.[27] His main concern in matters of style is for precision in ideas and their interconnection. Next, he applies himself, according to his recommendation in the excellent discourse he delivered upon being received into the Academie Française, to naming things in the most general terms; then he is concerned with harmony, which it is quite essential not to neglect. But this ought to be the last concern in matters of style.

It is of natural history and of style that he most enjoys conversing. In fact, I wouldn't be surprised if he preferred to talk of style. No one was more conscious of the metaphysics of style, unless it be Beccaria.[28] But Beccaria, though giving the precept, did not give examples as well as Buffon has. "Style is the man himself," he repeated to me often. "Poets don't have style, for they are constricted and enslaved by the meter of verses they compose. Whenever someone is praised to me, I always say: "Let's see his writing." I asked him how he found the style of M. Thomas. "Fair," he replied, "but too strained, too turgid." And Rousseau's style? "Much better; but Rousseau has all the defects of a bad education. He uses interjection, ready-made exclamation, and continual apostrophe."

"Give me, then, your principal ideas concerning style," I said. "They are in my discourse to the Academy. In two words, there are two things which go to creating style: invention and expression. Invention depends on patience. It is necessary to look at one's subject for a long time. Then little by little it just opens out and develops. You feel as if a small charge of electricity struck you in the head, and at the same time reached your heart. That is the moment of genius; that is when one experiences the pleasure of working, a pleasure so great that I would spend twelve hours, fourteen hours in study, completely delighted. In truth, I have given myself over more to such delight than I have been concerned with glory. Glory comes afterwards, if it will. And it comes almost always. But do you wish to augment the pleasure, and at the same time be inventive? When you prepare to deal with a subject, open no book; draw everything from your own brain. Consult authors only when you feel that you are unable to produce anything more yourself. This is the fashion I have always followed.

"Also, following this method, one truly enjoys reading great authors: one finds oneself on the author's level, or above it; one judges them, one puzzles them out, one reads them more quickly. So far as expression is concerned, it is always necessary to affix the image to the idea! It is even necessary that the image precede the idea in order to prepare the mind for it. One ought not always use the obvious word, for it is often trite. But you have to use the next closest term. In general, a comparison is ordinarily necessary if the force of the idea is to be felt. And to use a comparison myself, I imagine the style as a fretwork which must be trimmed, cleansed on every side, in order to get the form you desire. When you write, write what first comes to mind. This is generally best. Then leave what you have done for several days, or even longer. Nature does not produce things on demand; she only works bit by bit, after repose and with renewed force.[29] What is necessary is to occupy oneself continuously with the same object, keeping at it, not letting oneself get involved in several projects. When I am writing a work, I do not muse upon anything else."

"I exclude your profession, however," M. de Buffon said to me: "You often have several pleadings [to the court] to compose at the same time, and concerning matters of little interest. You have little time and can do nothing but speak from brief notes. In such cases, instead of correction, it is necessary to rely more on the eloquence of the words, that is enough for your listeners. Pardieu, Pardieu! The letter you wrote me (I cited the end of it at the beginning of this article, knowing I would have occasion to speak of it presently) offers a good parallel between the interpreter of nature and the interpreter of society. Put that in some speech and it would produce a superb effect. It would be curious to consider the foundations of opinion, and to show how they drift about in society."

I next asked M. de Buffon what the best fashion of training oneself would be. He replied that one should read only the works of great authors, but read them in all genres and sciences, for they are, as Cicero says, related, since views in one of these are applicable elsewhere, even though one may not be destined to use all of them. Thus, even for a jurisconsult, knowledge of military art and its principal operations would not be useless. "This is what I have done," the author of the <u>Histoire</u> <u>naturelle</u> told me. After all, the Abbé Condillac says it very well, at the beginning of his forth volume of the <u>Cours</u> <u>d'education</u> - if I remember correctly - that there is only one science, the science of Nature." M. de Buffon was of the same opinion, without citing the Abbé de Condillac, whom he did not like, having had previously polemic discussions with him.[30] But he thought that all our divisions and classifications are arbitrary; that mathematics itself is but an art which leads to the same end, that of applying its principles to Nature and of making it known; and we ought not be startled by that after all. The capital works in each genre are rare, and it would perhaps be possible to reduce the total to some fifty works which it would suffice to meditate on well.

Above all, it was the assiduous reading of the greatest geniuses which M. de Buffon recommended to me. He found very few of them in the world. "There are hardly five," he said to me, "Newton, Bacon, Leibnitz, Montesquieu, and myself." As regards Newton, he discovered a great principle, but he spent his whole life doing calculations to demonstrate it, and so far as style is concerned, he can hardly be of great use." He made more of Leibnitz than of Bacon himself. He claimed that Leibnitz fashioned things at the height of his genius, while with Bacon discoveries were born only after profound reflection. However, at the same time he said that what best showed the genius of Leibnitz was perhaps not in the collection of his works; to find his best it was necessary to search in the memoirs of the Academy of Berlin. In citing Montesquieu, he spoke of his genius and not of his style, which is not always perfect, being too curt and lacking development. "I knew him very well," he said to me, "and this defect was characteristic of his physique. The president was almost blind, and he was so keen that most of the time he forgot what he wanted to dictate,[31] so that it was necessary for him to contract his expression into the least possible space."

Finally, I was very eager to know what M. de Buffon would say to me of himself concerning my style. Here is the stratagem I devised.

He asked to see a sample of my style, and I was afraid for a moment: however, an extreme desire of hearing his observations and of forming myself with the help of his criticism made me forget my egotistical self-interests. I recited to him the only piece of mine which I could remember then. I saw with pleasure that he corrected only one single word, and that he criticized with rigor, but with reason, and he said to me with his customary frankness: "This is a page which I could not write better." Encouraged by this first success, I thought it would be pleasant to write another page about him, and present it to him. It was rash to dare to judge of genius in the presence of genius itself. I decided to compare M. de Buffon's "invention" with that of Rousseau, suspecting for whom the balance would tip without injustice. Thus, I secluded myself in my room that night, took <u>Emile</u> and the volume of the <u>Vues sur</u> [sic] <u>la nature</u>, and set myself to read alternatively one page of each work. I next meditated on the impressions which I felt interiorly, and reckoned up the different kinds of impressions. At the end of about an hour I came to work them out and write them down.[32] Next day I took what I had written to M. de Buffon. I can say that he was prodigiously satisfied with it. As I read it to him, he exclaimed in admiration, or at times corrected a few words. Finally, he spent five days in re-reading and retouching this piece himself. He continually called on me to ask if I agreed with such and such a change as he made. I struggled against the change sometimes, but almost always gave in. M. de Buffon, since then, put no limits on his affection for me. At times he would cry out: "There is a great conception, pardieu! pardieu! Better comparison could not be made than to compare Rousseau and myself in this fashion." At other times he entreated me to copy the piece well in my own hand, sign it, and allow him to send it to M. and Mme. Necker. Still again, he would urge me to have it inserted anonymously in the <u>Journal de Paris</u> or the <u>Mercure</u>. Wishing to have a bit of fun with this good and frank vainglory, I asked if it would not be good to send at the same time to the journals the inscription which his son had just dedicated to him at the base of the column which he had raised in his honor. "Save that for another time," he replied to me, "it is not good to divide readers' attention. That will be the subject of two letters."

Finally, not knowing what recompense to make me, nor how to show me his pleasure, here is what he said to me one day. I ought not say it, for I will thereby fall into self-esteem much more ridiculous and with much less basis than Buffon's. But the fidelity of my narration demands that I tell all. I should speak even against myself if this same narration demanded it. One morning I heard his house bell which he always rings three times, and a moment afterwards his <u>valet de chambre</u> came and said to me, "M. de Buffon asks for you." I went upstairs. He came to me, embraced me, and said: "Allow me to give you some advice." I did not know what he had in mind, but promised him that all that he wished to say to me would be received most gratefully. "You have two names," he said to me: "People sometimes call you by one, sometimes by the other, and sometimes they use both together. Believe me, use only one of them. Strangers then will not be able to mistake you."

He next spoke to me passionately of study, and of the happiness which it brings. He said to me that he always put himself outside society; that often he had sought out savants, believing that he would gain much by their conversation, but that he had seen that to glean a somewhat useful phrase it was not worth losing a whole evening; that work had become a necessity for him; that he hoped to indulge in it for the three or four years which yet remained to him of life, and that he had no fear of death whatsoever; that the idea of immortal renown consoled him; that if he had been of a mind to seek compensation for all the so-called sacrifices of his work, he would have found it in abundance in the estimation in which he was held in Europe, and in flattering letters from the principal European rulers.

This old man then opened a drawer, and showed me a magnificent letter from Prince Henri,[33] who had come to spend a day at Montbard. The Prince treated M. de Buffon with a sort of deference. Knowing that after dinner it was his custom to sleep, the Prince arranged his hours accordingly. He also had just sent to M. de Buffon a porcelain service set which he himself had designed. The design comprised swans in all their characteristic poses, in memory of the history of the Swan, which M. de Buffon had read to him during his visit. Finally, the Prince wrote these remarkable words to him: "If I had need of a friend, it would be you; of a father, once more, you; of an intellect to sharpen my perception, Eh! Who but you."

M. de Buffon showed me the next several letters from the Empress of Russia, written in her own hand, full of verve, wherein this great woman commended him in a way which was all the more pleasant to him, since it was obvious from the letters that she had read his works and that she understood them like a savant. She said to him: "Newton took the first step, you have taken the second." Indeed, Newton had discovered the law of attraction, and Buffon had demonstrated that of impulsion which, with the assistance of the former law, seems to explain all of Nature. She added: "You have not yet emptied your sack on the subject of man," making an allusion thereby to the system of generation, and Buffon congratulated himself on having been understood better by an Empress[34] than by an Academy. He also showed me some very thorny questions which the Empress proposed to him on the Époques de la nature, and shared with me the responses he made to them. In this elevated exchange of letters between power and genius, but where genius exercised the real power, I felt my soul touched, raised up. Glory appeared personified before my eyes. I imagined myself touching it, grasping it, and this admiration of sovereigns, forced to humiliate themselves thus before a genuine grandeur, touched my heart, as an act of homage well above all the honors they might have been able to bestow in their empire.

A few days later I left this good man, this great man, bearing in my heart a profound and deathless remembrance of all that I had seen and of all that I had heard. I recited to myself, as I left, these two beatiful lines from Voltaire's Oedipe:

> The friendship of a great man is a gift of the Gods;
> I read my duty and my destiny in his eyes.

It was fore-ordained that I would have the good fortune of seeing him once more. Leaving Semur, in order to return to Paris, the post brought me back by Montbard, contrary to my expectations. I was not able to refrain from sending my valet de chambre to see if there was any news of M. de Buffon, even though it was seven o'clock in the morning. He sent back word that he absolutely had to see me.

When I returned to him, I threw myself in his arms, and this good old man held me close to him for a long time with paternal tenderness. He wished to have breakfast with me, and supply my carriage with provisions. He also spoke to me for three hours more heatedly and actively than ever. He appeared to open his soul to me and allow me to probe it at my leisure. The love of work was scarcely forgotten in this conversation.

I consulted M. de Buffon about a project which I had conceived concerning legislation, which would take up, it is true, a great part of my life, and perhaps all of it. But what more beautiful monument could a magistrate leave? We reasoned together for a long time. My idea was to make a general review of all the rights of men, and of all their laws, comparing them, rating them, and then raising a new edifice. He approved my views, and encouraged me in my work. He enlarged my plan, and set its dimensions. He persuaded me, as I intended to do, to deal only with the leading principals of things, capita rerum, but to develop them well, though briefly, and to condense the work into one quarto volume, or two at the very most; and to work at it in four parts: 1) Universal moral principles, such as these ought to be at all times and in all places; 2) Universal legislation, which would consider the spirit of all the laws which exist in the universe. When I said to him there would be material for a nice book about the manner of drawing up a law, searching out all the possible circumstances wherein human reason could be exercised, he said to me that this would be the third part of my work; 3) concerning reforms that it would be necessary to introduce into the various systems of laws of the world; 4) finally, he suggested what would be a magnificent conclusion, which would be to end with a great chapter on necessity and on the abuse of juridical procedures. By this device one would take into consideration all objects which could possibly be the concern of legislation. This plan, however, immense in scope, appeared quite satisfactory to me, and I set myself the goal of executing it. I realize all that its execution will cost me; but a grand plan and a grand goal put happiness into the soul each day that one puts oneself to the task. M. de Buffon hid nothing for me, and I was well aware that I would have to work more than others, having in addition to fulfill the duties of my official position in life, which were sufficient to occupy anyone. But what superiority would not such a constant study give me, even for the completion of the duties of my state in life! He counselled me not to neglect them, and advised me that with patience and methodical application I should daily perceive some progress and vigor of my intelligence. He exhorted me to imitate him, that is to take a secretary solely for this work. Indeed, M. de Buffon always used many assistants. They furnished him with observations, experiences, memoranda, and he combined all these with the force of his genius. I found once a proof of his genius, in a small number of papers which he set aside in a carton. I saw a memoir on the magnet, on which he is working, sent him by the Comte de LaCépède, a young man full of ardor and of knowledge.

Buffon was right; there are a thousand things which must be left to assistants, or else one would be overwhelmed by work, and never arrive at one's goal. He said to me that at the time he was composing his greatest works he had a room full of cartons, which he has since burnt. He strengthened me in resolution not to consult books, to draw forth everything from within myself, to open books only when I could go no further on my own. Furthermore, among the books, he advised me to read only natural history, history, and accounts of voyages. His advice was certainly correct. Most men fall short of genius because they have neither the power nor the patience to consider things at a high level. They set out with too low an aim; yet every answer has to be contained in the origins of a phenomenon. When one knows the natural history of man, and then the natural history of a people,[35] one ought to be able to discover without effort what are their customs, and what are their laws. One will be able to discover almost the entire civil history of that people. But when one already knows their civil history, it ought to be even easier to discover and judge their laws and put them in perspective, either with their constitution, or with the events.

"I am not anxious about you so far as the first part is concerned," M. de Buffon said to me. "You will manage well so far as knowledge of universal moral principles is concerned. For this an upright soul and a penetrating and just spirit suffice. But when you come to discovering and classifying that innumerable multitude of institutions and laws - that requires great effort, one worthy of the utmost human courage." I could not refrain from making a delicate observation to him. "And as to religion, Monsieur, how are we to extricate ourselves from its tangle?" He replied to me: "There is a way of saying everything. You will note that religion is a separate object. You will wrap yourself in all the respect that one ought to show religion on account of the popular prejudice in its favor. It would be better to be understood by a small number of intelligent persons whose support would compensate for not being understood by the masses. For my part, I would treat Christianity and Mohammedanism with equal respect."

Figure 23 - From Pajou's bust of Buffon, 1775

And thus passed the hours in this conversation [filled with] glory and hope. I was not able to tear myself away from the bosom of this new father, which science and genius had given me. Eventually I had to leave; but not without remaining long in his close embraces nor without my reiterated promise to nourish myself continually on his works, which contain all of natural philosophy, and also to cultivate him with a filial assiduousness for the rest of his life.

With this I have finished that which I have to say of M. de Buffon. Since the details are only of significance for me, I have willingly stretched this out with a kind of veneration.[36]

Buffon in Retrospect: His Style and Glory

NOTES

*Hérault de Séchelles' Voyage à Montbar was published in Paris by Solvet, An IX (1801) Bound with it are several other short & miscellaneous literary pieces of the same author. The editor (apparently Solvet) notes that, for the most part, these pieces had already appeared in the Magasin encyclopédique. Solvet's text has been checked against two other and more recent editions, namely, those of Emile Dard and Hubert Juin. See Emile Dard, Hérault de Séchelles: Oeuvres litteraires (Paris: Librairie Academique; 1907), pp. 3-61. A more recent edition, with most useful biographical information on Hérault (cited in the "Introduction," above) is Hubert Juin, Hérault de Séchelles: Oeuvres litteraires et politiques (Paris: Editions Rencontre; 1970), pp. 69-100. Where the texts vary, I have followed Solvet's text. I have also taken the liberty, when necessary, of breaking down the elaborately complex sentences on which so many Eighteenth Century authors doted, and of re-paragraphing when it seemed appropriate. Spelling in the translation follows modern convention (e.g., "Montbard," rather than "Montbar") as does punctuation and the use of quotation marks; and titles of books cited by Herault are conventionally underlined, though they do not appear in italic type in the original. Where variant phrasing in the two editions have made no difference in the translation, footnotes specifying the variation have not been supplied.

[1] He perished on the scaffold several days before the 9th of Thermidor, calmly and with dignity announcing to those present: "Citizens, my name is Buffon." These words demonstrate his noble soul and his consciousness of the respect that his name ought to have inspired in everyone except assassins and hangmen. (A.L.M.) [This note, signed simply "A.L.M.," occurs in the Solvet edition.]

[2] [Gue-Jean Baptiste Target (Dec. 6, 1733-Sept. 9, 1806). French lawyer and politician. Elected to the Académie Francaise in 1785, he later became a deputy of the Third Estate of Paris to the Estates General, and president of that body on January 18, 1790. He became a judge in the court of cassation in 1797, was renamed to that post in 1800, and remained there until his death. Nouvelle biographie générale (hereafter NBG) (Paris: Firmin Didot Freres; 1865), Vol. 44, Cols. 876-78.]

[3] [Yves-Marie Desmarets, Comte de Maillebois (Aug. 1715- Dec. 14, 1791). The son of Jean-Baptiste-François Desmarets, Marquis de Maillebois and Marshall of France, Yves-Marie served in the Italian wars and was created lieutenant-general in 1748. Subsequently (1757) however, he was involved in a military scandal, connected with the battle of Hastembeck, was tried

before a military court, and imprisoned. Released several years later, in 1784 he was in Holland supporting popular elements against the Prussians. He bitterly opposed the principles of the Revolution in 1789. NBG (1860), Vol. 32, Col. 882.]

[4][Antoine-Leonard Thomas (Oct. 1, 1732–Sept. 17, 1785), French writer. One of seventeen children, he was educated at the Collège du Plessis in Paris. He subsequently became a professor at the Collège of Beauvais. A monograph on earthquakes gained him entrance to the Academy of Rouen in 1757. His Éloge du Marechal de Saxe (1759), his historical poem, Jumonville (1759), along with Éloges of d'Arguesseau (1760), Duquay-Trouin (1761), Sully (1763), Descartes (1765), and the Dauphin (1766), gained him a significant reputation, and he entered the Académie Française in 1766. Among his best known works were the Éloges of Descartes and of Marc-Aurele (given in 1770, published 1775), and an Essai sur le caractère, les moeurs, et l'esprit des femmes (1772), NBG (1866), Vol. 45, Cols. 222–226.]

[5][Suzanne Curchod, (1739–May, 1794), the daughter of a Swiss minister, became the wife of Jacques Necker (French Minister of finance) in 1764, after a friendship with Gibbon was terminated through his father's opposition. Her salon attracted such dignitaries as Buffon, Diderot, Thomas, Marmontel, Grimm, and d'Alembert. She greatly admired Buffon, and had a lively concern for Thomas. NBG (1863), Vol. 37, Cols. 590–592.]

[6][The Latin reads:

Excelsae Turri, Humilis Columna
Parenti suo, Filius Buffon, 1785.]

[7]In order to measure the strength and durability of trees, Buffon involved entire forests in his research. In order to obtain novel results concerning the development of heat, he put huge globes of metal in immense furnaces. In order to resolve several problems concerning the action of fire, he worked with torrents of flame and smoke. Finally, having brought together the foci of several mirrors into a common spot, he contrived the art which Proclus and Archimedes used, namely, that of setting fire to distant vessels (Vicq -d'Azyr, Disc. de recept. a l'Acad.)

[8][Solvet has "froid"; Dard has "frais."]

[9][Solvet has "carrele"; Dard has "carré."]

[10]Would this not be M. de Belle-Isle? A.L.M. [sic]

[11][Solvet has "savoyard"; Dard has "domestique."]

[12]M. de Buffon had always been very much occupied with himself, and put himself above everyone else. Since I knew that many women had been the objects of his attentions, I asked whether they had not caused him to squander his time. A person who knew him very well replied to me: "M. de Buffon constantly kept three things above all else before his eyes: his glory, his fortune, and his pleasure." He almost always reduced love to the physical side alone. Look at one of his discourses on the nature of animals, where, after a high-flown portrait of love, he abolishes it with a single stroke and debases it by claiming to prove that there is nothing more in the enjoyment of it than physical diversion, vanity, and self-esteem. There one finds his invocation to love. "They placed it next to Lucretia's declaration of love," he said to me once. Women were deadly vexed at him for his exertions in this area, or rather this abuse of reasoning. Mme. de Pompadour said to him at Versailles: "You are a little rascal . . . ." [The French text reads: "À Montbar, apres son travail, il faisait venir une petite fille, car il les a toujours beaucoup aimées; mais il se relevait exactement a cinq heures. Il ne voyait que des petites filles, ne voulant pas avoir de femmes qui lui depensassent son tems." This is obviously a sensitive issue. "Petite-fille" could mean "grand-daughter." But the rest of the text suggests-insinuates, perhaps--another context than grand-fatherly relaxation. And Buffon had no grandchildren of his own. Yann Gaillard (<u>Buffon</u>: <u>Biographie</u> <u>imaginaire et réelle</u>. Paris: Hermann; 1977, p. 55) suggests that Buffon might have been flattered by injurious allegations such as this one of Hérault's, and then goes on to suggest an alternative interpretation of this "calumny": "Qui sait si son grand âge n'éprouve pas quelques complaisance a l'évocation de ses relations supposées avec ses deux petites filles qui, d'après Hérault de Séchelles, Buffon reçevrait chaque jour à Montbard? Or, selon toute vraisemblance la racine de cette calomnie, c'est l'interêt tout paternel que Buffon porte a deux orphelines de la ville, qu' il fait venir quelquefois au chateau pour suivre les progrès de leur education, dont s'occupe l'autre Mme. Daubenton, la belle-soeur du medicin, mère de la jeune Betzy." Nowhere in the text does Hérault de Séchelles refer to <u>two</u> little girls. But, at least it does not seem that Buffon could have been indiscriminately pressing home some <u>droit de seigneur.</u>]

[13]So far as Buffon's flatterers, hangers-on, and devotees are concerned, I have an observation to make, which I have not come across anywhere. It is quite difficult for a great man to live without that sort of a group which naturally attaches itself to him, either out of curiosity, or admiration, out of a desire to imitate him as is the case with young people, or out

of vanity and a sense of self-importance, coming from being connected with a great man, when one is not able to be such a man oneself. So far as I am concerned, it is not revolting to see such a man love to be surrounded [by admirers]. I shall not say solely that this is a consolation for his exertion, a mitigation of his labors, a resource which constantly reminds him of his glory in the very midst of his ills and sufferings; I say further that this is even an encouragement to him in his studies which may provide him with a new facility. Admirers constantly remind you of the presence of your genius and your greatness. Furthermore, it is a fact that the presence of one's inferiors makes one's superiority stand out more. It is noticeable that the conversation becomes richer, freer, more plentiful. Manners are more relaxed and this free atmosphere adds much to it. Thus, far from finding a certain pettiness in the cortege which may surround a famous man, I often see in it an occasion and a means of being faithful to his renown.

14[Joseph-Marie Terray (Dec. 1715-February 18, 1778), French politician, friend of Madame de Pompadour, Comptroller-General of Finances under Louis XV from 1769. Active in the explusion of the Jesuits, he was rewarded with the Abbey of Molesmes in 1764. His financial measures appear to have amounted to a disguised form of bankruptcy. Unpopular as he was, his creation of a royal monopoly of grain led to the accusation that he and the King were partners in profiting from famine. He was replaced by Turgot after the accession of Louis XVI. NBG (1865), Vol. 44, Cols. 1011-1017; Michel Mourre, Dictionnaire d'histoire universelle (hereafter DHU). (Paris: Editions Universitaires; 1968), T. II, p. 2109.]

15His manner of conversing is not very consistent. He prefers to move from one topic to another. There is a reason for such conversational preference which may be alleged in favor of men of letters: First of all, they do not have any more, as they once did, that habit which philosophers used to have of conversing under the plane trees with their disciples, and of giving an account of their ideas. In the second place, their ideas are much more complex and premeditated than those of the ancient philosophers. New thoughts are needed; they are requested by both readers and auditors. The man of genius, unrelenting toward himself, only allows himself thus a small number of sentences, which he from time to time puts into his conversation, unless he should be seized, seduced, as it were, by the attraction of some sudden vision which dominates him, the ascendency of which he has not the power to elude [Hérault].

However long it already may be, we should not end our note concerning the conversation of this famous man without mentioning M. de Mont-Belliard, one of Buffon's most intimate friends. His name is automatically associated with that of the author of the Natural History by virtue of the significant role he played in this fine work. M. de Mont-Belliard offers a striking contrast to Buffon. Few men possessed what Buffon seemed to lack to the same degree as M. de Mont-Belliard. Nothing could be more spiritual, nothing more animated, nothing more engaging than his conversation. But as to the talent that constitutes the great writer, what a difference! "He has a pen of steel" said a woman of much wit. And she added: "But his strokes are far from those of the smooth brush of M. de Buffon!" [Editor, 1801].

[16]Besides, it ought to be admitted that his self-esteem never offended anyone. Here is a new trait, but it honors his character. Therefore, we do not hesitate to add it to the others scattered about already, perhaps in too great number, in this work:

Buffon held it as a general rule that children took after their mothers so far as their intellectual and moral characteristics were concerned. After having developed this topic in conversation, he thereupon applied it to himself, giving a pompous panegryic of his mother, who had indeed much wit, extensive knowledge, and a very well-organized mind. He loved to speak of her often.

[17]Since his earliest years, even since he was a schoolboy, he was passionately taken by geometry. This passion was such that he could never be without the Elements of Euclid, a copy of which he always carried with him; so that, while playing tennis with his comrades, he often secreted himself in a corner, or plunged into some deserted lane in order to open his book to try to resolve some problem which had been tormenting him. One day, carried along by his extraordinary taste for clockworks, he climbed up a belltower, and then climbed down on a knotted rope, skinning his hands painfully by slipping on the rope. He was not aware of the injury he had sustained, so preoccupied had he been with a geometrical proposition whch he could not fathom, the solution to which came to him all of a sudden while he was climbing down. [Guillaume-François-Antoine de L'Hôpital, Marquis de Sainte-Mesme (1661-Feb. 2, 1704). French geometrician, best known for his Analyse des infiniment petits pour l'intelligence des lignes courbes (1696), and the posthumously published Traité analytique des sections coniques (1707). NBG (1860), Vol. 31, Cols. 101-02; DHU, T. I, 1208.]

[18][Madeleine-Marie Blesseau entered Buffon's service May 1, 1769, "en qualite de femme de charge de ma maison." Humbert de Bazile, "Livre manuel contenant les charges annuelles," in Henri Nadault de Buffon, ed., Buffon: sa famille, ses collaborateurs et ses famileurs (Paris: Jules Renouard; 1863), p. 102. See also pp. 416-420. For a reference to Hérault's "Calumnies" concerning Mlle. Blesseau, see p. 420.]

[19] [For Ignace Bougot (Bougault), see Humbert-Bazile, in Henri Nadault de Buffon, ed., op. cit., pp. 405-415. The account of Pere Ignace's appearance with Buffon at the Académie (p. 408) sounds as if it were taken straight from Hérault's Voyage.]

[20] At that time, age had caused M. de Buffon to lose a bit of the charm of his youth. But it left him an impressive stature [une taille avantageuse], a noble air, a stately figure, and a countenance at once pleasant and majestic. In the eyes of Madame de Buffon, enthusiasm for [his] ability caused the disparity of their ages to disappear, and at their period of life, when felicity appears to restrict itself to replacing by amity and remembrance mixed with regrets a sweeter happiness which has escaped us, it had inspired those characteristics with a tender and constant passion. Never was a deeper admiration joined to a truer tenderness. These sentiments were shown in the glances, the manners, and the in the language [discours] of Madame de Buffon, and filled up his heart and his life. Each new work of her husband, each new award added to his glory, was for her a cause for rejoicing all the more sweet as it was selfless without any admixture of pride that could prompt her to claim the honor of sharing the esteem and the name of Buffon. She left behind only one son (he who has been mentioned in this work). Condorcet, Éloge de Buffon [Editor, 1801].

[21] [Marie-Françoise de Saint-Belin Malain (1732-1769), married to Buffon September 21, 1752, and mother of his son Georges-Louis-Marie (born 1764, guillotined in 1794). Dictionnaire de biographie française (Paris: Librairie Letouzey; 1956), T. 7, Col. 630. See also Humbert de Bazile, "La Comtesse de Buffon," in Henri Nadault de Buffon, op. cit., 179, footnote.]

[22] [For an alternate account of Buffon's religious posture, see Humbert de Bazile's "Memoirs" of Buffon, in Henri Nadault de Buffon, ed., op. cit., pp. 49-58.]

[23] We believe rather that the author of the Voyage à Montbar has tried to support his own opinions under the authority of a great man, and attributes to him a discourse which contradicts many fine pages in his works. It is only necessary to refer the reader to the eloquent invocation to the Supreme Being which concludes the first of the views of nature. Buffon's father, who had an almost religious respect for him, came one day to read this. He met his son, and in a transport of admiration, his first involuntary reaction was to throw himself on his knees. [Editor, 1801]

[24] It was thus that it took place. But it appeared that a sentiment other than what could be called a respect for the

institutions of society determined this last act of his private life. One need only note the singular circumstance of his general confession, which he made then in a loud voice and without inquiring as to whom were present. [Editor, 1801]

[25]M. de Buffon criticized these two lines of Racine:
"Le fer moissonna tout, et la terre humectée
But a regret le sang des neveux d'Erecthée."
He believed that the word "humectée" ought not precede a form of the verb "to drink." And it is true that nothing can be "moistened" before it has "drunk." But poetry, which is always delirious, has license to confound the sequence of events.
  He also criticized this fine line of the same writer:
  "Le jour n'est pas plus pur que le fond de mon coeur."
One cannot, he said, compare a day with the foundation of something. It is from this connection of words, to which too much attention cannot be paid, that perfection of style is often born. But, again, poetry need not be judged as prose. Mad. Necker, Mel. extr. de ses manuscr. [Editor, 1801].

[26]Independently of those who consulted him on their works there were few writers who did not hold it an honor to send him in homage their productions. But he had little time left for books which were sent him, ordinarily restricting himself to the table of contents in order to determine those which appeared most interesting. In the last fifteen years of his life there were few works which he read otherwise. Among the authors now dead, beyond those hereafter noted whose work he advises people to study, he had a particular liking for Fenelon and Richardson.

[27]He has also another manner of judging his works. When they are read to him, he asks his reader to put in other words certain sections of the work whose composition he has much labored on. Then, if the reader's paraphrase faithfully renders the sense which he was after, he leaves the section as it was. Whenever, on the contrary, the paraphrase diverges from the sense, he revises the passage, searching for that which made it opaque, and corrects it.

[28][Ceasare Bonesana, Marquis of Beccaria (Mar. 15, 1738-Nov. 28, 1794). Italian legal philosopher, and author in 1764 of the influential On Crimes and Punishments. DHU, I, 209.]

[29]Everytime that M. de Buffon felt feverish while writing he stopped working, for he then knew that the work had fatigued him, becoming conscious of it by the flushness of his skin. Then he took a walk and refreshed himself. "That happens to me most often," he said, "when I have an opinion, and come across great objections to it. People without ability never anticipate contradiction. They write without foreseeing it." Mad. Necker. Mel. extr. de ses manuscr. [sic] [Editor, 1801].

Hérault de Séchelles' Visit to Buffon   385

[30]Among a number of persons who arose to criticize the <u>Natural History</u> of M. de Buffon, L'Abbé Condillac, the most redoubtable of his adversaries, drew everyone's attention. His mind displayed all its force in the dispute. M. de Buffon, on the contrary, was almost a stranger to the dispute. But the author of the <u>Natural History</u> showed himself to be superior in the recognition of his faults. He called attention to them in the supplements with as much modesty as frankness, and thus demonstrated all that had the force of truth in him. Vicq-d'Azyr, <u>Disc. de recept.</u> [Editor, 1801].

[31][Solvet has "dicter"; Dard had "dire." (Montesquieu inherited from his uncle the title of "President à mortier" in the parliament of Bordeaux.)]

[32][Solvet has in an extensive footnote at this point Herault's "parallel between J.J. Rousseau and M. de Buffon." This material is contained in a separate section immediately after the "Visite" in Dard's edition. The "parallel" adds little to Herault's exposition, and has been omitted here.]

[33][Frederic-Louis-Henri, Prince of Prussia (Jan. 18, 1726-August 3, 1802). Third son of Frederick William I, and brother of Frederick II (the Great). <u>NBG</u> (1861), Vol. 24, Cols. 141-149.]

[34][Solvet has "Souveraine"; Dard has "femme."]

[35][Solvet has "l'histoire naturelle de l'homme, et ensuite l'histoire naturelle d'un peuple"; Dard has simply "l'histoire naturelle d'un peuple."]

[36][Solvet appends a brief obituary biography of Buffon at this point, in which he mentions that much of the material contained in his footnotes comes from a letter written to the <u>Journal de Paris</u> a few days after Buffon's death (April 7, 1788) by an unidentified person who "s'annonce avoir vecu dans la societe intime de Buffon. . . ."]

List of Illustrations

		PAGE
1	George Louis Leclerc, comte de Buffon, 1761 portrait by François Hubert Drouais. From R. Taton (ed.), *The Beginnings of Modern Science*, translated by A. Pomerans, (London: Thames and Hudson, 1964), p. 512. Reproduced by permission of J. E. Bulloz, Paris.	*Frontpiece*
2	Woodcut prefacing Daubenton's article, "Description du Cabinet du roi," *Histoire naturelle, générale et particulière*, original edition, (Paris: Imprimerie royale, 1749) Vol. III, p. 1. All subsequent *Histoire naturelle* plates are taken from this edition. (Courtesy Yale University Library.)	89
3	Woodcut by D. Sornigne prefacing the "Premier discours de la maniere d'etudier et de traiter l'histoire naturelle," *Histoire naturelle, générale et particulière* (Paris, 1749), Vol I, p. 1.	97
4	Woodcut facing title page of the "Second discours: histoire et theorie de la terre," *Histoire naturelle générale et particulière* (Paris, 1749), Vol. I, p. 64.	130

		PAGE
5	Woodcut by De Seve heading the "Second discours: Historie et Theorie de la Terre," *Histoire naturelle, générale et particulière* (Paris, 1749), Vol. I, p. 65.	134
6	Woodcut facing the "Preuves de la theorie de la terre: article 1," *Histoire naturelle générale et particulière* (Paris, 1749) Vol. I, p. 125.	150
7	Woodcut by De Seve heading article "Preuves de la theorie de la terre: article 1," *Histoire naturelle, générale et particulière* (Paris, 1749), Vol. I, p. 127.	151
8	Woodcut by J. Wigley of the Scroll-mounted Wilson screw-barrel microscope as manufactured around 1740 by John Cuff. From Henry Baker, *The Microscope Made Easy*, 3rd edition (London: Dodsley, 1744), p. 9. (Courtesy Yale University Library.)	164
9	Woodcut, probably by De Seve, prefacing "Histoire des animaux, chapitre premier: comparison des animaux et des vegetaux," *Histoire naturelle, générale et particulière* (Paris, 1749), Vol. II, p. 1.	170
10	Woodcut of hand-held version of Wilson screw-barrel microscope, from Henry Baker, *The Microscope Made Easy*, 3rd edition (London: Dodsley, 1744) facing page 1. (Courtesy Yale University Library.)	189
11	The John Cuff compound microscope similar to that depicted in Figure 9. From George Adams, *Essays on the Microscope*, 2nd edition, with Improvements and Additions by Frederick Kanmacher (London: Dillon and Keating, 1798), P. 7A. (Courtesy University of Notre Dame Library.)	190
12	Unsigned woodcut of male spermatic bodies, from "Histoire generale des animaux: experiments," *Histoire naturelle, générale particulière* (Paris, 1749), Vol. II, Pl. 1, facing page 181.	194

List of Illustrations

		PAGE
13	Unsigned Woodcut depicting male spermatic bodies, *Histoire naturelle, générale et particulière* (Paris, 1749), Vol. II, Pl. 2, facing p. 186.	195
14	Undersigned woodcut depicting male spermatic bodies, *Histoire naturelle, générale et particulière* (Paris, 1749) Vol. II, Pl. 3, facing p. 200.	197
15	Unsigned woodcut illustrating observation of female spermatic bodies, *Histoire naturelle, générale et particulière* (Paris, 1749), Vol. II, Pl. 4, facing p. 218.	198
16	Map of the New World by Robert de Vaugondy, woodcut by Guillaume Delahaye, from "Histoire et theorie de la Terre," *Histoire naturelle, générale et particulière* (Paris, 1749), Vol. I, p. 206. (Courtesy Yale University Library.)	259
17	Copy of oil portrait of Albrecht von Haller by Sigmund Freudenberger (1745-1807) from *Bulletin of the History of Medicine* 4 (1936), p. 650. (Reproduced by permission of the Johns Hopkins University Press.	294
18	Marble bust of Chretien-Guillaume Lamoignon de Malesherbes by Jean Antoine Houdon (1784). From H.H. Arnason, *The Sculptures of Houdon* (New York: Oxford University Press, 1975), Fig. 150. Reproduced by permission of Oxford University Press and Phaidon Press Limited. (Original in the Louvre).	328
19	Unsigned woodcut of Marie-Jean Hérault de Séchelles from Emile Dard, *Un Epicurien sous la Terreur: Hérault de Séchelles (1759-1794)*, (Paris: Perrin et Cie, 1907). (Courtesy Yale University Library.)	348
20	An anonymous engraving of the Tower of Saint-Louis and study at Buffon's estate in Montbard. From Fortunat Strowski, *Histoire des lettres*, part II, Vol. XIII of *Histoire de la nation française*, edited by Gabriel Hanotaux (Paris: Plon-Nourrit et Cie, 1923), p. 426. (Courtesy University of Notre Dame Library.)	355

		PAGE
21	Plaster bust of Buffon, 1781, by Jean Antoine Houdon. From H.H. Arnason, *The Sculptures of Houdon* (New York: Oxford University Press, 1975), Pl. 135. Reproduced by permission of Oxford University Press, Phaidon Press Limited, and the Glasgow Art Gallery and Museum.	*358*
22	Marble bust of Buffon by Jean Antoine Houdon, 1783. From H.H. Arnason, *The Sculptures of Houdon* (New York: Oxford University Press, 1975), Pl. 73. Used by permission of Oxford University Press and Phaidon Press Limited. (Original in the Hermitage, Leningrad.)	*359*
23	Marble bust of Buffon, circa 1775, by Augustin Pajou. From H.H. Arnason, *The Sculptures of Houdon* (New York: Oxford University Press, 1975), Pl. 46. Used by permission of Oxford University Press and Phaidon Press Limited. (Original in Museum National d'Histoire Naturelle, Paris.)	*377*

Index of Names and Subjects

Abstraction: source of error in metaphysics, 101, 127, 338-39; Malesherbes on, 344n4

Abstract Concepts: opposed to physical, 20-24; simple and complex, 171-72; have no existence, 172; Malesherbes' defense of, 334

Academy, Platonic: 298

Académie des sciences: on authority of Bible, 250

Actualism: and Uniformitarianism, 132, 133n4; and history of earth, 264

Aloisia Sigen: 278, 282n5. See also Nicolas Chorier

Analogy: and experiential knowledge, 56; and mathematical demonstration, 79; misleading character of, 101; in sciences 131, 185; and metaphysics, 343

Ancients & Moderns, 116-21, 218-20, 257, 298

Ancestral form: Kant's mention of, 1

Andes: Buffon's account of, 261

Angers: Buffon studies at, 5

Alchemists: use of hypotheses by, 301

Aldrovandi, Ulysse (1522-1605), 109-110, 218, 225

Animal: defined, 203

Animal kingdom: and relation to vegetable, 113

Apsides: motion of, 79-81

Archimedes (287-212 B.C.): ideas on infinity, 45

Aristotle: and Enlightenment science, 3, 185; and mammary gland, 92, 94-95n8, 115; on method, 118-208, 219; on classification, 119; theory of generation, 368

Armadillo: development of shell, 277

Arnauld, Antoine (1612-94): and "Port Royal Logic," 344n2

Art: as imitation of nature, 110-11

Atlantis: destruction by earthquake, 148

Attraction: and impulsion, 77, 79, 367; dependent on creator, 224

Average man: Statistical definition of, 73
Bacon, Francis (1561-1626): concept of natural history, 2; Buffon recommends, 38, 372
Baker, Henry (1698-1774): and Needham's Microscope, 166
Barrow, Isaac (1630-77): and Newton on calculus, 48
Bauhin, Casper (1550-1624): classification criticized, 302
Beccaria, Caesare Bonesana, Marquis de (1738-94): on style, 370; 383n28
Bernoulli, Daniel (1700-82): *On the Measure of Chance*, 66; and Buffon, 66; letter to Buffon, 73-74n2
Bernoulli, Jakob I (1654-1705): and probability theory, 25-26; and physical necessity, 26
Berthier, Guillaume - François (1704-82): editor of *Journal de Trévoux*, 213; and Buffon review, 214
Binomial theorem: Buffon's independent discovery of, 364
Blesseau, Marie-Madeleine: 351; relation to Buffon, 365, 382n18
Body: existence of doubtful, 287
Boerhaave, Hermann (1668-1738): scientific method praised, 38; respiration theory, 304; as metaphysician, 343
Bonnet, Charles (1720-1793): and preformation theory, 312

Botany: classification in, 102-107; no adequate system of, 103-104; need for system in, 257; utility of systems in, 301
Bossuet, Jacques Benigne (1627-1704); *Elevations* cited, 251-52; on truth, 251-52
Bougot, Père Ignace: Buffon's confessor, 365; 382n19
Bourguet, Louis (1678-1742); and Buffon's theory of the Earth, 258
Brown, Robert (1773-1885); use of simple microscope, 166-167; on Buffon-Needham observations, 167
Brownian Motion: and Buffon's organic molecules, 166-167; 209n5
Brownker, William (1620-1684): on infinite series, 47
Buffon, George-Louis-Marie (1764-94): monument to father, 360
Buffon, George-Louis Leclerc, comte de (1707-1788): biography and characteristics, 4-9, 355, 357, 360, 361, 362-64, 365, 367-69, 377n2, 379n12, 380-81n5, 382n20, 384n29; on methodology and principles of science, 8, 9-10, 11-27 *passim*, 37-38, 80, 81, 82-83, 109-10,111-14, 152, 153, 185, 225, 306-07; on contemporaries and style, 10, 11, 38, 89, 333, 336, 368, 372-74, 376, 383n25-27, 384n30; epistemology and concept of truth, 11, 19, 20-24, 53, 90, 95n9, 333-38; and Liebnizianism, 20-26, 131; controversy with

Linnaeus on classification, 22, 90, 91-93, 104-07, 216-17, 255-57; on mathematics, 22, 44-45, 46, 74n3, 258, 364; on position of man in nature and society, 22-23, 53, 95n9, 229, 230n2, 237-38; species concept of, 23-24, 94n5; generation theory of, 24, 368, 370; "Memoir sur le jeu de franc-carreau," 52; experimental work of, 92, 316, 360, 378-79n7; on fossils, 139-40, 244; and religion, 214, 223, 237, 250, 283, 352, 366-67, 368, 376-77, 383n23-24; cosmology of, 224, 265, 370; on naming, 256; destroys manuscripts, 362; *Treatise on the Magnet*, 370; on science of society, 371, 375
Burnet, Thomas (1635-1715): cosmology of, 13, 135, 226, 266
Calculus: Newton-Leibniz controversy over, 44
Calmar: Needham on seed of, 202-03
Cartography: eighteenth-century difficulties over, 267n2
Le Cat, Claude-Nicholas (1700 - 1768): opposes Buffon on theory of the Earth, 258
Catherine II, Empress of Russia (1729-1796): compares Buffon to Newton, 374
Cause: and effect, 36, 60, 102, 125, 152-53, 175; and sequence of events, 61-62; supernatural, 153; efficient, 176; final, 176-177, 322; in organisms, 183; secondary, 312

Cavalieri, Buonaventura (1598-1647): on the calculus, 47
Cellular tissue: animals and plants composed of, 270
Censorship: defended, 278; Buffon on, 367
Celsius, Anders (1701-44): on theory of earth, 258
Certitude: linked to mechanical philosophy, 11-12; moral, 51, 124; degrees of, 53-54; physical, 54, 123-24, 337; as property of physical truths, 241; Buffon's views on condemned, 287; of first principles, 289; as distinguished from evidence, 123-24; 338, 344n6. *See also* Evidence
Cesalpino, Andrea (1519-1603): as creator of systematic botany, 302-03
Chain of being: presupposition of Buffon's thought, 91; prevents systematization of nature, 102; links man and nature, 102, 237-38; 113, 203-04, 316; concept criticized, 237-38; produced by force of attraction, 315
Chance: calculation of, 59
Chapelle, Armand Boisbeleau de la (1676-1746): founder of *Bibliothèque raisonée*, 253
Châtelet, Gabrielle-Emilie Le Tonnelier de Breteuil, marquise du (1706 -1749): *Institutions de physique* (1740) as synthesis of Leibniz, Wolff and Newton, 20-22; on

reality of time and space, 20-21
Chorier, Nicolas (1612-1692): author of *Aloisia Sigen*, 278, 282n5
Clairaut, Alexis-Claude (1713-65): controversy with Buffon, 22; and lunar perturbations, 77; 81
Classification: imposes arbitrary divisions on nature, 100-107; single character, 103-06; metaphysical errors of, 106; anthropocentric nature of, 112, 114, 119, 218; Buffon's critique of, 22, 90, 91-93, 104-107, 216-17, 255-57; utility of, 255-57; weighting of characters in, 302-03
Clusius [Charles de l'Écluse] (1526-1609): classifications critized, 302
College de Godrans: Buffon attends, 4
Colson, John (1680-1760): and Newton's *Fluxions*, 41,43-44
Committee of Public Safety: Herault's role in, 350
Comet: role in formation of solar system, 135, 153-56, 224-25, 265-66, 286
Comparative anatomy: Aristotle on, 119-20
Condillac, Etienne Bonnot de (1715-80): 8, 10; and Buffon, 372, 384n30
Condorcet, Antoine-Nicolas, marquis de (1743-94): *Éloge* on Buffon, 5
Continents: Buffon's theory of formation, 148-49, 245
Corpus luteum (*corps glanduleux*): 187, 196-99, 205, 208n3, 273-74, 275, 276, 278, 316, 325n4;

Buffon's error in identifying, 321
Cotes, Roger (1682-1716): on inverse square law, 80; on lunar perturbations, 80
Cramer, Gabriel (1641-1724): and Swiss science, 4-5; Buffon visits, 5; edits work of Bernoulli brothers, 5; on St. Petersburg problem, 66; letter to Buffon on probability, 74-76
Creation: and theory of generation, 17, 312; mosaic account of, 131, 154, 225, 242-44, 247, 289; instantaneous denied by Buffon, 247-48;
Creator: Buffon on, 53; source of awe in nature, 101
Crystals: produced by organizing forces, 315
Cuff microscope: scroll-mounted Wilson, 165-66, 168n6; compound, 166, 170, 190
Dalempazius *see* de Plantade
Dalibard, Thomas-François (1703-1799): and Buffon's experiments on generation, 165, 188
Darwin, Charles (1809-1882): culmination of Enlightenment naturalism, 17; on connection of embryological development and species question, 18
Daubenton, Louis-Marie (1716-1800): tensions with Buffon, 7, 306-07; and experiments on generation, 165, 188, 196-99
Death: Buffon on causes of, 179-80

Deformities: present difficulties for Buffon's theory of generation, 271-72
Deluge: Noachian, 140-41, 248, 265; Buffon on, 223-24, 243, 247-50; and distribution of fossils, 264-65
Descartes, Rene (1596-1650): Mechanism and certitude linked, 11-12, 46; and mechanical philosophy, 184, 297; criticized, 185; and generation, 312
Description of nature (*Naturbeschreibung*): Kant's reference to, 1
Description: exactness necessary in, 108-109, 111
Diderot, Denis (1713-84): 367
Design: evident in simplicity of nature, 315
Design argument: and theory of generation, 278-79, 280, 322
Dualism: Buffon's views on, 220, 228, 287, 290
Dutch Newtonians: epistemology of, 12-14, 295
Earth, History of: Kant refers to, 1; Buffon on 157-58, 225, 242, 243-44, 258, 263; physical properties of, 136, 151-52, 305; direct creation of, 263
Earthquakes: and formation of mountains, 147-48
Effects: as means of knowing causes, 60, 175; *See also* Cause
Embryo: organization of, 272-73, 278, 279-80, 320
Empiricism: linked with historical scepticism, 12-14
Ennius, Quintus (239-169 B.C): 299-300

Epicureanism: irreligious character of, 278
Époques de la nature (Buffon, 1778), 7-8, 27, 362, 374
Error: sources of, 338-39
Erosion: and distribution of sediment, 146-47, 149
Ergot (blight): effects of, 206
Essences: unknowable, 299
Ethics: fundamental obligations of, 290
Ether: Newton on, 300
Eternalism: Buffon's cosmology tends to, 243-44
Euclid: Buffon's early study of, 364, 381n17
Evidence: no degree in, 54; as property of mathematical truths, 124, 154, 240, 337; Buffon's views on condemned, 287; distinguished from certitude, 338, 344n6. *See also* Certitude
External world: knowledge of, 287, 290
Existence: relational nature of, 108
Existence of God: Buffon's epistemology undermines, 241
Explanations: supernatural rejected, 224
Experience: foundation of physical and moral knowledge, 56
Experiment: importance of, 126
Extension: as abstract property of matter, 172
Faith: infallible truths of 290
Fenelon, François de Salignac de la Motte (1651-1715): *Aventures de Aristonous* quoted, 355; Buffon recommends, 383n26
Female, semen of. *See corpus luteum*

Fermat, Pierre de (1601-65): on calculation of infinite, 47
Fear of death: greatest of moral certainties, 58-59
Final causes: unknowable, 176-77; and organization of life, 322
First causes: knowledge of, 101-02, 175
First principles: as eternal truths, 289
Fixed air: Stephen Hales on, 39
Folkes, Martin (1690-1754): and Needham, 187
Fontenelle, Bernard Le Bovier de (1657-1757): on scientific method, 35; on formation of earth and animals, 247
Foramen ovale, 229
Forces: biological, 77, 229, 303; in organization of embryo, 272, 312; polarity of, 77; and natural theology, 279
Fossils: distribution of, 139-40, 244-45, 260; Buffon on origin of, 264
Frederick-Louis-Henri, Prince of Prussia (1726-1802): letter to Buffon, 374; 384n33
Function and structure: harmonious relation of, 323
Galileo Galilei (1564-1642): on laws of falling bodies, 152
Gambling: faulty logic of, 62-65
Garden, George (1649-1733): on preformation of embryo, 16
Gassendi, Pierre (1592-1655): opposition to Cartesianism, 297
Genera: definition of, 106

General views: role in natural history, 136
Generation: Descartes on, 15-16; preformation theory, 16-17; as marvel of nature, 55; and reality of species, 170; and organic molecule theory, 173, 269-81; asexual, 207; reviewed, 227; Buffon-Needham experiments on, 165-67, 187-204, 275; and variation, 319; and religion, 324
Geoffroy St. Hilaire, Isidore (1805-1861): periodization of Buffon's thought, 9
Geological change: causes of, 137-38, 147, 148; in remote ages, 140, 149
Geometry: reasoning in, 335-36; analytic nature of, 338
Germs: pre-existence of, 184, 207, 227, 269
Gesner, Conrad (1516-1565): on classification by organs of reproduction, 105; praised as metaphysican, 343
God: as non-deceiver, 290; proof of existence, 322
Graffian follicle. See *corpus luteum*
Grammar: science of, 342
Gravesande, Willem Jakob Van's (1688-1742), 19, 253
Gravitational attraction: opposed to inertia, 77; as general cause, 152; proof of, 152; universal diffusion of, 152-53; and internal molds, 178
Greek language: perfection of for science, 117
Gregorie, James [Jacques Gregori] (1638-75): on quadrature of hyperbola, 48, 49n9

Gueneau de Montbeillard, Philibert (1720-1785): Buffon's collaborator on generation experiments, 188; compared to Buffon, 381n5

Hales, Stephen (1677-1761): *Vegetable Staticks* translated by Buffon, 5; influence of, 35; empiricism of, 37; *Animal Staticks*, 39

Haller, Albrecht von (1708-1777): and prefaces to German edition of *Histoire naturelle*, 281; editor of *Göttingische Anzeigen*, 295; on preformation and epigenesis, 311

Hamm, Jan: discovers spermatozoa, 275, 316

Harris, John: definition of natural history, 2

Hartsoeker, Nicolas (1656-1725): discovers spermatozoa, 275, 316

Helvetius, Claude Adrien (1715-71), 31n42, 367

Hérault de Séchelles, Marie-Jean, (1760-1794): 283; character of, 349, 350, 351; *Declaration of Rights of Man*, 350; and "the terror," 350; on political reform, 375; on rights of man, 375; on Buffon's vanity, 380n13

Hickman, Nathan (1695-??): Buffon's tutor and travelling companion, 5

Himalayas: Buffon's account of, 261

Hippocrates (ca. 460-375 B.C.): theory of generation, 279, 315

Histoire naturelle, générale et particulière: Buffon's initial conception of, 6-7; continuations of after Buffon's death, 7; English translations delayed, 8; popularity in Enlightenment, 29n13; misleading character of title, 255

History of the Earth: determined from action of present causes, 264

History of nature: Kant's concept of, 1; Buffon on, 135

History: truths of, 336

Holland: topography of, 262-63

Homoeomeria (Anaxagoras): compared to Buffon's organic molecules, 270, 315

L'Hôpital, Guillaume-François-Antoine de, Marquis de Sainte-Mesme (1661-1704): Buffon's study of, 364

Hottentot: 272, 319

Human nature: historical view of, 3-4; Rousseau on, 3-4

Humbert-Bazile: and Herault's visit to Buffon, 351

Hume, David (1711-1776): on scientific Pyrrhonism, 15

Huyghens, Christiaan (1629-1695): mechanism and certitude linked, 11-12

Hypotheses: legitimacy of, 176-77, 295, 301, 304, 305; exclude final causes, 177; abuses of, 298; role in botany, 301-03; Buffon endorses, 289; Haller on, 295

Ideas: general and abstract, 171-72, 203-04, 289, 345

Imaginary goal: as necessary for science, 103

Imaginary numbers: problems arising from, 339

Imagination: effects on embryo, 278

Impulsion: opposed to attraction, 77, 79, 153, 224, 367; cause of, 154; Buffon as discover of, 374; and attraction explain all nature, 374. *See also* Gravitational attraction

Individual: as composed of similar parts, 171

Inertia: force opposed to gravity, 153. *See also* Impulsion

Infinite series: Cavaleri, Mercator, Brownker on, 47

Infinity: Buffon's concept of, 41-42, 44-48, 174; in ancient geometry, 47

Infusions: Buffon-Needham experiments on, 201-02; demonstrate existence of organic molecules, 206; bodies in 272, 276

Inheritance: and organic molecule theory, 317

"Initial Discourse" to the *Natural History*: as *Discourse on Method* for Enlightenment natural history, 6; English publication of, 93n3; critique of Linnaeus, 255-57

Instinct: more certain than reason, 102; as hypothetical entity, 303

Internal anatomy: value for classification, 96n11

Internal Molds: and Newtonian attraction, 165, 178, 182-83; as legitimate hypotheses, 177-78; reproduction of, 181, 183; and constancy of species, 202; defined, 205; as properties of nature, 270, 272; operation of, 272; in first man, 281; difficulties, 318-19; Buffon on, 370

Inverse square law: Buffon on, 79; Clairaut's modification of, 82

Jansenism: and *Nouvelles ecclésiastiques*, 235, 298

Jesuit-Jansenist Controversy, 237

Joblot, Louis: and bodies in infusions, 276

Journal de Trévoux: criticized by *Nouvelles ecclésiastiques*, 237

Jussieu, Bernard de (1699-1777), 6, 330

Kaestner, Abraham (1719-1800): and German edition of *Histoire naturelle*, 281, 295

Kamchatka, Sea of, 262

Kant, Immanuel (1724-1804): on history of nature, 1, 27; and Enlightenment critique of religion, 25; on teleology in nature, 132, 133n6; and Haller's hypotheticalism, 295-96; uses Buffon's work, 296, 296n5

Keill, John (1671-1721): attack on Burnet and Cartesian cosmology, 13

Kepler, Johann (1571-1630): laws of, 79-80, 152; and foundation for Newton's laws, 301

Kinckhuysen, Gerard (fl. 1645-1663), 43

Kingston, Evelyn Pierrpont, Second Duke of (1711-1773): Buffon's travelling companion, 5

Klein, Jacob Theodore (1685-1759): on classification, 255

Knowledge: relational character of, 55; derived from repetition of events, 55-60; origin of, 112-13, 218; and sensation, 171-72; degrees of, 337

Koenig, Samuel (1712-1757): du Chatelet's tutor, 31n43
Lacèpede, Bernard Germain, comte de (1756-1825): and completion of Buffon's *Histoire naturelle*, 7; memoir on magnet, 376
La Condamine, Charles-Marie de (1701-74): on topography of Brazil, 262
Laplace, Pierre Simon, Marquis de (1749-1827): extrusion of God from natural philosophy, 25
Laws: of nature 61, 125, 367; of society, 375
Leeuwenhoek, Antoni van 1632-1723): and spermatozoa, 188-189, 202, 275, 279; and microscope, 188; 271, 316
Leibniz, Gottfried Wilhelm, baron von (1646-1716): on time and space, 20-21; controversy with Clarke, 20; and controversy with Newton on calculus, 41, 43-44; theory of earth criticized, 247-48; predicts existence of polyps, 258; and mechanical philosophy, 312; Buffon recommends, 372
Lexicon Technicum: definition of natural history in, 2
Lieberkuhn, Johann Nathaniel (1711-1756): on organic matter, 270
Life: as product of organic molecules, 277; contrasted with vegetation, 316
Limestone: formed from fossils, 260
Linnaeus, Carolus (1707-1778): work compared to Buffon's, 6; Buffon's criticism of, 91-93, 105-06, 217-18, 340; classifications of, 105-07, 114-16, 257; theory of earth, 154, 158, 258; on distinction of man from animals, 299; Haller's defense of, 303
Locke, John (1632-1704): Buffon as disciple of questioned, 10, 20; on historical knowledge, 12-13
Lyell, Sir Charles (1797-1875): and geological actualism, 132
Magnetism: Buffon's study of, 361
Magnification: of simple microscope, 168
Maillebois, Yves-Marie Desmarets, Comte de (1715-91), 360, 378n3
Maillet, Benoit de (1656-1738): author of *Telliamed*, 223, 226n2; theory of earth, 258
Mairan, Jean-Jacques Dortous de (1678-1771): and Haller prefaces to *Histoire naturelle*, 281
Malpighi, Marcello (1628-94): praised as a metaphysician, 343
Mammary gland: and classification of quadrupeds, 92, 94-95n8, 107, 114, 115-116
Man: starting point of natural history, 22-23, 102, 220; ranked with animals, 102; natural history of, 227, 376; Buffon on eastern origin of, 262; creation of, 324-25
Materialism, 228
Mathematics: as axiomatic system, 53-54; inadequately represents nature, 84, 125-27; role of hypotheses in, 300. *See also* Evidence

Mathematical physics: development of in Europe, 298

Mathematical truths: definitional nature of, 53, 123, 239-42, 332-33, 335-37; Buffon's concept of condemned, 286; and physical truth, 336. *See also* Physical truth

Matter: organized and brute distinguished, 179 273; knowledge of, 185, 290; properties of, 185, 280; eternity of, 223; as mode of soul, 287; intentionality of, 322; inherent forces of, 323-24

Maupertuis, Pierre-Louis Moreau de (1698-1759): explanation of disappearing stars, 157; and Buffon's theory of generation, 273

Massuet, Pierre (1698-1776): edits *Bibliothèque raisonée*, 253

Mead, Richard (1673-1754): on nature of poisons, 206

Mechanical explanations: Buffon rejects adequacy of, 184-85

Mechanical philosophy: and embryological preformationism, 312

Mercator, Nicolaus (ca. 1619-87): on infinite series, 47, 49n9

Mettrie, Julien Offray de la (1709-1751): materialism of, 228, 280-81; 230n1

Metaphysics: source of errors in, 46, 338; and simplicity of physical law, 82; first principles of, 290; Buffon on, 122-27, 331, 342; defined, 340, 341; abuse of, 342-43; *See also* Truth

Method: ideal of, 108. *See also* Hypotheses

Methods of classification: utility of, 100-107. *See also* Classification; Linnaeus

Microscope: useless in classification, 105-06; used in Buffon-Needham experiments, 166-67; 168n6, 187-91; optical problems of, 189-91, 209n4

Milne-Edwards, Henri (1800-85), 167

Moderation: Haller recommends, 298

Money: calculation of value of, 64-72, 74-75

Monstrosity: difficulties presented by for Buffon's theory, 319

Montbard: description of Buffon's estate at, 356, 360-61

Montmort, Pierre Raymond (1678-1719): and St. Petersburg problem, 66-67

Montesquieu, Charles de Secondat, baron de (1689-1755): *Spirit of the Laws* criticized by Sorbonne, 367; Buffon on, 372

Moral certainty: and calculus of probability, 26-27, 51, 57-60, 67-68; intermediate position of, 51; and physical certitude, 56, 58-60. *See also* Certitude

Mornet, Daniel: on Buffon's popularity in Enlightenment, 29n13

Motion of planets: initiated by creator, 153

Mountains: gradual formation of, 147-48; 258, 260-62; Buffon's account of, 260-62

Musschenbroek, Pieter van (1692-1761): on necessity of natural laws, 14
Nadault de Buffon, Henri (1831- 90): on Herault's visit to Buffon, 351
Natural history: Bacon's definition of, 2; and history of nature, 2-3, 18, 27; and method, 14, 98, 100, 121-22, 136, 215-18, 232-33, 343; Buffon's concept of, 97, 132, 133n5; utility of, 97-100, 216; dangers in, 107; concerned with species, 111
Natural necessity: and theological understanding, 25-26
Natural philosophy: and need for hypotheses, 300-01
Naturbeschreibung: Kant's definition of, 2
Naturgeschichte: Kant's definition of, 2
Natural law: foundation of questioned, 13-14; ontological status of, 25-26; and moral truth, 290
Natural system of classification: Buffon on, 106-07, 112-13. *See also* Classification
Nature: concepts of, 3, 178-79; history of, 3; means of knowing, 55, 90-91; uniformity of, 100-101; order of, 136-37; autonomous organization of, 178, 270, 312, 323-24; superfecundity of, 179; harmony of, 263, 315; as substitute for Creator, 367; and literary style, 371
Necessities of life: value of, 65, 74-76
Necker, Mme Suzanne (1739-94): praise of Buffon, 360; 378n5

Needham, John Turberville (1713-1781): and experiments on generation, 165-66, 187-88, 196-99, 201-02, 316, 323-24
Newton, Isaac (1642-1727): Buffon's reading of *Principia*, 5; divine intervention in cosmology, 24-25; foregoes publishing *Fluxions*, 43; on lunar perturbations, 81; ignores force of magnetism, 83-84; on analogy, 131; on comets, 155; on hypotheses, 295, 300-01; and optical theory, 303-04; on size of earth, 305; on patience in natural science, 361; Buffon recommends, 324, 339, 372; on binomial theorem, 364
Newton-Leibniz controversy: dispute over calculus, 6, 41, 44
Nominalism: and Buffon's species concept, 91
Novel events: epistemological status of, 61
Number: defined by Buffon, 45-46
Observation: repetition of, as source of scientific truth, 220
Order of ideas: and connections of facts, 99
Organic: relation to inorganic, 171, 316
Organic matter: formation of, 266
Organic molecules: dissimilar character of, 165, 183; first assemblages of, 171, 184, 187, 205, 276; as substance of all organic matter, 171, 182, 202, 277; compose food, 182; comments on Buffon's theory of, 270-71, 316-17. *See also* Germs, Internal Mold

Organisms: composed of organic particles, 173; created directly by God, 269; as analogous to crystals, 270

Ontologia: Christian Wolff's treatment of time and space in, 21

Ontology: as subdivision of metaphysics, 340; as science of relations of ideas, 340-41

Ovary: relation to *corpus luteum*, 208n3. *See also Corpus luteum*

Ovid (43 B.C.-18 A.D.): on alternations of earth's surface, 134

Paradox: nature of, 339

Parasites (Internal): origin of, 206, 277

Parthenogenesis: 273

St. Petersburg Problem: Buffon's treatment of, 66-71; Gabriel Cramer on, 66; Montmort on, 66-67

Philosophy: neglect of in eighteenth century, 122

Physical certitude: as near infinite probability, 54, 123-24; product of repetition of observations, 56, 123; means of calculating, 57. *See also* Certitude, Truth, Physical Truth

Physical history: as true account of earth, 152

Physical knowledge: and succession of phenomena, 124-25

Physical truth: Buffon's concept of, 26-27; assessed by probability calculus, 26-27; 123, 240; based on repetition of events, 123-24; non-arbitrary character of, 123-24, 240; contrasted to mathematical truth, 123; and immanent causes, 131; Malesherbes on, 333, 337. *See also* Mathematical truths; Certitude

Physics: relation to mathematics, 125-27; 345n7

Planetary system: origin of, 154-60, 225, 317, 286

Plantade, François de (1670-1741): on homunculus hoax, 274-75

Plato (ca. 427-347 B.C.): Buffon's criticisms of, 338

Pliny, the Elder (23-79 A.D.): Buffon's praise of, 120, 121; classification of, 121

Pluche, Abbé Noel-Antoine (1688-1761): on knowledge of nature, 15; on formation of the embryo, 16-17

Polyp: regeneration of, 173; as composed of organized bodies, 173, 277; as intermediate form of life, 204; existence of predicted by Leibniz, 258

Pope, Alexander: (1688-1744), *Essay on Man*, 238, 250; compared to Buffon, 238

"Port Royal Logic:" grammatical principles of, 342

Positivism: Buffon's endorsement of, 36; and continental Newtonianism, 295

Preformation theory: and mechanical philosophy, 15-18; Abbe Pluche on, 16-17; and doctrine of creation, 16-17, 269; in regenerating organisms, 173-74; and concept of infinite series, 174; Buffon rejects, 207; Haller's changing views on, 311, 324; reinstated, 312; distinguished from

Index 403

preexistence, 313n3. *See also* Germ, Generation
Privation: source of metaphysical errors, 46
Probability: and natural necessity, 25-27; and physical truth, 26-27; degrees of, 53; of future events, 60; means of inferring causes, 81, 125; as inferior form of certitude, 306
Propositions: classification of, 333-34; as abstractions, 334-35
Psychology (Pneumatology): as subdivision of metaphysics, 340
Ptolemaic System: as fertile hypothesis, 305
Punctum Saliens: role in formation of chicken, 279
Pyrrhonism: and scientific empiricism, 14-15; Buffon accused of, 241
Ray, John (1628-1705): concept of classification, 255; controversy with Tournefort, 303
Regeneration: of polyp, 171, 173, 183
Religion: Enlightenment attacks on, 25; Buffon's theories and, 214, 223, 237, 250, 283, 352, 366-67, 368, 376-77, 383n 23-24; not threatened by Buffon--Needham experiments, 324; *see also* Creation; God; Faith; Design; Deluge
Reproduction: problem of insoluable, 175
Reproduction: parthenogenic, 273
Richardson, Samuel (1689-1761): Buffon's liking for, 383n26
Rivers: Buffon's account of, 266

Roche, Jacques Fontaine de la (1688-1761): editor of *Nouvelles ecclésiastiques*, 235
Rock, vitreous: location of, 260
Roger, Jacques: interpretation of Buffon's thought, 9, 51
Rousseau, Jean-Jacques (1712-78): *Discours sur l'inegalité* quoted, 3-4; historicist view of human nature, 4; visit to Buffon, 351, 361; Buffon criticizes, 364; style compared to Buffon's, 373; 370
Ruysch, Frederick (1638-1731): and generation theory, 273
Saint-Belin Malain, Marie Françoise de (1732-1769): wife of Buffon, 367, 382n21
Sallo, Denis de (1626-69): founder of *Journal des Savants*, 231
Sauvages, François Boissier de la Croix de (1706-1767): translator of Stephen Hales' *Haemastaticks*, 39n1
Saturn: formation of rings, 159
Scepticism: Pyrrhonist, 11; mitigated, 12; Buffon accused of, 241
Scheuchzer, Johann Jacob (1672-1733): on significance of fossils, 248, 264
Scholasticism: as useful method in philosophic debate, 333
Science: of man, 53; as knowledge of facts, 110; of society, 375

Sciences: abstract and concrete contrasted, 3, 127, 341; human and natural, 110, 337-38; general and particular approaches, 127; classification of, 341; mixed, 341. *See also* Method

Sea: description of bottom of, 137-38; former extent of, 140-41; mountains in, 143; Buffon on formation of, 258-60; evaporation of, 263-64; role in formation of continents, 286

Sediments: transport of, 143-44; horizontal positioning of, 145-46

Semen, female. *See corpus luteum*

Seminal bodies: vibration of, 193; non-animal nature of, 202-03; and relation to body parts, 207

Seminal fluid: observations on, 192-99; identity in male and female, 196-99, 201; in female, 273-74. *See also Corpus Luteum*; Spermatozoa; Generation

Society, science of: 375-76

Solar system: death of, 242

Sorbonne: condemns Buffon's work, 269, 278, 322; condemns Montesquieu's *Spirit of the Laws*, 367; Buffon's attitude toward, 367

Soul: distinct from matter, 228; knowledge of, 228; Buffon's views on condemned, 287-88; immortality of, 288, 290-91. *See Also* Dualism

Sectarianism, scientific: destructive consequences of, 297

Space: Newtonian absolute rejected, 21; Buffon denies infinity of, 45

Spallanzani, Abbé Lazzaro (1729-1799): attack on Buffon and Needham, 166; reinstates preformation theory, 312

Species: reality of, 23-24, 132, 170, 174; definition of, 23-24, 106, 174, 334; as collection of individuals, 24, 334; apparently unlimited, 101; as infinite series, 132, 170, 174; duration of, 173-74, 202, 207, 227, 280; internal mold constitutes unity of, 202; permanence of demonstrates God's existence, 280; as objects of universal ideas, 334-35

Spermatozoa: Buffon's observations on, 192-96, 275; animal nature of, 276-77, 316-17; Buffon disproves universal role of, 317

Spontaneous generation: 206, 270, 315-16; and existence of God, 278-79

Stahl, Georg-Ernst (1660-1734): Buffon praises his method, 38

Steno, Nicolas (1638-1686): and preformation theory, 316

Stones: figured (fossils), 264

Strata: positioning of, 141, 245

Style: Buffon on, 370-71, 372-73; Buffon's and Rousseau's compared, 373

Subtle matter: 38

Sun: rising of, as example of physical truth, 54, 56-57; as combustible star, 157; extinction of,

286. *See also* Comet; Planet
System-building mentality (*Esprit de système*): Buffon rejects, 38
Systems of classification: ideal requirements of, 102; as artificial signs, 108. *See also* Classification
Target, Gue-Jean-Baptiste (1733-1806), 357, 378n2
Teleology: and organization of embryo, 312, 320
Teleological forces: and necessity of creator, 321-22
Temperament: and life-forces, 299
Terray, Joseph-Marie (1715-1778), 380n14, 363
Theological truths: scientific explanation of, 226
Theophrastus (ca. 372-287 B.C.), 118
Theories of the Earth: Newtonian attacks on, 13; Buffon criticizes, 135, 149, 226; Buffon follows Leibniz on, 258
Thomas, Antoine-Leonard (1732-85): death of, 360; poetry of, 367; 370, 378n4
Tides: as continually-acting cause, 142-45; and distribution of sediments, 142-44
Time: Newtonian absolute criticized by Leibniz, 21; Newton's concept synthesized with Locke's, 31n44; Buffon denies infinity of, 45
Topsoil: organic nature of, 139
Tournefort, Joseph Pitton de (1656-1708): Buffon's praise of, 105; system preferred to Linnaeus', 217; use of floral parts in classification, 303; praised as metaphysician, 343
Toynbee, Arnold (1889-1975): on articulations of reality, 94n6
Transformism: de Maillet's theory of, 223
Trembley, Abraham (1700-1784): and regeneration of polyp, 203-04; influence on Haller, 311
Truth: does not exist as abstraction, 19-20, 122, 239, 289, 331; definition of, 122, 238-39, 331, 332, 333; as discovery of causes, 220; Buffon's concept of condemned, 238, 242, 286-87; moral, 242, 290; Bossuet on, 251-52; independent of convention, 290, 336, 337; as unchanging, 297; propositional character of, 333; Malesherbes on, 330, 333; and "truths," 333
Truths: mathematical status of, 62-63; mathematical and physical distinguished, 123, 332, 340
Tubules: as anatomical units, 270-71
Uniformitarianism: Buffon expresses principle of, 132, 141-42, 149. *See also* Actualism
Universal gravitation: *see* Gravitational attraction
Universe: creation of, 153. *See also* Planetary System
Vacuum: real existence of, 337
Valisnieri, Antonio (1661-1730): and search for human egg, 187
Valleys: Buffon on formation of, 260, 262

Variation: and permanence of species, 280; and embryological development, 312; difficulties for Buffon's generation theory, 318-320; proof of design in nature, 323
Vegetable Staticks (Hales): modifications by Buffon, 37
Verheyen, Phillippe (1648-1710): and existence of spermatozoa, 202
Vicq d'Azyr, Felix (1748-94): 379n7, 384n30
Volcanoes: insignificant as geological causes, 143, 260
Voltaire (Francois-Marie Arouet) (1694-1778): on *Journal des Savants*, 231; 278, 367; on friendship, 375
Voyage à Montbard: authenticity of, 351
Vues de la Nature: quoted, 131n5; Buffon on style of, 357
Wagers: calculation of, 59
Wallis, John (1616-1703): on calculation of infinity, 47
Whiston, William (1667-1752): cosmological theory of, 135, 226, 243, 244, 266
Wilson screw-barrel microscope: scroll mounted, 164-66, 168n6; and Buffon-Needham experiments, 165-66; hand-held version, 189
Wolff: Johann-Christian von (1679-1754): *Elementa matheseos universae* edited by Gabriel Cramer, 5; *Philosophia prima sive ontologia*, 20-21; and Mme du Châtelet, 20, 21; on space and time, 20-21; on influence in France, 31n43
Woodward, John (1665-1728): on crust of the earth, 135; 226, 266; on significance of fossils, 248
Zinck, Georg Heinrich (1692-1768): probable translator of German edition of *Histoire naturelle*, 281
Zoological classification: arbitrary character of, 108, 114; Linnaeus' system criticized, 114-16; Greeks on, 117-20

www.ingramcontent.com/pod-product-compliance
Lightning Source LLC
Chambersburg PA
CBHW051242300426
44114CB00011B/851